The Institute of Mathematics
and its Applications
Conference Series

The Institute of Mathematics
and its Applications
Conference Series

Previous volumes in this series were published by
Academic Press to whom all enquiries should be addressed.
Forthcoming volumes will be published by
Oxford University Press throughout the world.

NEW SERIES

1. *Supercomputers and parallel computation* Edited by D. J. Paddon
2. *The mathematical basis of finite element methods*
   Edited by David F. Griffiths
3. *Multigrid methods for integral and differential equations*
   Edited by D. J. Paddon and H. Holstein

Multigrid methods for integral and differential equations

Based on lectures at a Summer School/Workshop
held at Burwalls Conference Centre, University of Bristol, England,
September 1983

Edited by

D. J. PADDON
University of Bristol

and

H. HOLSTEIN
University College of Wales, Aberystwyth

CLARENDON PRESS · OXFORD · 1985

Oxford University Press, Walton Street, Oxford OX2 6DP
Oxford New York Toronto
Delhi Bombay Calcutta Madras Karachi
Kuala Lumpur Singapore Hong Kong Tokyo
Nairobi Dar es Salaam Cape Town
Melbourne Auckland
and associated companies in
Beirut Berlin Ibadan Nicosia

Oxford is a trade mark of Oxford University Press

Published in the United States
by Oxford University Press, New York

British Library Cataloguing in Publication Data
Multigrid methods for integral and differential equations : based on lectures at a summer school/ workshop held at Burwalls Conference Centre, University of Bristol, England, September 1983.— (The Institute of Mathematics and its Applications Conference series. New series; 3)
1. Equations—Numerical solutions
I. Paddon, D.J. II. Holstein, H. III. Institute of Mathematics and its Applications IV. Series
512.9'4 QA218
ISBN 0–19–853606–2

Library of Congress Cataloging in Publication Data
Main entry under title:
Multigrid methods for integral and differential equations.
(The Institute of Mathematics and Its Applications conference series : new ser., 3)
Includes indexes.
1. Integral equations—Numerical solutions—Addresses, essays, lectures. 2. Differential equations—Numerical solutions—Addresses, essays, lectures.
3. Numerical grid generation (Numerical analysis)—Addresses, essays, lectures. I. Paddon, D. J.
II. Holstein, H. (Horst) III. Institute of Mathematics and Its Applications. IV. Series.
QA431.M75 1985 515.3'5 85–11567
ISBN 0–19–853606–2

Printed in Great Britain by
St Edmundsbury Press Ltd.
Bury St Edmunds, Suffolk

PREFACE

These proceedings consist of papers whose authors addressed a small but enthusiastic group of numerical analysts on the Multigrid Method, during a Summer School/Workshop held at the Burwalls Conference Centre, University of Bristol, England, in September 1983. Most of the papers were substantially revised after the meeting, and bring together the authors' views and research up to the effective closing date of the proceedings (July 1984).

During the first part of the meeting, participants had the privilege of hearing the guest speakers, all of whom are leading authorities in this relatively new and exciting branch of numerical analysis: Achi Brandt, Wolfgang Hackbusch, Pieter Hemker, Ulrich Trottenberg and Pieter Wesseling. Each presented two papers on the theory and practice of the Multigrid Method. The second part of the meeting consisted of the presentation of contributed papers. Throughout the meeting, participants responded warmly to the excellent quality of the papers, and to the open and friendly discussions led by the guest speakers.

Multigrid Methods came into prominence in the late 1970's. The guest speakers have made major contributions to its development and its acceptance as a significant computational tool. Even though computers have shown orders of magnitude performance increases during the last decade, the demand for efficient, robust methods for solving field equations remains open. It is in this direction that the Multigrid Method has made notable advances, and has made accessible new areas of computation, e.g. three-dimensional simulation. In many cases, the Multigrid Method allows a realisation of optimal algorithms, whose computational work grows linearly with the number of unknowns in a system of equations. The Multigrid Method may itself be regarded as a model numerical process, capable of influencing future research into optimal algorithms.

These proceedings fall into two sections.

Section 1. This starts with a witty introduction by A. Brandt, from which we may gain insight into the deep-felt philosophical convictions held by a principal founder of the Multigrid Method. It is followed by the paper of W. Hackbusch, which gives a substantial review of the Multigrid Method for integral equations, together with a considerable innovative content. A very wide range of problems is shown to be amenable to the integral equation approach. P. Hemker's contribution gives a state-of-the-art guide to the design and implementation of programs for the Multigrid Method, applied to the solution of difference equations derived from general elliptic equations of the convection-diffusion type. Scalar and vector codes are considered. The paper of P. Sonneveld, P. Wesseling and P.M. de Zeeuw gives a most lucid treatment of conjugate gradient and Multigrid as methods for the acceleration of iterative convergence. The efficiency and robustness of the methods are compared for a variety of difficult test problems derived from the convection-diffusion equation.

Section 2. This section contains the contributed papers, and represents a spectrum of research ranging from new theory (J.F. Maitre and F. Musy; S. McCormick) to applications of Multigrid in engineering design (K.E. Barrett et al.) and in particular, to three-dimensional simulation (T. Scott). Hybrid techniques, in which Multigrid plays a key role, are described in papers on Spectral Methods (T.N. Phillips, T.A. Zang and M.Y. Hussaini) and the Hierarchical Finite Element Method (A.W. Craig and O.C. Zienkiewicz).

Both finite difference (B. Favini and G. Guj) and finite element approaches are exemplified in this section. An outstanding contribution is made by J. Ruge and K. Stüben to the understanding of the Algebraic Multigrid Method. Here the Multigrid approach is extended to apply to a fairly wide class of sparse matrix systems, irrespective of their origin from a finite difference or finite element discretisation, or indeed from any geometric grid structure at all. While there are computational overheads compared to the most efficient geometric multigrid solvers when applicable, the authors exhibit remarkably successful results on a variety of problems, using a truly "pitch black box" Algebraic Multigrid routine.

The Summer School/Workshop and subsequent communication with the authors has given the editors a great deal of pleasure. We warmly express our gratitude to the authors for their cooperation and friendship while producing these proceedings, and for the opportunity to present for publication this collection of stimulating papers.

We would like to thank Roger Moses and the staff of the Burwalls Conference Centre for creating an atmosphere so conductive to the meeting and discussions that took place. We would also like to thank Catherine Richards and her staff at the Institute of Mathematics and its Applications for the careful typing and efficient yet cheerful way in which they coordinated the production of these proceedings. We are grateful to the staff of the OUP for publishing this volume, and in particular to Anthony Watkinson, for the help, encouragement and patience he has shown throughout its preparation.

Derek Paddon

Horst Holstein

ACKNOWLEDGEMENTS

The Institute thanks the authors of the papers, the editors, Dr. D. Paddon, AFIMA (University of Bristol) and Dr. H. Holstein, AFIMA (University College, Wales) and also Mrs. Janet Parsons, Miss Denise Wright, Miss Pamela Irving and Miss Karen Jenkins for typing the papers.

CONTENTS

LIST OF CONTRIBUTORS

K.E. Barrett, Department of Mathematics, Coventry (Lanchester) Polytechnic, Coventry CV1 5FB, UK

A. Brandt, Department of Applied Mathematics, The Weizmann Institute of Science, Rehovot, Israel

D.M. Butterfield, Department of Mathematics, Coventry (Lanchester) Polytechnic, Coventry CV1 5FB, UK

A.W. Craig, Civil Engineering Department, University College of Swansea, Singleton Park, Swansea SA2 8PP, UK

S.E. Ellis, Department of Mathematics, Coventry (Lanchester) Polytechnic, Coventry CV1 5FB, UK

B. Favini, Dipartmento di Meccanica e Aeronautica, Università di Roma, 'La Sapienza', Italy

G. Guj, Dipartmento di Meccanica e Aeronautica, Università di Roma, 'La Sapienza', Italy

W. Hackbusch, Institute für Informatik und Praktische Mathematik, Christian-Albrechts-Universität Kiel, Olshausenstr. 40, D-2300, Kiel 1, Germany

P.W. Hemker, Centre for Mathematics and Computer Science, Department of Numerical Mathematics, Kruislaan 413, 1098 SJ Amsterdam, The Netherlands

M.Y. Hussaini, Institute for Computer Applications in Science and Engineering, Mail Stop 132C, NASA Langley Research Centre, Hampton, Virginia 23665, USA

C.J. Judd, Department of Mathematics, Coventry (Lanchester) Polytechnic, Coventry CV1 5FB, UK

J.F. Maitre, Département de Mathématiques Informatique Systèmes, Ecole Centrale de Lyon, B.P. 163-69131, Ecully, Cedex, France

S. McCormick, Department of Mathematics, Colorado State University, Fort Collins, Colorado 80523, USA

F. Musy, Départment de Mathématiques Informatique Systèmes, Ecole Centrale de Lyon, B.P. 163-69131, Ecully, Cedex, France

T.N. Phillips, Institute for Computer Applications in Science and Engineering, Mail Stop 132C, NASA Langley Research Center, Hampton, Virginia 23665, USA

J. Ruge, Gesellschaft für Mathematik und Datenverarbeitung, Postfach 1240, D-5205 St. Augustin 1, West Germany

T. Scott, UKAEA, Winfrith, Dorchester, Dorset DT2 8DH, UK

P. Sonneveld, Department of Mathematics and Informatics, Delft University of Technology, Julianalaan 132, 2628 BL Delft, The Netherlands

K. Stüben, Gesellschaft für Mathematik und Datenverarbeitung, Postfach 1240, D-5205 St. Augustin 1, West Germany

J.H. Tabor, Department of Mathematics, Coventry (Lanchester) Polytechnic, Coventry CV1 5FB, UK

P. Wesseling, Department of Mathematics and Informatics, Delft University of Technology, Julianalaan 132, 2628 BL Delft, The Netherlands

T.A. Zang, NASA Langley Research Center, Mail Stop 132C, Hampton, Virginia 23665, USA

P.M. de Zeeuw, Centre for Mathematics and Computer Science, Department of Numerical Mathematics, Kruislaan 413, 1098 SJ. Amsterdam, The Netherlands

O.C. Zienkiewicz, Civil Engineering Department, University College of Swansea, Singleton Park, Swansea SA2 8PP, UK

INTRODUCTION - LEVELS AND SCALES

A. Brandt

*(Department of Applied Mathematics,
The Weizmann Institute of Science, Israel)*

Open your eyes and you see multi-level processes all around. They have always been here. The organization and operation of military forces is an obvious example: soldiers are grouped in squads, which in turn are grouped in sections, grouped in platoons, then in companies, battalions, regiments, brigades, divisions, corps and armies. Civil society, less strict and more complicated, still operates in a variety of hierarchical structures: geographical, economical, political, judicial, educational, and so forth.

Such hierarchies are necessary because there are very many, sometimes millions, interdependent decisions to be made; they cannot be made by one governor, because of their multitude and their complicated interdependence, nor can they be decided by many independent administrators without being coordinated with each other. The hierarchical structure effectively deals with this situation by exploiting the fact that each decision has a certain "scale". The location of a new hospital, for example, is a decision which has the scale of a district. It strongly affects the district served by it, marginally affects neighbouring districts, and very weakly affects others. The decision can therefore be made by a certain administration assigned to the district, in some coordination with neighbouring administrations. The decision cannot effectively be negotiated at a too low administrative level, e.g., at the level of individual families living in the district; the relevant information (concerning needs, constraints, etc.) must be gathered into one point of decision. Similarly, the general outline of principal throughways should be decided at the inter-state level, while local back roads should be regulated at the district or town levels.

The two kinds of roads should be connected, of course. The effective way to manage the inter-dependence between the different levels is based on the assumption that global decisions should only marginally be affected by local ones (otherwise the latter would not be local). Thus, ideally, a two-level hierarchical structure should operate in the following way. The global government first gathers some general figures summarized at the local level, representing sum totals of local needs, important overall constraints, etc. Based on these it prepares preliminary global plans. These global plans give the local governments the framework for devising their own, more detailed, plans. In course of doing that, the local government may realize that some, usually marginal, aspects of the global plans do not quite fit the local situation and, therefore, need some adjustments or corrections. So, at a second round, the global government again gathers information summarized at the local level, now representing sum totals of needed *corrections* ("defect corrections", in the language of numerical analysts). Since in practice this process is seldom fully recognized, let alone fully effectively organized, more such rounds may be needed. When more levels of government are involved, the process is applied recursively, in a variety of manners.

For very much similar reasons, iterative hierarchical procedures, similar to the process we have just described, are very efficient in solving large and complicated problems on computers. Such procedures were naturally introduced to solve problems where the hierarchical structure was already explicit in the problem itself. A good example is the field of production planning, notably in the Soviet Union, where hierarchic divisions into sectors and pyramidal management naturally led to the introduction of iterative "aggregation/disaggregation" (a/d) algorithms, starting in the mid sixties (Dudkin and Yershov (1965)) and growing in the seventies into extensive Russian literature on iterative a/d procedures for large linear programming problems (see Vakhutinsky, Dudkin and Ryvkin (1979)). Multi-level approaches have in fact quite naturally emerged in all branches of computer technology, as in the structured organization of computer hardware (see for example Stone (1972)), the top-down structured design of software (Yourdon, Swann (1978)), the pyramidal data structures (trees, heaps, etc.) and many of the most efficient algorithms in computer science, such as fast sorting (sorting n numbers in $O(n \log n)$ operations) the "divide and conquer" class of algorithms, etc. (See, e.g., Aho, Hopcroft and Ullmann (1974) . Most of these procedures are not iterative, though.)

Also, for very much the same reasons, multi-level algorithms have come forth as the most efficient algorithms in solving the very large algebraic systems arising in discretising partial differential boundary value problems, especially those describing steady-state physical systems. Here the fully hierarchical structure is not at all generally explicit in the problem itself, so it took some effort, and interesting historical development, to realise it. Fully effective multi-level algorithms were first developed as *direct*, not iterative, solvers, treating very special situations where it was algebraically clear enough how to construct recursively hierarchical solvers. I refer here to the fast solvers based on fast Fourier transforms (FFT) and/or reduction methods, especially the cyclic odd-even reduction, both of which are clearly recursive, but non-iterative, multi-level processes (see Buzbee, Golub and Nielson (1970), Hockney (1970) and Temperton (1979)). The total reduction method also belongs to this class (see Trottenberg (1977)). The solution of n equations by this kind of solvers require at most $O(n \log n)$ computer time and storage, but the class of problems for which this full efficiency is attained is quite limited: separable problems, essentially meaning constant-coefficient elliptic equations with constant-coefficient boundary conditions on rectangular domains. This class has been substantially enlarged by using these fast solvers, and various combinations of them, *iteratively*. Thus, for example, if the coefficients are not constant but sufficiently uniform, the iterative application of a constant-coefficient fast solver can be very effective (the number of iterations depending on the uniformity of the coefficients, but not on the meshsize). Extensions to domains of arbitrary shapes have been obtained by "capacitance matrix methods", using the fast direct solvers iteratively, with conjugate-gradient acceleration, each solution typically costing the equivalent of some 15 applications of the fast solver (see e.g. Proskurowski (1981)). The nested-dissection approach to elimination ordering (George and Lin (1981)) is another powerful multi-level approach, more general but less efficient than the FFT and reduction solvers.

Meanwhile, steady-state PDE problems and the solution of their discretised equations were examined from two other points of view, which jointly led to the realisation that each such problem contains a natural hierarchy of levels, not immediately explicit, but very powerful and much more general than the mechanical hierarchy exploited by the above methods. First, studying reasons for slow convergence of various relaxation solvers, it became clear that relaxation is a "local" process which cannot efficiently treat "global" or "smooth" solution components. A smooth error component shows relatively small "residuals", i.e., small errors in the individual difference equations. The smooth error can be much larger than shown by the residuals because, throughout a large region, these residuals have the same sign, so they reinforce each other. Now, since the relaxation corrections are based on the individual residuals, they are necessarily small compared with the actual error, if the latter is smooth. Thus, in order to get a correction comparable to a smooth error, information concerning the residuals throughout sufficiently large regions must be summed up to one point of decision, very much as in the case of global social decisions mentioned above. The size of the regions over which residuals should be summed up - so that a correct picture about the error magnitude is obtained - must be comparable to the *scale* of the error, i.e., to the typical distance over which the error substantially changes. As long as each relaxation step (each step of correction) works on a much smaller scale, convergence must be slow.

A second, complementary viewpoint evolved from examining the nature of the discretisation error, i.e., the difference between the true solution of the differential problem and the exact solution of the discretised equations. The relative magnitude of this error is clearly determined by the relative magnitudes of the discretisation meshsize and the solution scale. A smooth solution, which is a large-scale solution, can thus be approximated on a coarse grid. The same is obviously also true for a smooth *error*. Thus, it became clear, exactly those errors that are slow to converge by relaxation processes on some fine grid can be approximated on a coarser grid, where the meshsize is comparable to their scale and hence their convergence need not be slow.

A natural hierarchy of levels emerges, based on viewing the solution to each boundary value problem as a linear combination of components with different scales. Each component is most effectively controlled by grids with meshsize comparable to its scale, and efficient multi-level control can thus be realised as a multi-grid processing.

A two-grid process, for example, is fully analog to the ideal operation of the two-level hierarchical government described above. The problem is first represented on the coarser grid, e.g., by averaging its equations to the scale of that grid. The (approximate) solution to the resulting coarse-grid problem, once computed, is then interpolated to the fine grid, serving there as a first approximation, a framework, to be next improved by fine-grid processes, such as relaxation. This fine-grid processing finds the fine features of the solution which were invisible to the coarser grid, and also, as a result, encounters some residuals of global (smooth) errors, which it cannot efficiently reduce. (These are smooth errors caused by aliasing, i.e., by the previous coarse-grid processing having misinterpreted coarse-grid traces of the fine features. Now that those fine features have been removed from the error by the fine processing, that aliasing error becomes the dominant one.) So, in the next round, the residual problem is approximately transferred, by some averaging, to the coarse grid, where it can efficiently be solved and its solution is then interpolated back to the fine grid and added as a correction to the previous fine-grid solution.

The process just described is the two-level "full multigrid" (FMG) algorithm. It can be used recursively in a variety of manners in case more levels are involved.

The number of levels that should be used depends on the ratio between the size of the domain and the finest scale one wants to see in the solution. Between these two scales as many should be introduced as practical. Namely, the ratio between successive scales (successive meshsizes) should be as small as possible, as long as this does not substantially increase the total number of gridpoints. The ratio 1:2 between successive meshsizes is very convenient: the total number of gridpoints is still dominated by their given number on the finest grid, but successive meshsizes are close enough to effectively treat any solution component. In fact, with such a ratio, the experience so far showed that suitable FMG algorithms could solve all test problems "to the level of truncation error" (i.e., to the point where the error in approximating the differential solution is dominated by the discretisation error, not by the error in solving the discrete system) in just few (less than 10) "work units", where the work unit is the amount of operations involved in *expressing* the given (finest-grid) system of discrete equations. The 1:2 ratio is also most convenient in programming the inter-grid transfers. Note that in this respect the multigrid processing is different from multi-level social structures. Its levels are chosen much more tightly, to achieve maximum efficiency.

Note also that the decomposition of the PDE solution into components of different scales is only implicit; it is used above to motivate and explain the validity and strength of the multigrid process; but the actual multigrid algorithm does not use any such decomposition. It only transfers equations (or residual equations) from fine grids to coarse grids, and solutions (or corrections) from coarse to fine. Decomposition in terms of Fourier components, in particular, can be used as a powerful tool to *analyse*, and even exactly predict, the performance of multigrid algorithms, but the algorithms themselves do not employ such decompositions, and their efficiency extends far beyond the cases where the Fourier analysis is rigorously valid. (Incidentally, it is important to realize that in some cases Fourier analysis is the wrong tool to separate local effects from global ones. For example: a local discontinuity gives rise to global high-frequency Fourier components. Other tools should then be used to understand quantitatively, and optimise, the multigrid performance.)

The FMG solvers (solving discretised PDEs to the level of truncation errors) have been developed to the point that they are today even faster than the FFT and reduction solvers mentioned before. (The multigrid Poisson solver described in Barkai and Brandt (1983) is the fastest we know.) More importantly, of course, these FMG solvers are much more general. They solve with the same efficiency (i.e., in few work units) complicated nonlinear systems on general domains. Moreover, it is possible to integrate into each application of an FMG solver, for small extra computer work, various processes of *local* mesh refinement, mesh optimisation and *local* coordinate transformations, making it very effectual in treating singularities, unbounded domains, curved boundaries, boundary layers, discontinuities, etc.

Furthermore, multigrid solvers can be directly applied to "higher" problems - such as optimisation, optimal design and optimal control problems, or system identification problems - whose solution would normally be accomplished through solving a *sequence* of boundary value problems. An important principle indeed is always to try to multigrid the given, *original* problem, instead of merely using fast multigrid solvers to a sequence of intermediate subproblems. The *original* problem (e.g, the optimisation problem itself) should first be solved on a coarser grid, then relaxed on the finer grid (including for example, local optimisation of parameters, in case some of the functions to be optimised do have local scales), then brought back to the coarser grid, etc. The entire solution of the *original* problem may thus cost only few work units. Multigridding the original problem is especially advantageous in case that problem is autonomous (i.e., having solution-dependent but not directly space-dependent coefficients, as for example most fluid dynamic problems), while the subproblems are nonautonomous (having coefficients spatially depending on the solution of a previous subproblem).

Sometimes it is still required to solve a sequence of subproblems. A typical example is the interactive design of a certain structure, where between two solutions of the system the structure is changed in some specific parts and/or in some of its global parameters. The new solution can then be obtained from the old one by a remarkably short multigrid processing, in which the finer grids are relaxed only around the changed parts. This technique may allow the design of a large structure to be done mostly in core memory, since for several design steps only the currently designed parts, and some neighbouring parts, should fully be kept in memory while the rest of the structure may be represented by coarser grids (using the FAS version of multigrid). Re-solving by such techniques should be so efficient as to allow the designer to introduce some changes to a large structure and immediately view the new solution on the screen.

Similarly, in evolution problems with implicit time differencing, the solution of a new system of equations seems to be required at each time step. Each of these systems could be solved very efficiently by an FMG algorithm, costing the equivalent of just few *explicit* time steps. But, here again, multigrid techniques may be even more effective if they are designed in terms of the original evolution problem itself. For example, one can often drastically reduce the overall work by exploiting the fact that the solution is a superposition of pure convection and smooth changes, or changes which are smooth throughout most of the domain, hence requiring fine-grid interactions only at some small subdomains.

The full efficiency of multigrid solvers, as described here, is not easy to obtain, though. It sometimes depends on the correct treatment of each feature of the problem at each multigrid stage. Many things can easily go wrong, such as relaxing a certain boundary condition in a way which conflicts with the interior smoothing; or improper fine-to-coarse transfers of boundary conditions and their residuals; or treating at relaxation conditions which seem local but are not; or using a relaxation scheme which is not as powerful a smoother as it should, and can, be; or wrong order or type of interpolations; or any inadequate treatment of any difficulty, from structural singularities (e.g., reentrant corners) to discontinuities in the solution or in the equations, anisotropies, non-ellipticity, etc., etc. A single mistake at any of these may substantially degrade the whole performance, not to mention plain programming bugs, which, due to the corrective nature of the algorithm, may well disguise themselves in the innocent form of slow convergence. To obtain full efficiency it is therefore necessary to construct the algorithmic concepts and the actual programs in a gradual, systematic way, using available knowhow (see Hackbusch and Trottenberg (1982), Brandt (1984)).

The class of partial differential equations that can be solved by multigrid solvers has been ever extending, from second order equations to arbitrary orders, from linear to nonlinear, from smooth coefficients to strongly discontinuous ones (Alcouffe, Brandt, Dendy and Painter (1981)), from definite to indefinite problems, from scalar equations to general systems (Brandt (1984) §3.8), and from elliptic type to other types.

Unlike evolution problems, where properties like hyperbolicity and parabolicity are all important, the only feature that matters concerning the type of the differential operator in (stationary) boundary value problems is whether it is nicely (isotropically) elliptic or not. If it does not have a good ellipticity measure, it does not make any difference whether it is anisotropic elliptic (like $\varepsilon\partial^2/\partial x^2+\partial^2/\partial y^2$) or semi-elliptic, weakly elliptic (e.g., having elliptic singular perturbation), hyperbolic or any other nonelliptic type. All these types have the same basic difficulty and can be treated essentiallly by the same approach. The difficulty is that the solution scales discussed above are *not isotropic*. (Incidentally, one dimensional problems, such as $du/dx = f$, are elliptic, unlike their two-dimensional counterparts. Their use as models for treating non-ellipticity is thus erroneous.) In other words, at each point in the domain of the problem there passes a "characteristic line" (sometimes just a characteristic *surface*; and in case of non-scalar PDEs there may pass *several* such lines or surfaces at each point, corresponding to different solution components) such that a "global" error component (in the sense that its residuals are small compared with its size) can change rapidly in directions perpendicular to the characteristic lines. Such global but rapidly changing components are called *characteristic components*. As global components, their convergence by relaxation is inefficient, but as rapidly changing components they cannot be well approximated by coarser grids.

The key for efficient multigrid treatment of such anisotropic problems is to distinguish clearly between two very different situations, depending on whether or not one wishes to approximate characteristic components far from boundaries, where "far" is meant relative to the component's smaller scale, i.e., the scale of its rapid change perpendicular to the characteristic. To so approximate characteristic components by the discrete system, the grid must (roughly) be aligned with the characteristic lines throughout the domain. Fast multigrid solvers can then be based on these aligned gridlines, either by relaxing these gridlines simultaneously, or by coarsening the grid only along these lines and not in the perpendicular directions. (The latter approach is preferred when the characteristics are surfaces rather than lines.) If, on the other hand, the grid is not consistently aligned with the characteristics, then characteristic components cannot be approximated far from boundaries (except accidentally, in some regions of accidental alignment), and it is then unwise to attempt to have fast algebraic convergence for such components. This convergence, which is meaningless in terms of approximating the *differential* solution, is harder to obtain: costlier and more complicated relaxation schemes and/or intergrid transfers must be developed, and even then fast convergence is not always guaranteed. It is absurd, we believe, to invest most of your computer resources and programming effort to get fast algebraic convergence exactly for those components whose algebraic solution is generally (out-side accidental regions of alignment) no approximation to the differential solution.

This fashion in treating anistropy, as well as some other developments, has led to the recognition that fast algebraic convergence should not be the main objective of multigrid solvers. The objective is of course to get the desired accuracy, in terms of solving the *differential* problem, for minimal computer (and also human) resources. This is obtained by FMG solvers which, especially in cases of anisotropic equations without corresponding grid alignment, do not necessarily employ uniformly good smoothers, hence nor do they attain uniformly good algebraic convergence rates. Working with such solvers requires of course certain modifications in the traditional "smoothing factor" approach for measuring the effectiveness of relaxation (see Brandt (1984), §20.3.1), as well as new approaches for a priori predicting, and a posteriori judging, the overall success of the FMG solver, (see Brandt (1984), §7.4, §7.5 and §1.6). An important advantage of these approaches for performance evaluation is that the performance becomes less sensitive to the precise treatment of all problem features at all algorithmic stages.

These approaches also allow the evaluation of important schemes which deliberately avoid any algebraic convergence. Such for example is the "double discretization" scheme (Brandt (1984), §10.2), which employs less accurate, stable discretizations in relaxation processes, and others, more accurate but not necessarily stable ones, in the fine-to-coarse transfers of residuals, thus combining easier local stability with higher global accuracy.

Another interesting development has recently led to the extension of multigrid-like techniques to cases where no grid is actually present. It started, in a way, with the development of usual multigrid algorithms for diffusion problems which were isotropically elliptic but in which the diffusion coefficients were strongly discontinuous. It turned out (Alcouffe, Brandt, Dendy and Painter (1981)) that to obtain the usual multigrid efficiency in such cases, the coarse-to-fine interpolation of corrections should be based on the difference equations themselves, rather than being the standard polynomial interpolation. This later led to the recognition that interpolation can be based solely on the algebraic equations, without even using the geometry of the grid. In a similar manner, the fine-to-coarse transfer of residual equations can completely be based on the *adjoint* (transposed) algebraic system. Since these intergrid transfers are what give the coarse grid a definite meaning, it was further realized that even the choice of the coarse-grid *variables* can be freed from its traditional geometric context, and purely "algebraic multigrid" (AMG) schemes were introduced (Brandt (1983), Brandt (1984), §13.1, Brandt, McCormick and Ruge (1984), Ruge and Stüben (1984)). In these schemes the selection of coarser levels is based on the principle that each variable of any level should have a sufficiently strong "total algebraic connection" to variables of the next-coarser level. The entire processing is thus made in terms of the given (sparse) algebraic system of equations, with no reference to their geometric origin. AMG algorithms can thus be used as very efficient "black box" solvers for important classes of matrix equations. (For some other classes the current AMG solvers are not suitable.) The typical multigrid efficiency is obtained by AMG even for cases where it would be very difficult to construct conventional (geometric) multigrid algorithms, such as cases of finite element discretisation on arbitrary, irregular triangulations, or even cases where topologically rectangular grids are used, but with highly and non-uniformly stretched coordinates (Lagrangian discretizations in particular) or with peculiarly distributed physical coefficients. In addition, AMG solvers can be applied to many large algebraic systems which are not at all derived from continuous problems, such as the geodetic problem treated in Brandt, McCormick and Ruge (1983) .

The scope of multi-level computations has thus been extended very much. To state it most broadly, consider any matrix equation $Ax = b$. (That the system is linear is not really essential; but it simplifies the following statements.) Denote by $\tilde{x}$ the evolving computed approximation to x, and define the error vector $e = x - \tilde{x}$ and the *normalized* residual vector $r = (r_i) = (a_i e/\|a_i\|)$, where a_i is the i-th row of A and $\|\cdot\|$ is the ℓ_2 norm. For a suitable relaxation scheme it can be shown (Brandt (1983), Theorem 3.4) that the decrease in $\|e\|$ per sweep can be slow only when $\|r\| << \|e\|$. Since r is properly normalized, for most error components $\|r\|$ is comparable to $\|e\|$. Hence, convergence can be slow only for *special* types of error components. Slowly converging errors can therefore be approximated by far fewer parameters, that is, by a much smaller algebraic system — a coarser level. To exploit this fact one of course needs some characterisation for those vectors e for which $\|r\| << \|e\|$. In case the matrix A approximates a differential operator L, those vectors e approximate functions v for which $\|Lv\| << |L|\|v\|$. This usually implies smoothness of v, or, when L is anisotropic, at least smoothness in characteristic directions. In other cases other characterisations can be found, so the general rule which emerges is that slow convergence should always be avoidable.

This corresponds, more or less, to the "golden rule of computational physics", which states that the amount of computational work should be proportional to the amount of real physical changes in the computed system. Stalling numerical processes must be wrong. Indeed, multi-level processing is a general vehicle to effect this rule. So, whenever you have stalling computations — either in the form of slowly converging iterative procedures, or in the form of computational grids, in space and/ or in time, which almost everywhere tend to excessively over-resolve the scales of real physical changes — try to think in terms of multi-level techniques.

Multi-level methods are now in the process of being introduced into a variety of new fields, including various systems of tomography, image processing and pattern recognition; statistical physics; queueing theory; network simulation and design; geodesy; multivariate interpolation; large transportation problems and linear programming. The aggregation/ disaggregation methods developed earlier for linear programming can now be improved by an AMG-type approach, because the latter provides a more mathematical basis for defining the levels, hence "tighter" hierarchies, exploiting implicit levels not necessarily recognized by the real-life systems. Applications to image processing and pattern recognition are in a sense not quite new, either. There is a strong evidence (see Campbell and Robson (1968), Terzopoulos (1963), Wilson and Bergen (1979) that the human vision processes themselves, in our brain, are multi-levelled. Thus indeed, when you open your eyes, you see multi-level processes all around — social, biological as well as physical (hierarchical structure of matter — resulting from some primordial evolution toward "efficiency"?), and your seeing itself is multi-levelled. Which is fine, of *coarse*.

REFERENCES

Aho, A.V., Hopcroft, J.E. and Ullmann, J.D., (1974) The Design and Analysis of Computer Algorithms. Addison Wesley.

Alcouffe, R.E., Brandt, A., Dendy Jr., J.E. and Painter, J.W. (1981) The multi-grid methods for the diffusion equation with strongly discontinuous coefficients. *SIAM J. Sci. Stat. Comput.*, **2**, 430-454.

Barkai, D. and Brandt, A., (1983) Vectorized multigrid Poisson solver for the CDC Cyber 205. *Applied Math. and Computations,* **13**, 215-229.

Brandt, A., (1984) Multigrid Techniques: 1984 Guide, with Applications to Fluid Dynamics. A monograph, 187 pages. Weizmann Institute of Science, Rehovot, Israel, February 1984. Appeared as Lecture Notes in Computational Fluid Dynamics, von Karman Institute for Fluid Dynamics, Rhode-Saint-Genèse, Belgium. Available as GMD-Studien Nr. 85 from GMD-FlT, Postfach 1240, D-5205, Sankt Augustin 1, W. Germany.

Brandt, A., (1983) Algebraic multigrid theory: The symmetric case. Preliminary Proc. Int. Multigrid Conf., Copper Mountain, Colorado, April.

Brandt, A., McCormick, S. and Ruge, J., (1984) Algebraic multigrid (AMG) for sparse matrix equations. "Sparsity and Its Applications" (D.J. Evans, ed.), Cambridge University Press.

Brandt, A., McCormick, S. and Ruge, J., (1983) Algebraic multigrid (AMG) for automatic multigrid solution with application to geodetic computations. Report, Colorado State University, Fort Collins, Colorado.

Buzbee, B.L., Golub, G.H. and Neilson, C.W., (1970) On direct methods for solving Poisson's equations. *SIAM J. Num. Anal.*, **7**, 627-656.

Campbell, F.W.C. and Robson, J. (1968) Application of Fourier Analysis to the visibility of gratings. *J. Physiol. (Lond.)* **197**, 551-566.

Dudkin, L.M. and Yershov, E.B., (1965) Interindustries input-output models and the material balances of separate products. *Planned Economy,* **5**, 54-63.

George, J.A. and Lin, J.W.H., (1981) Computer Solution of Large Sparse Positive Definite Systems. Prentice Hall, Englewood Cliffs, N.J.

Hackbusch, W. and Trottenberg, U., (eds.) (1982) Multigrid Methods, Lecture Notes in Math. **960**, Springer-Verlag.

Hockney, R.W., (1970) The potential calculation and some applications. *Meth. in Comput. Phys.* **9**, 135-211.

Proskurowski, W., (1981) Capacitance matrix solvers — a brief survey "Elliptic Problem Solvers" (M. Schultz, Ed.), Academic Press, 391-398.

Ruge, J. and Stüben, K., (1984) Efficient solution of finite difference and finite element equations by algebraic multigrid (AMG). This proceedings.

Stone, H.S., (1972) Introduction to Computer Organization and Data Structures, McGraw-Hill, New York.

Swann, G.H., (1978) Top-Down Structured Design Techniques. Petrocelli Books.

Temperton, C., (1979) Direct methods for the solution of the discrete Poisson equation: some comparisons. *J. Comp. Phys.*, **31**, 1-20.

Terzopoulos, D., (1983) Multilevel computational processes for visual surface reconstruction. *Computer Vision, Graphics and Image Processing* **24**, 52-96.

Trottenberg, U., (1977) Reduction methods for solving discrete elliptic boundary value problems — an approach in matrix terminology. "Fast Elliptic Solvers" (U. Schumann, ed.). Advanced Publications, London.

Vakhutinsky, I.Y., Dudkin, L.M. and Ryvkin, A.A., (1979) Iterative aggregation — a new approach to the solution of large-scale problems. *Econometrica*, **47**, 821-841.

Wilson, H.R. and Bergen, J.R., (1979) A four mechanism model for spatial vision. *Vision Res.* **19**, 19-32.

Yourdon, E., Structured Design. Yourdon, Inc.

MULTIGRID METHODS OF THE SECOND KIND

W. Hackbusch
(Christian-Albrechts-Universität Kiel, Germany)

1. INTRODUCTION

1.1 Summary

Fredholm integral equations of the second kind,

$$u = Ku + f, \quad (K : \text{integral operator}) \tag{1.1.1}$$

give rise to large linear systems

$$u_h = K_h u_h + f_h \quad (h : \text{discretisation parameter}) \tag{1.1.2}$$

with full $n_h \times n_h$ matrices K_h. Direct methods such as Gaussian elimination need $O(n_h^3)$ operations, whereas the 'multigrid iteration of the second kind', described in this paper, requires only Cn_h^2 operations with relatively small constant C.

The characteristic feature of the multigrid iteration is the simultaneous use of discrete equations (1.1.2) for different parameters h, 2h, 4h, etc. The linear multigrid process and variants are developed in §2. Numerical results are given for genuine integral equations as well as the integral equation method applied to elliptic boundary value problems.

The convergence analysis as well as the numerical results show that the multigrid iteration of the second kind is even faster than multigrid methods developed for elliptic boundary value problems. In the latter case the convergence rate is bounded by a constant, whereas in the former convergence is the faster the smaller is the discretisation parameter h.

It is an important characteristic of multigrid methods that they can be extended to nonlinear equations

$$u_h = \mathcal{K}_h(u_h), \quad (\mathcal{K}_h : \text{nonlinear}). \tag{1.1.3}$$

The nonlinear multigrid method is described and analysed in §3.

In the case of linear integral equations, the term $K_h u_h$ describes the evaluation of the integral and is obtained from the usual matrix multiplication of K_h times u_h. The algorithm requires nothing more than the evaluation of $K_h u_h$; in particular, the entries of K_h are not needed. Therefore, the multigrid methods of the second kind apply to any linear (or nonlinear) equation (1.1.2) of the second kind, provided that $u_h \to K_h u_h$ can be evaluated and K_h satisfy some theoretical properties. For instance, $u_h \to K_h u_h$ may involve the solution of an elliptic boundary value problem or a parabolic initial-boundary value problem, etc. Accordingly, we report results on a variety of quite different problems: nonlinear elliptic problems (§3.3), elliptic eigenvalue problems (§3.5.2) (e.g. the Steklov eigenvalue problem (§3.5.3)), elliptic control problems (§3.5.1), parabolic boundary control problems (§3.6.2), time-periodic parabolic equations (§3.6.1), and integro-differential equations (§3.5.4).

1.2 Contents

This paper is organised as follows.

1.3 Historical Comments

Multigrid methods were first developed for elliptic boundary value problems (for references see Brandt (1977), Hackbusch and Trottenberg (1982)). However, a two-grid iteration closely related to a multigrid iteration for integral equations of the second kind (which we call multi-grid iteration of the second kind) is already described in Brakhage (1960). Abramov (1962) and Šišov (1962) also describe algorithms for integral eigenvalue problems that approach the two-grid idea. The later monograph of Lučka (1980) should also be mentioned. Atkinson (1973 and 1976) took up the idea of Brakhage and developed automatic programs for linear integral equations.

Independently of the author, P. Hemker and H. Schippers formulated the multigrid iteration of the second kind (Schippers (1979), Hemker and Schippers (1981) and Wolff (1979)). The author's first report (1978) on this subject appeared in 1981 (Hackbusch (1981b)). The extension to nonlinear problems is almost of the same kind as used for the elliptic multigrid methods (Brandt (1977)).

Important applications to problems in fluid mechanics are described in the thesis of Schippers (1982c), which contains his papers (Schippers (1980a), (1980b), (1982a) and (1982b)).

1.4 Notation

The norm of a Banach space U is denoted by $\|\cdot\|_U$. In §2.2 we shall introduce finitely dimensional spaces U_ℓ ($\ell = 0,1,\ldots$). Their norms are also written $\|\cdot\|_U$, to avoid the notation $\|\cdot\|_{U_\ell}$. Accordingly, the norms of V, V_ℓ ($\ell = 0,1,\ldots$) are denoted by $\|\cdot\|_V$. We shall consider Hölder spaces $C^s(D)$ and Sobolev spaces $H^s(D)$. For grid functions defined on a grid $D_\ell \subset D$ there are discrete counterparts denoted by $C^s(D_\ell)$ (cf. §2.2) and $H^s(D_\ell)$ (cf. §2.8.1).

The associated operator norm of a bounded linear mapping $K : U \to V$ is defined by

$$\|K\|_{V \leftarrow U} := \sup \{\|Ku\|_V / \|u\|_U : 0 \neq u \in U\} .$$

Analogously, $\|K\|_{U \leftarrow U}$ is the operator norm of $K : U \to U$. The same notation is used for mappings $K_\ell : U_\ell \to V_\ell$ or $K_\ell : U_\ell \to U_\ell$ respectively.

2. MULTIGRID METHODS FOR INTEGRAL EQUATIONS

2.1 Fredholm's Integral Equation of the Second Kind

Fredholm's integral equation of the second kind is

$$u(x) = \int_D k(x,y)u(y)dy + f(x) \text{ for all } x \in D \qquad (2.1.1)$$

with given kernel function k defined on D × D and given right hand side f. A short notation for Eq. (2.1.1) is

$$u = Ku + f \ , \qquad (2.1.2)$$

where $(Ku)(x) = \int_D k(x,y)u(y)dy$ defines the integral operator K.

Let U be the underlying Banach space involved in (2.1.2): for $f \in U$ a solution of (2.1.2) is sought. Under the usual assumption that $K: U \to U$ is compact, the estimate

$$\|(I - K)^{-1}\|_{U \leftarrow U} \leqslant C \qquad (2.1.3)$$

holds, whenever $\lambda = 1$ is not an eigenvalue of K.

In the following, we assume that there is some Banach subspace $V \subset U$ such that image(K) $\subset V$, or

$$\|K\|_{V \leftarrow U} \leqslant C. \qquad (2.1.4)$$

The elements of V must have a higher order of differentiability than those of U, as in the following examples. Eq. (2.1.4) expresses the smoothing effect of K.

<u>Example 2.1.1</u> Suppose $k \in C^s(D \times D)$ with bounded D, where C^s is the Hölder space of order s (C^s is the space of s-times continuously differentiable functions for integral $s \geqslant 0$; if $s = n + t$, $n \in \mathbb{Z}$, $0 < t < 1$, the nth derivatives are Hölder continuous with exponent t). Then K satisfies (2.1.4) with $U = C^\circ(D)$, $V = C^s(D)$.

The assumption on k can be weakened. Weakly singular kernels are included, as shown:

<u>Example 2.1.2</u> For $D = [0,1]$ and $k(x,y) = 1/(x-y)^\lambda$ with $0 \leqslant \lambda \leqslant 1$, inequality (2.1.4) is fulfilled with $U = C^\circ(D)$ and $V = C^{1-\lambda}(D)$.

U and V may also be chosen as Sobolev spaces $U = H^s(D)$ and $V = H^t(D)$ with $t > s$.

<u>Example 2.1.3</u> Let $D = [0,1]$ and $k(x,y) = 1/\sqrt{2-x-y}$. Inequality (2.1.4) holds for all pairs $U = H^s(D)$, $V = H^{s+\frac{1}{2}}(D)$ with constants $C = C(s)$ as long as $-1 < s < \frac{1}{2}$, where $H^{-t}(D)$ is the dual of $H^t(D)$ for $t > 0$.

In all the previous examples V is compactly embedded in U and Note 2.1.4 applies.

Note 2.1.4 If (2.1.4) holds for V compactly embedded in U, the operator $K : U \to U$ is compact.

2.2 *Discretisation*

There are various kinds of discretisation methods, e.g. kernel approximations, projection methods (in particular Galerkin and collocation methods), and Nyström's method (Baker (1977)). Taking the trapezoidal rule as quadrature formula, we obtain the following very simple discretisation.

Example 2.2.1 (Nyström's method with repeated trapezoidal rule) Suppose $D = [0,1]$, $k \in C^0(D \times D)$ and

$$h_\ell = 1/n_\ell \, , \; n_\ell = 2^\ell n_0 \quad (\ell = 0,1,\ldots) \tag{2.2.1}$$

with some $n_0 \in \mathbb{N}$. The desired solution u_ℓ is a grid function defined on

$$D_\ell = \{0, h_\ell, 2h_\ell, \ldots, 1 - h_\ell, 1\} \tag{2.2.2}$$

satisfying the system

$$u_\ell(x) = h_\ell \sum_{y \in D_\ell} w(y) k(x,y) u_\ell(y) + f(x), \; x \in D_\ell \tag{2.2.3}$$

of $n_\ell + 1$ equations. The weights are $w(0) = w(1) = \frac{1}{2}$ and $w(y) = 1$ elsewhere.

The system (2.2.3) or any other discretisation with discretisation parameter h_ℓ will be denoted by

$$u_\ell = K_\ell u_\ell + f_\ell . \tag{2.2.4}$$

Grid functions u_ℓ and f_ℓ belong to U_ℓ, which is the discrete analogue of U. Similarly, a discrete counterpart V_ℓ of V is needed. For instance, the discrete space $V_\ell = C^s(D_\ell)$ corresponding to $C^s(D)$ has the norm

$$\|v_\ell\|_{C^0(D_\ell)} = \max_{x \in D_\ell} |v_\ell(x)| \qquad \text{if } s = 0,$$

$$\|v_\ell\|_{C^s(D_\ell)} = \max \{\|v_\ell\|_{C^0(D_\ell)}, \; |v_\ell(x) - v_\ell(x')|/|x-x'|^s : x,x' \in D_\ell, \; x \neq x'\}$$

$$\text{if } 0 < s \leq 1,$$

etc. The norm of $U_\ell (V_\ell)$ is written as $\|\cdot\|_U$ ($\|\cdot\|_V$). They give rise to matrix norms $\|\cdot\|_{U \leftarrow U}$, $\|\cdot\|_{V \leftarrow U}$, etc.

Under usual assumptions (Anselone (1971)) the system (2.2.4) is solvable if h_ℓ is sufficiently small. Further, $(I - K_\ell)^{-1}$ is uniformly bounded. The counterpart of (2.1.3) is the stability assumption

$$\|(I - K_\ell)^{-1}\|_{U \leftarrow U} \leq C_S \qquad (\ell = 0,1,\ldots) \qquad (2.2.5)$$

with C_S independent of ℓ.

The analogue of (2.1.4) is the regularity assumption

$$\|K_\ell\|_{V \leftarrow U} \leq C_R \qquad (\ell = 0,1,\ldots). \qquad (2.2.6)$$

Example 2.2.2 Under the conditions of Example 2.1.1, the discretisation of Example 2.2.1 satisfies the regularity assumption (2.2.6) with $C_R = \max\{\|k(\cdot,y)\|_{C^s([0,1])} : 0 \leq y \leq 1\}$.

Instead of Note 2.1.4 we have

Note 2.2.3 Assume (2.1.3) and (2.2.6), and let V be compactly embedded in U. Suppose that there are interpolations $P_\ell : V_\ell \to V$ and restrictions $R_\ell : U \to U_\ell$ with $\|P_\ell\|_{V \leftarrow V} \leq C$, $\|R_\ell\|_{U \leftarrow U} \leq C$, satisfying $R_\ell P_\ell = I$. Under the consistency condition

$$\|(P_\ell K_\ell R_\ell - K)v\|_U \to 0 \ (\ell \to \infty) \text{ for all } v \in U,$$

the stability condition (2.2.5) follows for sufficiently large ℓ.

A restriction

$$r: U_\ell \to U_{\ell-1} \qquad (2.2.7)$$

is needed in the multigrid process. The restriction r may be weighted, but whenever $D_{\ell-1} \subset D_\ell$ and $U_\ell = C^s(D_\ell)$, $V_\ell = C^t(D_\ell)$, the trivial injection $r = r_{inj}$,

$$(r_{inj}\, u_\ell)(x) = u_\ell(x), \qquad x \in D_{\ell-1}, \qquad (2.2.8)$$

can be used.

The relative consistency condition states that the expression $(r\,K_\ell - K_{\ell-1} r) v_\ell$ is small for smooth v_ℓ; more precisely:

$$\|r\,K_\ell - K_{\ell-1} r\|_{U \leftarrow V} \leqslant C_C\, h_\ell^{\kappa} \tag{2.2.9}$$

where κ is called the consistency order. κ depends on the discretisation (e.g. on the order of the quadrature formula) and on the choice of U and V.

Example 2.2.4 Let K_ℓ be the discretisation of Example 2.2.1, assume $\|k(\cdot,y)\|_{C^\alpha([0,1])} \leqslant C$ and choose $r = r_{inj}$. Then for all $s \geqslant 0$, $0 \leqslant t \leqslant \alpha$ the estimate

$$\|r\,K_\ell - K_{\ell-1} r\|_{C^t(D_{\ell-1}) \leftarrow C^s(D_\ell)} \leqslant C(t,s) h_\ell^{\min(2,\alpha,s)}$$

holds. The appropriate choice of U_ℓ and V_ℓ will be $U_\ell = C^0(D_\ell)$ and $V_\ell = C^{\min(2,\alpha)}(D_\ell)$ corresponding to $t = 0$, $s = \kappa = \min(2,\alpha)$.

The multigrid iteration requires a prolongation

$$p : U_{\ell-1} \rightarrow U_\ell, \tag{2.2.10}$$

which is assumed to be U-stable:

$$\|p\|_{U \leftarrow U} \leqslant C_p . \tag{2.2.11}$$

The interpolation error of p should be small enough to satisfy

$$\|I - p\,r\|_{U \leftarrow V} \leqslant C_I\, h_\ell^{\kappa} . \tag{2.2.12}$$

Example 2.2.5 Let $r = r_{inj}$, $h_{\ell-1} = 2h_\ell$, p be a piecewise linear interpolation operator, $U_\ell = C^0(D_\ell)$, $V_\ell = C^\alpha(D_\ell)$. Then (2.2.11) holds with $C_p = 1$, while the interpolation error is $\|I - p\,r\|_{U \leftarrow V} \leqslant C_I(\alpha) h_\ell^{\min(\alpha,2)}$ with $C_I(\alpha) = 1$ for $0 \leqslant \alpha \leqslant 1$, $C_I(\alpha) = \frac{1}{2}$ for $\alpha > 1$, $\kappa = \min(2,\alpha)$.

Roughly speaking, the maximal κ satisfying (2.2.9) and (2.2.12) is the minimum of the interpolation order of p, the order of the quadrature method and the differentiability order of k.

We conclude with estimates of the relative discretisation errors (error of $u_{\ell-1}$ relative to u_ℓ).

<u>Proposition 2.2.6</u> Let $f_{\ell-1} = rf_\ell$. Regularity (2.2.6) and consistency (2.2.9) imply

$$\|ru_\ell - u_{\ell-1}\|_U \leq C_R C_C h_\ell^\kappa \|u_\ell\|_V . \qquad (2.2.13)$$

The similar statement

$$\|u_\ell - pu_{\ell-1}\|_U \leq (C_p C_R C_C + C_I) h_\ell^\kappa \|u_\ell\|_V \qquad (2.2.13')$$

requires (2.2.11) and (2.2.12). We can estimate $\|u_\ell\|_V$ from

$$\|u_\ell\|_V \leq C_R C_S \|f_\ell\|_U + \|f_\ell\|_V \leq (C_R C_S \|I\|_{U \leftarrow V} + 1)\|f_\ell\|_V . \qquad (2.2.14)$$

<u>Proof</u> Use $u_{\ell-1} - ru_\ell = (I - K_{\ell-1})^{-1}(rK_\ell - K_{\ell-1}r)u_\ell$ and $pu_{\ell-1} - u_\ell = p(u_{\ell-1} - ru_\ell) - (I - pr)u_\ell$. □

2.3 *Multigrid Iteration (First Variant)*

<u>2.3.1 Algorithm</u>

Formally, the multigrid iteration is the same as for elliptic problems $L_\ell u_\ell = f_\ell$ with L_ℓ replaced by $I - K_\ell$. However, the smoothing step is much simpler than in the elliptic case, where a careful choice of the smoothing iteration is essential. Here we use Picard's iteration $u_\ell \to K_\ell u_\ell + f_\ell$ and perform only one ($\nu = 1$) iteration step. The choice $\nu \geq 2$ is possible (cf. §2.8.1), but has a quite different justification than in the elliptic case. The convergence of the Picard iteration (i.e. $\rho(K_\ell) < 1$) is not required.

Two-grid iteration (TGM) for solving $u_\ell = K_\ell u_\ell + f_\ell$	(2.3.1)
$u'_\ell := K_\ell u_\ell^j + f_\ell$ (smoothing step)	(2.3.1a)
$u_\ell^{j+1} := u'_\ell - p(I - K_{\ell-1})^{-1} r(u'_\ell - K_\ell u'_\ell - f_\ell)$ (coarse-grid correction)	(2.3.1b)

Here r is the restriction (2.2.7) and p the prolongation (2.2.10). By (2.3.1a), the error $u_\ell^j - u_\ell$ (u_ℓ: solution of (2.2.4)) of the jth iterate is mapped into $u'_\ell - u_\ell = K_\ell(u^j_\ell - u_\ell)$. Because of the regularity assumption (2.2.6), this error is smoothed. Replacing the exact solution $(I - K_{\ell-1})^{-1}d_{\ell-1}$, $d_{\ell-1} := r(u'_\ell - K_\ell u'_\ell - f_\ell)$ in (2.3.1b) by two steps of the same iteration at level $\ell-1$, we obtain the first variant of the multigrid iteration. It is written as a quasi-ALGOL procedure. The second parameter u is input as u^j_ℓ and afterwards output as u^{j+1}_ℓ, while $f = f_\ell$ is the inhomogeneous term of Eq. (2.2.4). Therefore, the procedure performs one step of the multigrid iteration at level ℓ.

Multigrid iteration (MGM′) for solving $u_\ell = K_\ell u_\ell + f_\ell$ (2.3.2)

```
procedure MGM (ℓ,u,f); integer ℓ; array u,f;                 (2.3.2a)

if ℓ = 0 then u := (I - K_0)^-1 f else

begin integer i; array d,v;

        u := K_ℓ*u + f;                                        (2.3.2b)

        d := r(u - K_ℓ*u - f);                                 (2.3.2c_1)

        v := 0; for i := 1,2 do MGM (ℓ- 1,v,d);                (2.3.2c_2)

        u := u - p*v                                           (2.3.2c_3)

end;
```

Algorithm (2.3.2) is a so-called 'W-cycle' since two iterations are performed in $(2.3.2c_2)$. Three iterations yield no further improvement, while one iteration does not give the convergence results described in §2.3.3.

2.3.2 Computational Work

The work of one multigrid iteration is dominated by the multiplication $K_\ell * u_\ell$, since K_ℓ is a full matrix. Therefore, the work taken by r, p and vector additions can be neglected. Assume that there are numbers $n_\ell = \dim(U_\ell) + O(1)$ (cf. Example 2.2.1) such that

$$u_\ell \mapsto K_\ell u_\ell \text{ requires } \le C_K n_\ell^2 + O(n_\ell) \text{ operations} \tag{2.3.3a}$$

$$u_\ell \mapsto rK_\ell u_\ell \text{ requires} \le C_{rK} n_\ell^2 + O(n_\ell) \text{ operations} \tag{2.3.3b}$$

while

$$n_{\ell-1} \leqslant C_N\, n_\ell. \tag{2.3.4}$$

Note 2.3.1 For a usual matrix multiplication the constant in (2.3.3a) is given by $C_K = 2$ (1 addition, 1 multiplication). In the case of a weighted restriction r the constant C_{rK} equals C_K. If r is the trivial injection r_{inj} (cf. (2.2.8)), $K_\ell u_\ell$ has to be evaluated only partially. One needs $2n_{\ell-1}n_\ell \leqslant 2C_N n_\ell^2$ operations; hence $C_{rK} = C_K C_N$. The standard choice (2.2.1) of n_ℓ satisfies (2.3.4) with $C_N = \frac{1}{2}$. Therefore, depending on r, (5a) or (5b) hold:

$$C_K = 2, \quad C_{rK} = 2, \quad C_N = \tfrac{1}{2}, \tag{2.3.5a}$$

$$C_K = 2, \quad C_{rK} = 1, \quad C_N = \tfrac{1}{2}. \tag{2.3.5b}$$

By the following observation one can save some matrix multiplications:

Note 2.3.2 The two multigrid iterations in $(2.3.2c_2)$ are started with $u_{\ell-1} = 0$. Whenever the starting iterate is zero, the smoothing step (2.3.2b) requires no matrix multiplication.

Thanks to Note 2.3.2, two multigrid iterations at level ℓ starting with $u_\ell^0 = 0$ require $C_\ell^*\, n_\ell^2 + O(n_\ell)$ operations, where C_ℓ^* satisfies the inequality $C_\ell^* \leqslant C_K + 2C_{rK} + 2C_N^2\, C_{\ell-1}^*$. Since the work of (2.3.2a) can be neglected, we obtain $C_\ell^* \leqslant (C_K + 2C_{rK})/(1 - 2C_N^2)$. One iteration (2.3.2) applied to $u_\ell^0 \neq 0$ takes $C'_{MGM}\, n_\ell^2 + O(n_\ell)$ operations, where $C'_{MGM} \leqslant C_K + C_{rK} + C_N^2 C_\ell^*$. This proves

Proposition 2.3.3 Suppose $C_N < \sqrt{\tfrac{1}{2}}$. One step of the multigrid iteration (2.3.2) requires $C'_{MGM}\, n_\ell^2 + O(n_\ell)$ operations, where

$$C'_{MGM} = [(1 - C_N^2)C_K + C_{rK}]/(1 - 2C_N^2). \tag{2.3.6}$$

The constants from (5a) and (5b) yield

$$C'_{MGM} = 7 \text{ (case of (5a))}, \quad C'_{MGM} = 5 \text{ (case of (5b)}. \tag{2.3.7}$$

The last number expresses that one multigrid iteration costs as much as $2\frac{1}{2}$ Picard iterations. A more precise analysis is needed to answer the following question. Let a fixed finest step size $h = 1/n$, $n = n_{\ell_{max}}$ be given. The choice of ℓ_{max} determines $n_o = n/2^{\ell_{max}}$. What is the optimal n_o? Taking into account the work of the exact solution in (2.3.2a), one finds $n_o \simeq \sqrt{3n/2}$. However, this choice gives only a very slight improvement: $C'_{MGM} = 5$ from (2.3.7) is replaced by $5 - \sqrt{6/n}$. The same calculation shows that the three-grid iteration ($\ell_{max} = 2$) is cheaper than the two-grid iteration ($\ell_{max} = 1$) as soon as $n > 10$.

2.3.3 Convergence Analysis

The two-grid iteration (2.3.1) can be written as

$$u_\ell^{j+1} = M_\ell^{TGM} u_\ell^j + N_\ell^{TGM} f_\ell$$

with the iteration matrix

$$M_\ell^{TGM} = [I - p(I - K_{\ell-1})^{-1} r(I - K_\ell)]K_\ell \; . \tag{2.3.8}$$

The iteration converges if and only if $\rho(M_\ell^{TGM}) < 1$ (ρ: spectral radius). As $\rho(A) \leqslant \|A\|_{U\leftarrow U}$ it suffices to prove that $\|M_\ell^{TGM}\|_{U\leftarrow U} < 1$.

Theorem 2.3.4 Assume (2.2.5), (2.2.6), (2.2.9), (2.2.11), (2.2.12) with some $\kappa > 0$. Then the contraction number of the two-grid iteration (2.3.1) with respect to $\|\cdot\|_u$ is given by

$$\|M_\ell^{TGM}\|_{U\leftarrow U} \leqslant C_{TGM}\, h_\ell^\kappa \tag{2.3.9}$$

with $\quad C_{TGM} \leqslant (C_I + C_p C_S C_C) C_R$.

Proof The splitting

$$M^{TGM} = \{(I-pr) + p(I-K_{\ell-1})^{-1}[(I-K_{\ell-1})r - r(I-K_\ell)]\}K_\ell$$

$$= [(I-pr) + p(I-K_{\ell-1})^{-1}(rK_\ell - K_{\ell-1}r)]K_\ell$$

implies

$$\|M_\ell^{TGM}\|_{U\leftarrow U} \leq [\|I-pr\|_{U\leftarrow V} + \|p\|_{U\leftarrow U}\|(I-K_{\ell-1})^{-1}\|_{U\leftarrow U}\|rK_\ell - K_{\ell-1}r\|_{U\leftarrow V}]\|K_\ell\|_{V\leftarrow U}$$

$$\leq (C_I + C_pC_SC_C)C_R\, h_\ell^\kappa . \qquad \square$$

Inequality (2.3.9) implies convergence for sufficiently small h_ℓ. Convergence cannot be guaranteed for all levels, since possibly $C_{TGM}\, h_1^\kappa \geq 1$.

The multigrid iteration (2.3.2) can be represented by

$$u_\ell^{j+1} = M'_\ell u_\ell^j + N'_\ell f_\ell .$$

Standard arguments (Hackbusch (1982), p. 202) show:

Lemma 2.3.5 The iteration matrix M'_ℓ of the multigrid algorithm (2.3.2) is recursively defined by $M'_1 = M_1^{TGM}$ and

$$M'_\ell = [I - p(I-M'^{\,2}_{\ell-1})(I-K_{\ell-1})^{-1}r(I-K_\ell)]K_\ell \qquad (2.3.10)$$

$$= M_\ell^{TGM} + p\, M'^{\,2}_{\ell-1}(I-K_{\ell-1})^{-1}r(I-K_\ell)K_\ell \qquad (2.3.10')$$

$$= M_\ell^{TGM} + p\, M'^{\,2}_{\ell-1}[r - (I-K_{\ell-1})^{-1}(rK_\ell - K_{\ell-1}r)]K_\ell \qquad (2.3.10'')$$

Let r be bounded by

$$\|r\|_{U\leftarrow U} \leq C_r . \qquad (2.3.11)$$

In all previous examples $C_r = 1$ holds. Usually, the estimate $\|K_\ell\|_{U\leftarrow U} \leq \|K_\ell\|_{V\leftarrow U} \leq C_R$ is valid. We assume

$$\|K_\ell\|_{U\leftarrow U} \leq C_K . \qquad (2.3.12)$$

C_K may be small or large depending on the problem. Eq. (2.3.10") yields the inequality

$$\|M'_\ell\|_{U\leftarrow U} \leqslant \|M_\ell^{TGM}\|_{U\leftarrow U} + C_p\|M'_{\ell-1}\|^2_{U\leftarrow U}[C_rC_K + C_SC_CC_Rh_\ell^\kappa]. \tag{2.3.13}$$

If h_1 is small enough there exists C* with

$$C_pC_{TGM}(1 + C^*h_{\ell-1}^\kappa)^2[C_rC_K + C_SC_CC_Rh_\ell^\kappa] \leqslant C^* \quad \text{for } \ell \geqslant 2. \tag{2.3.14}$$

This proves

Theorem 2.3.6 (Convergence of the multigrid iteration). Assume (2.2.5), (2.2.6), (2.2.9), (2.2.11), (2.2.12) with $\kappa > 0$, (2.3.11), and (2.3.12). Let h_1 be sufficiently small. Then C* exists such that

$$\|M'_\ell\|_{U\leftarrow U} \leqslant C_{TGM}\,(1 + C^*h_\ell^\kappa)h_\ell^\kappa\;. \tag{2.3.15}$$

Hence, asymptotically, two-grid and multigrid iterations have the same contraction numbers. However, if h_1 is too large, there is no C* satisfying (2.3.14), and $\|M'_\ell\|_{U\leftarrow U}$ diverges rapidly:

$$\|M'_\ell\|_{U\leftarrow U} = O([C_pC_rC_K]^{2^\ell-1}) \to \infty.$$

To gain a better understanding we shall suppose rp = I, which holds for $r = r_{inj}$ (cf. (2.2.8)) whenever p is an interpolation.

Lemma 2.3.7 If rp = I, a further representation of M'_ℓ is

$$M'_\ell = M_\ell^{TGM} + p\, M'^{\,2}_{\ell-1}\, r\, (M_\ell^{TGM} - K_\ell). \tag{2.3.16}$$

For the usual values $C_p = C_r = 1$ Eq. (2.3.16) yields

$$\|M'_\ell\|_{U\leftarrow U} \leqslant \|M_\ell^{TGM}\|_{U\leftarrow U} + \|M'_{\ell-1}\|^2_{U\leftarrow U}(\|M_\ell^{TGM}\|_{U\leftarrow U} + C_K) \tag{2.3.17}$$

with C_K from (2.3.12).

Note 2.3.8 Assume rp = I, $C_r = C_p = 1$, and define

$$\zeta = \sup\{\|M_\ell^{TGM}\|_{U\leftarrow U} : \ell \geqslant 1\} \text{ (usually } \zeta = \|M_1^{TGM}\|_{U\leftarrow U}).$$

The multigrid convergence, or more precisely, the inequalities $\|M'_\ell\|_{U\leftarrow U} < 2\zeta < 1$ and (2.3.15) are ensured by the condition

$$4\zeta(\zeta + C_K) \leqslant 1. \tag{2.3.18}$$

Obviously, it is hard to satisfy this condition if $C_K >> 1$.

2.4 Second Variant of the Multigrid Iteration

2.4.1 Algorithm

Hemker and Schippers (1981) proposed a modification of the multigrid iteration which is applicable also for large C_K. It is based on the equivalent formulation of the coarse-grid correction in the two-grid iteration (2.3.1) by

$$u_\ell^{j+1} = u'_\ell + pr(u_\ell^j - u'_\ell) - p(I - K_{\ell-1})^{-1}[(I - K_{\ell-1})r(u_\ell^j - u'_\ell) + rd_\ell], \tag{2.4.1}$$

where $d_\ell = u'_\ell - K_\ell u'_\ell - f_\ell$. The restriction r in front of $u_\ell^j - u'_\ell$ may also be replaced by another $\hat{r} \neq r$. The evaluation of the square bracket requires a further multiplication by $K_{\ell-1}$.

Although (2.4.1) and (2.3.1b) are equivalent, the multigrid versions differ.

Multigrid iteration MGM" for solving $u_\ell = K_\ell u_\ell + f_\ell$ (2.4.2)

```
procedure MGM(ℓ,u,f); integer ℓ; array u,f;
if ℓ = 0 then u := (I - K0)^-1 f else                      (2.4.2a)
begin integer i; array v,d;
      d := u - Kℓ*u - f;                                    (2.4.2b1)
      u := u - (I - pr)d;                                   (2.4.2b2)
      d := ((I - Kℓ-1)*r - r*Kℓ)*d;                         (2.4.2b3)
      v := 0; for i := 1,2 do MGM(ℓ-1,v,d);                 (2.4.2b4)
      u := u - p*v                                          (2.4.2b5)
end;
```

The smoothing step is hidden in statement (2.4.2b3)

2.4.2 Computational Work

Note 2.3.2 remains valid. Instead of Proposition 2.3.3 we obtain

Proposition 2.4.1 Suppose $C_N < \sqrt{1/2}$. One step of the multigrid iteration (2.4.2) requires $C''_{MGM} n_\ell^2 + O(n_\ell)$ operations, where

$$C''_{MGM} = (C_K + C_{rK})/(1 - 2C_N^2) \tag{2.4.3}$$

where C_K, C_{rK} from (2.3.3). The constants (2.3.5a) or (2.3.5b) yield

$$C''_{MGM} = 8 \text{ (case (2.3.5a))}, \quad C''_{MGM} = 6 \text{ (case (2.3.5b))}. \tag{2.4.4}$$

Note 2.4.2 In special cases the work can be less than indicated by (2.4.4). Assume (2.3.5b). If $r = r_{inj}$ and if K_ℓ is obtained by the trapezoidal rule (2.2.3), the components of $rK_\ell v_\ell$ are the sum of $\frac{1}{2}K_{\ell-1}rv_\ell$ and $n_\ell/2$ further terms. Therefore, the evaluation of $(K_{\ell-1}r - rK_\ell)v_\ell$ in (2.4.2b$_3$) requires only $2n_{\ell-1}n_\ell \leqslant 2C_N n_\ell^2$ operations. One concludes that $C''_{MGM} = 5$, as in (2.3.7).

2.4.3 Convergence Analysis

Again, the two-grid iteration matrix M_ℓ^{TGM} is given by (2.3.8), whereas the multigrid iteration matrix M''_ℓ of algorithm (2.4.2) equals $M''_1 = M_1^{TGM}$ for $\ell = 1$ and

$$\begin{aligned} M''_\ell &= I - \{I - pr + p(I - M''^2_{\ell-1})(I - K_{\ell-1})^{-1}[(I - K_{\ell-1})r - rK_\ell]\}(I - K_\ell) \\ &= M_\ell^{TGM} + pM''^2_{\ell-1}[r - (I - K_{\ell-1})^{-1}(rK_\ell - K_{\ell-1}r)K_\ell] \end{aligned} \tag{2.4.5}$$

for $\ell \geqslant 2$. Note that (2.4.5) differs from (2.3.10") only in that the initial term r in the square brackets has replaced rK_ℓ in (2.3.10"). Instead of (2.3.13) one obtains

$$\|M''_\ell\|_{U\leftarrow U} \leqslant \|M_\ell^{TGM}\|_{U\leftarrow U} + C_p\|M''_{\ell-1}\|^2_{U\to U}(C_r + C_S C_C C_R h_\ell^\kappa). \tag{2.4.6}$$

Theorem 2.4.3 Assume (2.2.5), (2.2.6), (2.2.9), (2.2.11), (2.2.12) with $\kappa > 0$, and (2.3.11). Let h_1 be sufficiently small. Then C* exists such that (2.3.15) holds for M''_ℓ. In contrast to Theorem 2.3.6 the upper bound of h_1 and the constant C* do not depend on C_K (cf. (2.3.12)).

The situation becomes simpler in

Corollary 2.4.4 If $rp = I$, a further representation of M''_ℓ is

$$M''_\ell = M_\ell^{TGM} + pM''^{2}_{\ell-1} r (M_\ell^{TGM} - I) . \tag{2.4.7}$$

If in addition $C_p = C_r = 1$, the multigrid iteration converges when

$\|M_\ell^{TGM}\|_{U\leftarrow U} \lesssim 1/[2(\sqrt{2} + 1)] \approx 0.207$ for $\ell \geq 1$.

Proof M''_ℓ satisfies (2.3.17) with $C_K := 1$. Note 2.3.8 with $C_K = 1$ yields the desired result. An analysis of these and further versions are given in Mandel (1983). □

2.5 Third Variant of the Multigrid Iteration

2.5.1 Algorithm

The difficulty of the first variant was due to the presence of the factor K_ℓ in the second term of (2.3.10"). This factor is not avoided in the second version but the dominant term rK_ℓ from (2.3.10") is replaced by r in (2.4.5). In the case of the first variant (2.3.2) the multiplication by K_ℓ yields a smoothing which has not been exploited hitherto. Since the defect d from (2.3.2c_1) is already smooth, the first of the two iterations (2.3.2c_2) can be performed without the smoothing step (2.3.2b). Therefore, the resulting third variant is even cheaper than the first one. The resulting procedure MGM" has a further parameter sm. If its value is true (false), the smoothing is (not) performed. One step $u_\ell^j \rightarrow u_\ell^{j+1}$ is defined by means of MGM(.,., true). The value sm = false is used only inside the algorithm at lower levels.

Multigrid iteration MGM''' for solving $u_\ell = K_\ell u_\ell + f_\ell$ (2.5.1)

```
procedure MGM (ℓ, u, f, sm); integer ℓ; array u,f; boolean sm;
if ℓ = 0 then u:= (I - K_0)^-1 f else                  (2.5.1a)
begin array v, d;
      if sm then u:= K_ℓ*u + f;                         (2.5.1b)
      d:= u - K_ℓ*u - f;                                (2.5.1c_1)
      v:= 0; MGM(ℓ-1,v,d,false); MGM(ℓ-1,v,d,true);     (2.5.1c_2)
      u:= u - p*v                                       (2.5.1c_3)
end;
```

2.5.2 Computational Work

Proposition 2.5.1 Suppose $C_N < \sqrt{1/2}$. One step of the multi-grid iteration (2.5.1) (with sm = true) requires $C'''_{MGM}\, n_\ell^2 + O(n_\ell)$ operations, where

$$C'''_{MGM} = (C_K + C_{rK})(1 - C_N^2)/(1 - 2C_N^2) \tag{2.5.2}$$

(cf. (2.3.3), (2.3.4)). The constants (2.3.5a) or (2.3.5b) yield

$$C'''_{MGM} = 6 \text{ (case (2.3.5a))}, \quad C'''_{MGM} = 4.5 \text{ (case (2.3.5b))}. \tag{2.5.3}$$

2.5.3 Convergence Analysis

The two-grid iteration matrix (for all variants) can be written as $M_\ell^{TGM} = A_\ell^{TGM} K_\ell$ with

$$A_\ell^{TGM} = I - pr + p(I - K_{\ell-1})^{-1}(rK_\ell - K_{\ell-1}r)$$

(cf. (2.3.8)). The proof of Theorem 2.3.4 showed

$$\| A_\ell^{TGM} \|_{U\leftarrow V} \leqslant (C_I + C_p C_S C_C) h_\ell^\kappa = \alpha_\ell^{TGM}. \tag{2.5.4}$$

The iteration matrix of algorithm (2.5.1) is $M'''_\ell = A'''_\ell K_\ell$ with

$$\begin{aligned} A'''_\ell &= A_\ell^{TGM} + pA'''_{\ell-1} K_{\ell-1} A'''_{\ell-1}(I - K_{\ell-1})^{-1} r(I - K_\ell) \\ &= A_\ell^{TGM} + pA'''_{\ell-1} K_{\ell-1} A'''_{\ell-1}[r - (I - K_{\ell-1})^{-1}(rK_\ell - K_{\ell-1}r)] \end{aligned} \tag{2.5.5}$$

(cf. (2.3.10') and (2.3.10")). We need an estimate of $\| (I - K_{\ell-1})^{-1} r(I - K_\ell) \|_{V\leftarrow V}$.

Lemma 2.5.2 The constant $C_S^* := C_R C_S \| I \|_{U\leftarrow V} + 1$ satisfies

$$\| (I - K_\ell)^{-1} \|_{V\leftarrow V} \leqslant C_S^*. \tag{2.5.6}$$

Proof. Use (2.2.14). □

Lemma 2.5.3 The inequalities (2.5.6) and (2.5.7),

$$\|r\|_{V\leftarrow V} \leq C_r^*, \tag{2.5.7}$$

imply $\|(I - K_{\ell-1})^{-1} r(I - K_\ell)\|_{V\leftarrow V} \leq \text{const.}$

$(I - K_{\ell-1})^{-1} r(I - K_\ell)$ equals $r - (I - K_{\ell-1})^{-1}(rK_\ell - K_{\ell-1}r)$. A natural estimate of the latter difference is

$$\|rK_\ell - K_{\ell-1}\, r\|_{V\leftarrow V} \leq C_C^*\, h_\ell^\kappa \tag{2.5.8}$$

(cf. Example 2.2.4 with $V_\ell = C^\kappa(D_\ell)$).

Lemma 2.5.4 (2.5.6), (2.5.7) and (2.5.8) imply

$$\|(I - K_\ell)^{-1} r(I - K_\ell)\|_{V\leftarrow V} \leq C_r^* + C_S^*\, C_C^*\, h_\ell^\kappa \;. \tag{2.5.9}$$

Set β_ℓ := const in the case of Lemma 2.5.2 or $\beta_\ell := C_r^* + C_S^*\, C_C^*\, h_\ell^\kappa$ in the case of the Lemma 2.5.4. The norm $\alpha_\ell := \|A_\ell'''\|_{U\leftarrow V}$ can be estimated by

$$\alpha_\ell \leq \alpha_\ell^{TGM} + C_p\, \alpha_{\ell-1}^2\, C_R\, \beta_\ell.$$

Thus we obtain the inequality

$$\|M_\ell'''\|_{U\leftarrow U} \leq \|A_\ell'''\|_{U\leftarrow V}\, \|K_\ell\|_{V\leftarrow U} \leq \alpha_\ell C_R =: \zeta_\ell$$

with

$$\zeta_\ell \leq \zeta_\ell^{TGM} + C_p\, \zeta_{\ell-1}^2 \beta_\ell, \quad \zeta_\ell^{TGM} := \alpha_\ell^{TGM}\, C_R, \quad \alpha_\ell^{TGM} \text{ from (2.5.4).}$$

This recursion formula is similar to (2.4.6), if β_ℓ is defined by the right side of (2.5.9). Therefore, this variant (2.5.1) as well as the second variant are both applicable for large values of $C_K = \sup\{\|K_\ell\|_{U\leftarrow U} : \ell \geq 1\}$.

2.6 Numerical Results

2.6.1 Comparison of the three variants

Consider the integral equation

$$u(x) = \int_0^1 \lambda \cos(\pi xs)\, u(s)ds + f(x), \quad x \in D = [0,1], \qquad (2.6.1)$$

with f such that $u(x) = e^x\cos(7x)$ is a solution (Atkinson (1973 and 1976), Hemker - Schippers (1981)). The parameter λ will be chosen as 1, 10 and 100. We discretise Eq. (2.6.1) by the trapezoidal rule of Example 2.2.1 and use the step sizes $h_\ell = 1/n_\ell$ of (2.2.1). Since this quadrature formula is of second order, the appropriate prolongation p is the piecewise linear interpolation

$$(p\, v_{\ell-1})(\nu h_\ell) = v_{\ell-1}(\nu h_\ell) \qquad (\nu \text{ even}),$$

$$(p\, v_{\ell-1})(\nu h_\ell) = \tfrac{1}{2}[v_{\ell-1}(\nu h_\ell - h_\ell) + v_{\ell-1}(\nu h_\ell + h_\ell)] \qquad (\nu \text{ odd}).$$

The restriction can be chosen as $r = r_{inj}$ from (2.2.8). Then, conditions (2.2.5), (2.2.6), (2.2.9), (2.2.11), (2.2.12) are satisfied with $\kappa = 2$.

The observed convergence rates of the two-grid iterations are listed in Table 2.6.1 for λ = 1. The 'observed convergence rate' is defined by $[\|u_\ell^5 - u_\ell\|_U / \|u_\ell^0 - u_\ell\|_U]^{1/5}$ (we do not claim that these numbers are always very close to the true rate). The corresponding convergence rates of the multigrid iterations (2.3.2), (2.4.2) and (2.5.1) are almost independent of the number of levels. For instance, the two-grid iteration with $n_0 = 64$, $n_1 = 128$ has nearly the same rate (namely 0.00004) as any of the three multigrid variants with $n_0 = 1$, $n_1 = 2, \ldots, n_7 = 128$. Hence, the numbers of Table 2.6.1 also hold for the multigrid case (n_1 replaced by n_ℓ).

Table 2.6.1

Observed rate of convergence of the two-grid iteration ($n_1 = 2n_0$) for Equation (2.6.1) with λ = 1.

n_1	2	4	8	16	32	64	128
rate*	1.5-1	3.1-2	9.6-3	2.5-3	6.3-4	1.6-4	4.0-5

* 1.5-1 means $1.5_{10}{-1}$

Table 2.6.2

Observed multigrid convergence rates for Eq. (2.6.1) with λ = 10.

	n	$n_0=1$	$n_0=2$	$n_0=4$	$n_0=8$	$n_0=16$	$n_0=32$	$n_0=64$
1st variant	2	7.2-1	-	-	-	-	-	-
	4	6.7+0	4.5-1	-	-	-	-	-
	8	6.8+1	1.2+0	1.2-1	-	-	-	-
	16	7.3+4	4.4+0	1.2-1	3.3-2	-	-	-
	32	4.7+10	1.6+2	2.5-2	1.4-2	8.3-3	-	-
	64	2.0+22	1.7+5	9.2-3	2.7-3	2.4-3	3.6-3	-
	128	2.5+45	4.0+11	5.1-4	5.7-4	5.5-4	5.5-4	5.6-4
2nd variant	2	7.2-1	-	-	-	-	-	-
	4	7.0-1	4.5-1	-	-	-	-	-
	8	5.8-1	1.7-1	1.2-1	-	-	-	-
	16	4.2-1	4.0-2	2.1-2	3.3-2	-	-	-
	32	2.2-1	7.9-3	8.0-3	7.5-3	8.3-3	-	-
	64	6.1-1	2.1-3	2.1-3	2.1-3	2.1-3	3.6-3	-
	128	4.7-3	5.3-4	5.3-4	5.3-4	5.3-4	5.3-4	5.6-4
3rd variant	2	7.2-1	-	-	-	-	-	-
	4	4.2-1	4.5-1	-	-	-	-	-
	8	1.9-1	1.5-1	1.2-1	-	-	-	-
	16	6.2-2	3.6-2	2.1-2	3.3-2	-	-	-
	32	5.8-3	8.8-3	7.9-3	7.5-3	8.3-3	-	-
	64	2.1-3	2.0-3	2.0-3	2.0-3	2.1-3	3.6-3	-
	128	5.3-4	5.3-4	5.3-4	5.3-4	5.3-4	5.3-4	5.6-4

Table 2.6.3

Observed multigrid convergence rates for Eq. (2.6.1) with λ = 100.

	n	n_o=1	n_o=2	n_o=4	n_o=8	n_o=16	n_o= 32	n_o=64
1st variant	2	9.3+0	-	-	-	-	-	-
	4	7.7+3	5.0+0	-	-	-	-	-
	8	1.5+7	1.1+3	1.3+0	-	-	-	-
	16	1.7+17	1.0+7	8.5+1	4.1-1	-	-	-
	32	1.0+36	1.7+16	1.7+4	5.7+0	1.2-1	-	-
	64	1.0+74	8.2+33	4.6+10	2.5+2	3.8-1	3.2-2	-
	128	-	8.0+69	8.6+22	9.1+6	7.3-1	2.9-2	2.0-2
2nd variant	2	9.3+0	-	-	-	-	-	-
	4	3.7+2	5.0+0	-	-	-	-	-
	8	6.8+5	8.5+0	1.3+0	-	-	-	-
	16	5.7+11	3.5+1	1.3+0	4.1-1	-	-	-
	32	3.2+23	1.4+3	1.4+0	2.4-1	1.2-1	-	-
	64	1.0+47	1.8+6	1.8+0	6.7-2	4.3-2	3.2-2	-
	128	-	3.3+12	2.9+0	1.0-2	9.2-3	8.9-2	2.0-2
3rd variant	2	9.3+0	-	-	-	-	-	-
	4	1.2+2	5.0+0	-	-	-	-	-
	8	1.5+4	1.9+1	1.3+0	-	-	-	-
	16	2.1+8	3.1+2	1.1+0	4.1-1	-	-	-
	32	4.5+16	8.5+4	8.1-1	2.3-1	1.2-1	-	-
	64	1.9+33	6.2+9	8.0-1	7.9-2	4.3-2	3.2-2	-
	128	3.8+66	3.7+19	5.7-1	1.4-2	9.6-3	8.9-3	2.0-2

Table 2.6.1 shows that the rates are proportional to $h_\ell^\kappa = h_\ell^2$ as predicted by Theorem 2.3.4. Theorem 2.3.6 as well as Theorem 2.4.3 and the corresponding statement in §2.5.3 are also confirmed, since two-grid and multigrid rates can hardly be distinguished. It can be checked that conditions (2.2.5), (2.2.6), (2.2.9), (2.2.12) are needed. If, e.g., the prolongation p is chosen as unsymmetric, piecewise constant interpolation (of order 1), the inequality (2.2.12) is fulfilled at most for $\kappa = 1$ and consequently the convergence rates are observed to be $O(h_\ell)$. Later (§3.6.2), numerical results will demonstrate that in the case of Example 2.1.3 ($\kappa = 1/2$) the rates are proportional to $h_\ell^{1/2}$. In the present example, where the kernel is very smooth, one should employ higher order quadrature formulae. Simpson's rule is of fourth order. To obtain a rate $O(h_\ell^4)$ one has to choose p as cubic interpolation ($\kappa = 4$).

As stated in Theorem 2.3.6, the success of the first variant (2.3.2) depends on the smallness of the bound C_K of $\|K_\ell\|_{U \leftarrow U}$. Obviously, this number is proportional to the factor λ in Eq. (2.6.1). Tables 2.6.2 and 2.6.3 contain the observed rates of the multigrid variants for $\lambda = 10$ and $\lambda = 100$ respectively. The finest step size 1/n is determined by $n = n_\ell$. The coarsest grid width is $1/n_0$ on level 0 (cf. (2.2.1)). As predicted by Theorem 2.3.6, the first variant (2.3.2) diverges if the coarsest step size h_0 is not small enough ($h_0 \geq 1/2$ for $\lambda = 10$, $h_0 \geq 1/8$ for $\lambda = 100$).

The numbers of Tables 2.6.2 and 2.6.3 in the diagonal entries ($n = 2n_0$) correspond to the two-grid case. As mentioned above, all three variants are equivalent as two-grid iterations and their rates coincide. The two-grid rates of both Tables 2.6.2, 2.6.3 show that Theorem 2.3.4 holds even in the case of divergence.

The numerical results illustrate that there may be two different reasons for divergence. The first variant diverges if $\|K_\ell\|_{U \leftarrow U}$ is too large and all three variants fail if the two-grid iteration with some coarsest grid size n_0 has a convergence rate of about 1.

The second and third variants yield very similar convergence rates; however, the latter variant is cheaper. It is remarkable that the third multigrid version (2.5.1) converges for $\lambda = 100$, $n_0 = 4$, $n_\ell \geq 32$, although the two- and three-grid iterations ($n \leq 16$) diverge.

2.6.2 The Integral Equation Method for Elliptic Problems

Let $\Omega \subset \mathbb{R}^2$ be a connected domain with boundary Γ. Consider the Dirichlet problem

$$\Delta\phi = 0 \text{ in } \Omega, \qquad \phi = g \text{ on } \Gamma. \tag{2.6.2}$$

A harmonic function in $\mathbb{R}^2 \setminus \Gamma$ is given by the 'double layer potential'

$$\phi_d(\zeta) = \frac{1}{2\pi} \int_\Gamma \mu(z) \cos(n_z, z - \zeta)|z - \zeta|^{-1} d\Gamma_z, \quad \zeta \notin \Gamma, \tag{2.6.3}$$

with arbitrary 'doublet distribution' μ. The term $\cos(n_z, z - \zeta)$ is the cosine of the vector $z - \zeta \in \mathbb{R}^2$ and the outward normal n_z at $z \in \Gamma$. The integral $\phi_d(\zeta)$ can also be evaluated at $\zeta \in \Gamma$, provided that Γ is sufficiently smooth. The limits $\phi_d^+(\zeta)$ and $\phi_d^-(\zeta)$ of ϕ_d at $\zeta \in \Gamma$ from the outer and inner side, respectively, differ by $\mu(\zeta)$ if μ is continuous at ζ:

$$\phi_d^{\pm}(\zeta) = \phi_d(\zeta) \mp \mu(\zeta)/2, \qquad \zeta \in \Gamma.$$

The solution of the Dirichlet problem (2.6.2) is ϕ_d if $\phi_d^-(\zeta) = g(\zeta)$ on Γ, i.e. $g(\zeta) = \phi_d(\zeta) + \mu(\zeta)/2$. From this condition we obtain the integral equation

$$\mu(\zeta) = -\frac{1}{\pi} \int_\Gamma \mu(z)\cos(n_z, z - \zeta)\ |z - \zeta|^{-1} d\Gamma_z + 2g(\zeta) \tag{2.6.4}$$

for the unknown μ. Given a representation $\gamma(t)$, $0 \leqslant t \leqslant 1$, of the boundary Γ, the function $u(t) := \mu(\gamma(t))$ is the solution of

$$u(s) = -\frac{1}{\pi} \int_0^1 u(t)\cos(n_{\gamma(t)}, \gamma(t) - \gamma(s))\,|\gamma(t) - \gamma(s)|^{-1} \|d\gamma/dt\|_2 dt$$
$$+ 2g(\gamma(s)). \tag{2.6.4'}$$

The regularity (2.1.4) is proved by Schippers (1982b):

<u>Proposition 2.6.1</u> Assume $\Gamma \in C^{2+\alpha}$ with $0 < \alpha < 1$, i.e. there is a parameterisation of Γ by $\gamma(t)$, $0 \leqslant t \leqslant 1$, with $0 < \|d\gamma/dt\| < \infty$, such that the periodic extension of γ to $\mathbb{R}$ satisfies $\gamma \in C^{2+\alpha}(\mathbb{R})$. Then the integral operator associated with Eq. (2.6.4) is a bounded mapping from $L^\infty(\Gamma)$ (or $C^0(\Gamma)$) into $C^{1+\alpha}(\Gamma)$:

$$\|K\|_{C^{1+\alpha}(\Gamma) \leftarrow L^\infty(\Gamma)} \leqslant C, \quad \alpha \in (0,1).$$

Problem (2.6.4') can be discretised by the following collocation method ('panel method'). Define $h_\ell = 1/n_\ell$ as in (2.2.1). For $0 \leqslant j < n_\ell$ let $u^{(j)}$ be the function with value 1 in $[jh_\ell, (j+1)h_\ell]$ and 0 otherwise. The entries $K_{\ell,ij}$ of K_ℓ are defined by

$K_{\ell,ij} = (Ku^{(j)})((i + 1/2)h_\ell)$. The components of f_ℓ are $f_{\ell,i} = f((i + 1/2)h_\ell)$ with $f(t) = 2g(\gamma(t))$. The ith component of the grid function is regarded as the value of u_ℓ at the 'grid point' $s = (i + 1/2)h_\ell$ $(0 \leq i < n_\ell)$. The (weighted) restriction of u_ℓ is

$$(ru_\ell)((i + 1/2)h_{\ell-1}) = [u_\ell((2i + 1/2)h_\ell) + u_\ell((2i + 3/2)h_\ell)]/2.$$

The operator p is defined as piecewise constant interpolation:

$$(p\, u_{\ell-1})((2i + 1 \pm 1/2)h_\ell) = u_{\ell-1}((i + 1/2)h_{\ell-1}).$$

Inequality (2.2.12) holds with $\kappa = 1$ if we choose $U_\ell = C^0(D_\ell)$, $V_\ell = C^1(D_\ell)$. But $\kappa = 2$ can be obtained by the choice $U_\ell = C^{-1}(D_\ell)$ with norm

$$\|u_\ell\| = \max\{|u_\ell(x + h_\ell/2)v_\ell(x + h_\ell/2) + u_\ell(x - h_\ell/2)v_\ell(x - h_\ell/2)|$$

$$: x = (2i + 1)\, h_\ell,\ \|v_\ell\|_{C^1(D_\ell)} \leq 1\}.$$

An exterior boundary value problem arising from the non-circulatory potential flow around an aerofoil can be described by the same integral operator (Schippers (1982c)). The flow is represented by the potential ϕ satisfying

$$\begin{aligned} &\Delta\phi = 0 \text{ in } \mathbb{R}^2 \setminus \bar{\Omega} \quad (\text{exterior of } \Omega), \\ &\partial\phi/\partial n = 0 \quad \text{on } \Gamma \\ &\phi(\zeta) \to \langle U,\zeta\rangle \quad \text{for } \|\zeta\| \to \infty \end{aligned} \qquad (2.6.5)$$

where $\langle\cdot,\cdot\rangle$ is the scalar product in $\mathbb{R}^2$. $U \in \mathbb{R}^2$ is the velocity of the undisturbed flow. The solution of Eq. (2.6.5) is given by the double layer potential (2.6.3) with μ satisfying the integral equation

$$\mu(\zeta) = -\frac{1}{\pi}\int_\Gamma \mu(z)\cos(n_z,\ z - \zeta)\,|z - \zeta|^{-1}d\Gamma_z - 2\langle U,\zeta\rangle.$$

The following numerical results are taken from Schippers (1982c). He considers the problem (2.6.5) for the Kármán-Trefftz aerofoil, which is defined in the complex plane by

$$z(s) = c(e^{is} - \sqrt{1-\rho^2} + i\rho)^k/(e^{is} - \delta + i\rho)^{k-1}, \quad 0 \leq s \leq 2\pi, \tag{2.6.6}$$

with $c = 2L(\delta + \sqrt{1-\rho^2})^{k-1}/(2\sqrt{1-\rho^2})^k$, δ : thickness,

ρ : camber, k : trailing edge angle, L : length.

The conformal mapping

$$z \mapsto w = z(1 - \tilde{z}/z)^{1-1/k}, \quad \tilde{z} \in \Omega \text{ fixed},$$

removes the corner ('trailing edge', cf. Fig. 2.6.1).

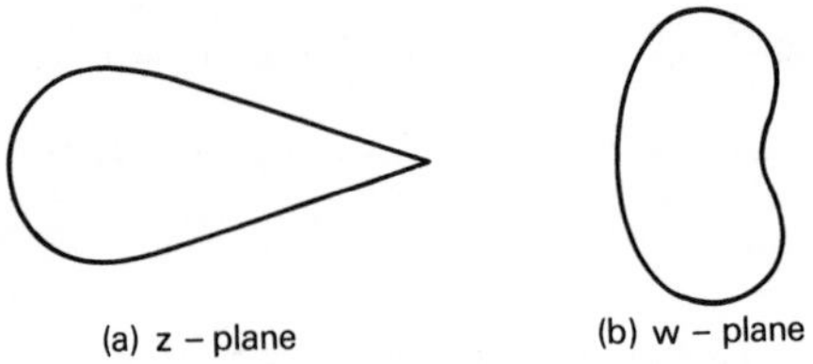

Fig. 2.6.1 Kármán-Trefftz aerofoil in z- and w- coordinates.

The resulting boundary Γ is described by $w = w(s)$, $0 \leq s \leq 2\pi$. The parameters in (2.6.6) are $\rho = 0$, $L = 1$, $\delta = 0.05$, $k = 1.9$ and $k = 1.99$. Further data are $U = (1,0)^T$, $\tilde{z} = 1.95$.

Following Schippers (1982c), the observed convergence rates of the multigrid iteration (2.3.2) (first variant) with coarsest grid size $h_0 = 1/8$ are listed in Table 2.6.4. The numbers confirm that the convergence rate is proportional to h_ℓ^κ, $\kappa = 2$.

The circulatory flow problem will be discussed in §2.8.3.

Table 2.6.4

Multigrid convergence rates for non-circulatory potential flow problem.

k \ n	16	32	64	128	256
1.90	.26	.091	.020	.0045	.0011
1.99	.43	.26	.095	.022	.0054

2.7 *Nested Iteration*

2.7.1 Algorithm

Taking $u_\ell^0 = 0$ as starting guess and iterating i times at level ℓ, we obtain the result u_ℓ^i with an iteration error $\|u_\ell^i - u_\ell\|_U \leq O(h_\ell^{i\kappa})\|u_\ell^0 - u_\ell\| = O(h_\ell^{i\kappa})$. In the case i = 1, the error has the same order as the discretisation error (cf. (2.2.13)) but not necessarily the same size. Therefore, it may be safer to perform i = 2 iterations.

Instead of using the multigrid iteration only at level ℓ, one can apply the following nested iteration

$$\tilde{u}_0 := (I - K_0)^{-1} f_0; \tag{2.7.1a}$$

<u>for</u> k := 1 <u>step</u> 1 <u>until</u> ℓ <u>do</u>

<u>begin</u> $\tilde{u}_k := \tilde{p}\,\tilde{u}_{k-1}$; (2.7.1b)

<u>for</u> j := 1 <u>step</u> 1 <u>until</u> i <u>do</u> MGM$(k,\tilde{u}_k,f_k)$ (2.7.1c)

<u>end</u>;

yielding approximations $\tilde{u}_k$ for <u>all</u> levels k = 0,1,..., ℓ. The operator $\tilde{p}$ may be the same prolongation as in the multigrid algorithm, but it can also be chosen as interpolation of higher order than p. The usual choice of i is i = 1, as explained below.

The advantage of algorithm (2.7.1) is that the knowledge of the approximations $\tilde{u}_0, \tilde{u}_1,\ldots,\tilde{u}_\ell$ allows us (1) to estimate the discretisation error of $\tilde{u}_\ell$, (2) to decide a posteriori whether the computation should be continued at level $\ell + 1$, (3) to perform Richardson's extrapolation (Baker (1977), p 155 ff, Schneider (1977)). The work of the nested iteration with i = 1 is less than the work of two multigrid iterations at level ℓ.

<u>Note 2.7.1</u> Assume (2.3.4): $n_{\ell-1} \leq C_N\, n_\ell$ with $C_N < \sqrt{1/2}$. The work of (2.7.1a) and (2.7.1b) can be neglected. The total computational work of algorithm (2.7.1) is bounded by $i/(1 - C_N^2)$ times the work of one MGM-iteration at level ℓ. For i = 1 the nested iteration takes $C\, n_\ell^2 + O(n_\ell)$ operations with C given in Table 2.7.1:

Table 2.7.1

The factor C in the formula for the computational work of the nested iteration (2.7.1).

multigrid variant	first	second	third
case (2.3.5a)	9.33	10.67	8.00
case (2.3.5b)	6.67	8.00	6.00
case of Note 2.4.2	-	6.67	-

2.7.2 Accuracy

Proposition 2.7.2 Assume

$$\|u_k - \tilde{p}\, u_{k-1}\|_U \leq C_D h_k^{\kappa} \quad \text{(relative discretisation error)} \tag{2.7.2a}$$

$$\|\tilde{p}\|_{U \leftarrow U}\, h_{k-1}^{\kappa}/h_k^{\kappa} \leq \text{const} \quad (\tilde{p} = \tilde{p}_{k \leftarrow k-1}) \tag{2.7.2b}$$

$$\|M_k^{MGM}\|_{U \leftarrow U} \leq C_{MGM}\, h_k^{\kappa} \tag{2.7.2c}$$

for $1 \leq k \leq \ell$, where M_k^{MGM} is the respective iteration matrix M_k', M_k'' or M_k''' of the multigrid algorithm (2.3.2), (2.4.2), (2.5.1). Then all results $\tilde{u}_k$ $(0 \leq k \leq \ell)$ of the nested iteration (2.7.1) satisfy

$$\|\tilde{u}_k - u_k\|_U \leq C_D h_k^{\kappa}\, C_{MGM} h_k^{\kappa} + O(h_k^{3\kappa}), \quad 0 \leq k \leq \ell. \tag{2.7.3}$$

The relative discretisation error of (2.7.2a) is already mentioned in (2.2.13′). Proposition 2.7.2 shows that the iteration error of $\tilde{u}_k$ is much smaller than the (relative) discretisation error. Note that condition (2.7.2c) implies multigrid convergence. In that case we know from Theorem 2.3.6 that $\|M_\ell^{MGM}\|_{U \leftarrow U} \leq C_{TGM}\, h_k^{\kappa} + O(h_k^{2\kappa})$. Thereby, (2.7.3) may be replaced by

$$\|\tilde{u}_k - u_k\|_U \leq C_D h_k^{\kappa}\, C_{TGM} h_k^{\kappa} + O(h_k^{3\kappa}). \tag{2.7.3′}$$

2.7.3 Nested Iteration with Nyström's Interpolation

If a quadrature formula is used as discretisation method, one can replace the interpolation $\tilde{p}$ in (2.7.1b) by Nyström's interpolation (Nyström (1928)):

$$\tilde{u}_k := K_{k,k-1}\tilde{u}_{k-1} + f_k, \qquad (2.7.4a)$$

where the rectangular matrix $K_{k,k-1}$ is defined by

$$(K_{k,k-1}u_{k-1})(x) = h_{k-1} \sum_{y \in D_{k-1}} w(y)k(x,y)u_{k-1}(y), \quad x \in D_k. \qquad (2.7.4b)$$

$K_{\ell+1,\ell}\, u_\ell + f_{\ell+1}$ is the right hand side of (2.2.3) evaluated at $x \in D_{\ell+1}$ instead of $x \in D_\ell$. If $D_{k-1} \subset D_k$ and $f_k(x) = f_{k-1}(x)$ on D_{k-1}, the restriction of $K_{k,k-1}u_{k-1} + f_k$ to D_{k-1} equals $K_{k-1}u_{k-1} + f_{k-1}$.

The interpolation (2.7.4) is much more expensive. It requires

$$C_K n_k n_{k-1} \leqslant C_K C_N n_k^2 \qquad (2.7.5)$$

operations, whereas standard interpolations $\tilde{u}_k = \tilde{p}\,\tilde{u}_{k-1}$ take work of order $O(n_k)$. Nonetheless, it pays to apply Nyström's interpolation, since the resulting approximation $\tilde{u}_k$ is already smoothed. The first iteration of the following multigrid algorithm (if i = 1 there is only one iteration) can be performed <u>without</u> a smoothing step. Combining the modified nested iteration (i = 1) with the third multigrid variant (2.5.1), we obtain the following algorithm:

Nested iteration with Nyström's interpolation	(2.7.6)
$\tilde{u}_0 := (I - K_0)^{-1} f_0;$	(2.7.6a)
<u>for</u> k := 1 <u>step</u> 1 <u>until</u> ℓ <u>do</u>	
<u>begin</u> $\tilde{u}_k := K_{k,k-1}\tilde{u}_{k-1} + f_k;$	(2.7.6b)
MGM$(k,\tilde{u}_k,f_k,$<u>false</u>$)$; <u>comment</u> MGM from (2.5.1)	(2.7.6c)
<u>end</u>;	

<u>Note 2.7.3</u> The nested iteration (2.7.6) requires Cn_ℓ^2 operations with

$$C = 6.67, \quad \text{case (2.3.5a)},$$
$$C = 4.67, \quad \text{case (2.3.5b)}.$$

The respective numbers are C = 8 and C = 5.33 if the first multigrid variant (2.3.2) is used in (2.7.6c).

It is remarkable that 14/3 n_ℓ^2 operations (equivalent to 2 1/3 Picard iterations) are sufficient for producing results $\tilde{u}_0, \tilde{u}_1, \ldots, \tilde{u}_\ell$ to an accuracy beyond the discretisation error.

We conclude with numerical results. Let $\tilde{u}_k$ be the results of the nested iteration (2.7.6) for the problem from §2.6.1 with $\lambda = 10$. A comparison of the iteration error $\|\tilde{u}_k - u_k\|_\infty$ with discretisation error $\|u_k - u\|_\infty$ is given in Table 2.7.2:

Table 2.7.2

Results of nested iteration (2.7.6) applied to problem (2.6.1).

k	0	1	2	3	4	5	6
h_k	1/2	1/4	1/8	1/16	1/32	1/64	1/128
iteration error	0.0	4.8-1	8.2-2	6.7-3	1.4-4	6.3-6	6.0-7
discretisation error	1.8+0	4.9-2	1.5-2	4.9-3	1.7-3	5.9-4	2.1-4

2.8 Comments and Modifications

2.8.1 More than one Picard Iteration

Multigrid algorithms for elliptic equations use ν steps of some smoothing iteration, where ν must be sufficiently large (usual values are $\nu = 2$, $\nu = 3$). One can try to replace the smoothing step (2.3.2b) by

$$\underline{\text{for}}\ \ i = 1 \text{ step } 1\ \underline{\text{until}}\ \nu\ \underline{\text{do}}\ u := K_\ell * u + f; \tag{2.8.1}$$

As known from §§2.3-2.5, the choice of $\nu > 1$ is dangerous if $\|K_\ell\|_{U\leftarrow U} >> 1$. Even if $\|K_\ell\|_{U\leftarrow U} \approx 1$ the analysis given above shows no advantage in taking $\nu > 1$. However, there are cases where $\nu > 1$ is reasonable.

In Example 2.1.3 the operator K does not map from $L^2(D)$ into $H^\gamma(D)$ for fixed $\gamma \in (1/2, 1)$, but K^2 does so. Another example follows.

Example 2.8.1 Assume $D = [0,1]$ and let $k(x,y)$ be sufficiently smooth in the two triangles $0 \leq x \leq y \leq 1$ and $0 \leq y < x \leq 1$, but discontinuous across the diagonal $x = y$. Then K maps from $C^s(D)$ into $C^{s+1}(D)$, $s \geq 0$, but not into $C^t(D)$ with $t > s + 1$. As a consequence $K^2 : C^0(D) \to C^2(D)$ holds.

Assume that the multigrid iteration (2.3.2) (with $\nu = 1$) is applied. In the case of Example 2.1.3 one has to choose $U_\ell = L^2(D_\ell)$, $V_\ell = H^{½}(D_\ell)$

(discrete analogues of $L^2(D)$, $H^{\frac{1}{2}}(D)$), which yields $\kappa = 1/2$ as maximal exponent in the $O(h^\kappa)$ right hand sides of (2.2.4), (2.2.12). For Example 2.8.1 we have $U_\ell = C^0(D_\ell)$, $V_\ell = C^1(D_\ell)$, and $\kappa = 1$. The possibly better orders of the interpolation p and of the quadrature formula cannot be exploited. The multigrid convergence rate is only $O(h_\ell^\kappa)$ with $\kappa = 1/2$ or $\kappa = 1$ respectively.

The situation improves if we perform $\nu \geq 2$ smoothing iterations (cf. (2.8.1)). The regularity condition (2.2.6) can be replaced by

$$\|K_\ell^\nu\|_{V \leftarrow U} \leq C_R.$$

Now the respective choices of V_ℓ in the previous examples are $H^\gamma(D_\ell)$, $\gamma \in (1/2, 1)$, and $C^2(D_\ell)$. The multigrid convergence rates become $O(h_\ell^\gamma)$ and $O(h_\ell^2)$ respectively, instead of $O(h_\ell^{\frac{1}{2}})$ and $O(h_\ell)$.

2.8.2 Other Smoothing Iterations

A general iteration can be written as

$$u_\ell \longmapsto A_\ell^{-1}(B_\ell u_\ell + f_\ell) \text{ with } A_\ell - B_\ell = I - K_\ell. \tag{2.8.2}$$

We always used Picard's iteration: $A_\ell = I$, $B_\ell = K_\ell$. In the case of a weakly singular kernel $k(x,y) = k_1(x,y)f(x-y)$ with $f(s) = |s|^{-\lambda}$ or $f(s) = \log|s|$ etc. one might think of employing a Jacobi-like iteration, since the main diagonal and the neighbouring diagonals contain entries of larger size than outside the diagonal band. A natural choice of A_ℓ and B_ℓ would be

$$A_\ell = I - K_\ell^B, \quad B_\ell = K_\ell - K_\ell^B, \tag{2.8.3a}$$

where the entries of K_ℓ^B are

$$K_{\ell,ij}^B = \left\{ \begin{array}{ll} K_{\ell,ij} & \text{if } |i-j| \leq r_\ell, \\ 0 & \text{otherwise} \end{array} \right\}. \tag{2.8.3b}$$

The number r_ℓ determines the band width of A_ℓ. For periodic functions the difference i-j in $|i-j| \leq r_\ell$ is to be understood modulo $\dim(U_\ell)$. It has to be ensured that A_ℓ is stable: $\|A_\ell^{-1}\|_{U \leftarrow U} \leq$ const.

The smoothing iteration (2.8.3) can be regarded as Picard's iteration applied to a modified integral equation. Set $k^{\delta}(x,y) = k(x,y)$ if $|x-y| \leq \delta$, $k^{\delta} = 0$ elsewhere, and define the operator K^{δ} by means of the kernel k^{δ}. The integral equation $u = Ku + f$ can be written as

$$u = (I - K^{\delta})^{-1} [(K - K^{\delta})u + f]. \tag{2.8.4}$$

Iteration (2.8.2) with A_{ℓ}, B_{ℓ} from Eq. (2.8.3) is Picard's iteration applied to the discretisation of Eq. (2.8.4), if $\delta = r_{\ell}h_{\ell}$.

A somewhat modified splitting of $I - K_{\ell}$ may be better than (2.8.3). Define the integral operator $\tilde{K}$ by the kernel $\tilde{k}(x,y) = k(x,y)$ for $|x-y| \geq \delta$, $\tilde{k}$ smoothly continued for $|x-y| \leq \delta$. $K^{\delta} := K - \tilde{K}$ can be used in (2.8.4). The corresponding discrete version (2.8.3) involves a matrix B_{ℓ} with improved smoothing property.

The previous iterations can also be applied to integral equations of the first kind as done by Oskam and Fray (1982). In this case Eq. (2.8.4) becomes $u = (K^{\delta})^{-1}[(K - K^{\delta})u + f]$. Another approach to equations of the first kind will be proposed in §3.5.5.

Not only the diagonal entries of K_{ℓ} can become large. The next subsection reports such an example and describes an efficient smoothing iteration.

2.8.3 Application of the Integral Equation Method to the Calculation of Circulatory Flow around an Aerofoil

Consider the exterior boundary value problem (2.6.5) in the case of an aerofoil with trailing edge ζ_t (cf. Fig. 2.6.1a, Fig. 2.8.1). Again the double layer potential (2.6.3) is used. At both sides of the edge ζ_t the flow speed must vanish ('Kutta condition'). As a consequence

$$d\mu(\zeta)/d\tau \to 0 \text{ as } \zeta \to \zeta_t, \quad \zeta \in \Gamma,$$

must hold for the doublet distribution μ, where $d/d\tau$ represents derivation in the tangential direction. The resulting integral equation for μ is

$$\mu = K\mu - 2 \langle U,\zeta\rangle + \beta (\mu^{-} - \mu^{+}), \tag{2.8.5}$$

where $\beta(\zeta) = \frac{1}{\pi} \arg(\zeta_t - \zeta)$ and K is the integral operator associated with Eq. (2.6.4). The quantity μ^+ (μ^-) is the limit of $\mu(\zeta)$ along the upper (lower) side of the trailing edge (cf. Fig. 2.8.1). Schippers (1982a and 1982c) discretises the problem (2.8.5) by the panel method of §2.6.2. The parametrisation of Γ is chosen such that $\gamma(0) = \gamma(1) = \zeta_t$. For $s \in (0,\frac{1}{2})$ the boundary points $\gamma(s)$ and $\gamma(1-s)$ are lying on the upper and lower part of Γ having the same x-coordinates. Quantities μ^+ and μ^- are identified with the first and last components $u_{\ell,1}$ and u_{ℓ,n_ℓ} of the discrete grid function (cf. Fig. 2.8.1).

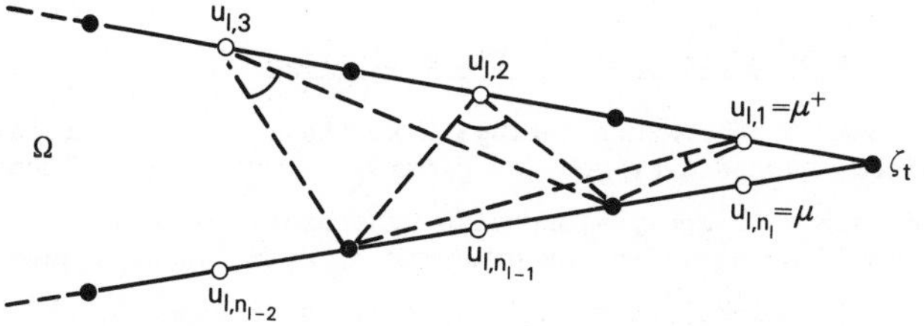

Fig. 2.8.1 Neighbourhood of the trailing edge

The matrix K_ℓ of the resulting discrete equation

$$u_\ell = K_\ell u_\ell + f_\ell \tag{2.8.6}$$

has large entries $K_{\ell,i,n_\ell-i-1}$ in the cross-diagonal. The size of $K_{\ell,i,j}$ is determined by the angle at the ith grid point with sides intersecting the endpoints of the jth interval. Fig. 2.8.1 depicts the respective angles involved by $K_{\ell,i,n_\ell-1}$ for $i = 1,2,3$. Obviously the angle is extreme for $i = 2$ (in general for $i + j = n_\ell + 1$).

It turns out that the Picard iteration does not yield a fast converging multigrid iteration. This is in accordance with the following result of Schippers (1982c):

<u>Note 2.8.2</u> If a trailing edge is present, the integral operator K is not bounded as a mapping from $L^\infty(\Gamma)$ into $C^0(\Gamma)$.

Schippers (1982a and 1982c) proposed to smooth Eq. (2.8.6) either by block-Jacobi iteration (called 'paired Jacobi iteration'), where the ith block consists of $u_{\ell,i}$ and $u_{\ell,n_\ell+1-i}$ $(1 \leq i \leq n_\ell/2)$, or by the corresponding 'paired Gauss-Seidel relaxation'. In the former case, the matrix A_ℓ from (2.8.2) is given by

$$A_{\ell,ij} = \begin{cases} 1 - K_{\ell,ii} & \text{if } j = i, \\ -K_{\ell,i,n_\ell+1-i} & \text{if } j = n_\ell+1-i, \\ 0 & \text{otherwise.} \end{cases}$$

The multigrid convergence is satisfactory if paired Jacobi iterations are used for smoothing. Even better results are obtained for paired Gauss-Seidel iterations (cf. Schippers (1982c and 1982d)).

2.8.4 Construction of Coarse-Grid Discretisation

If one chooses a quadrature formula like the trapezoidal rule, one obtains a nicely nested hierarchy of grids D_ℓ. A different situation arises if one wants to apply a Gaussian quadrature formula. The first possibility would be to define each matrix K_ℓ by a Gaussian quadrature rule with n_ℓ grid points, where $n_0 < n_1 < \ldots < n_\ell$. Since $D_{\ell-1} \not\subset D_\ell$, the restriction r cannot be injection. E.g., ru_ℓ may be the evaluation of the piecewise linearly interpolated grid function u_ℓ at the points of $D_{\ell-1}$. Similarly, $pu_{\ell-1}$ can be defined.

A second possibility is explained below. For a fixed (finest) level ℓ let K_ℓ be a given discretisation using the grid D_ℓ. Choose $D_0 \subset D_1 \subset \ldots \subset D_{\ell-1} \subset D_\ell$, and define r and p suitably, and set

$$K_{\ell-1} := r\, K_\ell\, p, \qquad K_{\ell-2} := r\, K_{\ell-1}\, p, \ \ldots \ . \tag{2.8.7}$$

E.g., r may be the trivial injection (2.2.8), while p is a piecewise interpolation.

Proposition 2.8.3 Let $K_0, K_1, \ldots, K_{\ell-1}$ be constructed by (2.8.7). The two-grid iteration matrix (2.3.8) equals

$$M_k^{TGM} = [I + p(I - K_{k-1})^{-1} r\, K_k][I - pr]K_k, \quad k = 1,2,\ldots,\ell.$$

The multigrid iteration matrices M_k' and M_k'' from (2.3.10) and (2.4.5), respectively, become

$$M'_k = M_k^{TGM} + p\, M'^{\,2}_{k-1}[r - (I - K_{k-1})^{-1}\, r\, K_k(I - pr)]K_k,$$

$$M''_k = M_k^{TGM} + p\, M''^{\,2}_{k-1}[r - (I - K_{k-1})^{-1}\, r\, K_k(I - pr)K_k].$$

The relative consistency condition (2.2.9) follows from $r\,K_\ell - K_{\ell-1}r = r\,K_\ell[I - pr]$, while the regularity condition (2.2.6) at level ℓ implies the same inequality at the coarse-grid levels $k < \ell$ if $\|p\|_{U\leftarrow U} \leq 1$, $\|r\|_{V\leftarrow V} \leq 1$.

2.8.5 Further Modifications

All multigrid variants involved $\gamma = 2$ iterations of the coarse-grid equation (cf. (2.3.2c_2), (2.4.2b_4), (2.5.1c_2)). It does not pay to perform $\gamma = 3$ iterations since the multigrid rate is already sufficiently close to the two-grid rate (cf. Theorem 2.3.6 etc.).

In particular, if $\|K_\ell\|_{U\leftarrow U} < 1$, the choice $\gamma = 1$ is possible. However, if $\|K_\ell\|_{U\leftarrow U} \approx 1$, the contraction numbers $\|M'_\ell\|_{U\leftarrow U}$ and $\|M''_\ell\|_{U\leftarrow U}$ will be of the size of $\|M_0^{TGM}\|_{U\leftarrow U}$ (instead of $\|M_\ell^{TGM}\|_{U\leftarrow U}$)! A compromise

$$\gamma = 2 \quad \text{if} \quad \ell \leq \ell^*, \quad \gamma = 1 \quad \text{if} \quad \ell > \ell^*.$$

Then, $\|M''_\ell\|_{U\leftarrow U} \approx \|M_{\ell^*}^{TGM}\|_{U\leftarrow U}$ can be expected for $\ell > \ell^*$. If $\|K_\ell\|_{U\leftarrow U} >> 1$, the first variant cannot be used with $\gamma = 1$.

The convergence behaviour described in Theorems 2.3.6 and 2.4.3 is preserved by the choice

$$\gamma = 1 \quad \text{if } \ell \text{ is odd}, \quad \gamma = 2 \text{ if } \ell \text{ is even}. \tag{2.8.8}$$

Note 2.8.4 Assume (2.3.5b). The nested iteration (2.7.1) with odd ℓ and with MGM from (2.3.2) modified by (2.8.8) requires $Cn_\ell^2 + O(n_\ell)$ operations, where $C = 82/15 \approx 5.47$. The corresponding number for the analogous modification of the nested iteration (2.7.6) is $C = 52/15 \approx 3.47$.

For the sake of convenience only, the ratio $h_{\ell-1}/h_\ell = 2$ is proposed (cf. (2.2.1)). One can, instead, use a sequence $\{h_\ell\}$ with increasing quotient $h_{\ell-1}/h_\ell$ (McCormick (1980)).

A multigrid approach to the 3-dimensional panel method is studied by Nowak (1982).

3. NONLINEAR MULTIGRID METHODS OF THE SECOND KIND

3.1 The Equation of the Second Kind

The equation of the second kind is the fixed point equation

$$u = K(u), \tag{3.1.1}$$

where K: $U^O \subset U \to U$ is a (nonlinear) operator. Eq. (3.1.1) is assumed to have at least one isolated solution $u^* \in U^O$. If K is affine (i.e. $K(u) = Ku + f$), we regain the linear problem (2.1.2).

The first multigrid treatment of equation (3.1.1) is described in Hackbusch (1979a). For a brief survey see Hackbusch (1980c).

Example 3.1.1 The nonlinear Fredholm integral equation of the second kind

$$u(x) = \int_D k(x,y,u(y))dy + g(x), \quad x \in D,$$

leads to Eq. (3.1.1) with $K(u) = \int_D k(.,y,u(y))dy + g$. However, there are numerous other problems that can be formulated by Eq. (3.1.1). Examples will be given in §§3.3, 3.5, 3.6.

As in §2 we assume that there be a sequence of discrete problems

$$u_\ell = K_\ell(u_\ell) \ , \quad \ell = 0,1,\ldots, \tag{3.1.2}$$

where K_ℓ: $U_\ell^O \subset U_\ell \to U_\ell$ are nonlinear mappings. At least one isolated solution $u_\ell^* \in U_\ell^O$ is supposed to exist. The derivative of K_ℓ is denoted by

$$K_\ell(v_\ell) := K'_\ell(v_\ell) \quad \text{for} \quad v_\ell \in U_\ell^O$$

and is assumed to be Lipschitz continuous:

$$\|K_\ell(v_\ell) - K_\ell(w_\ell)\|_{U\leftarrow U} \leq C_K \|v_\ell - w_\ell\|_U \text{ for all } v_\ell, w_\ell \in U_\ell^O, \ \ell \geq 0. \tag{3.1.3}$$

The derivative at $v_\ell = u_\ell^*$ (u_ℓ^* : solution of Eq. (3.1.3)) is denoted by

$$K_\ell := K_\ell(u_\ell^*) = K'_\ell(u_\ell^*). \tag{3.1.4}$$

We shall require that the matrix M_ℓ^{TGM} from (2.3.8) defined with K_ℓ and $K_{\ell-1}$ from (3.1.4) satisfies

$$\|M_\ell^{TGM}\|_{U\leftarrow U} \leq C_{TGM}\, h_\ell^\kappa, \quad M_\ell^{TGM} := [I - p(I - K_{\ell-1})^{-1} r(I - K_\ell)]K_\ell. \tag{3.1.5}$$

Consequently, the linear two-grid iteration applied to the linearised equation has a contraction number $\leq C_{TGM}\, h_\ell^\kappa$. By Theorem 2.3.4, estimate (3.1.5) follows from (2.2.5), (2.2.6), (2.2.9), (2.2.11), (2.2.12).

Example 3.1.2 Discretise the nonlinear integral equation of Example 3.1.1 by the trapezoidal rule as in Example 2.2.1. $K_\ell(v_\ell)$ is given by Eq. (2.2.3) with $k(x,y)$ replaced by $k_u(x,y,v_\ell(y))$. Relations $k_u \in C^1(D \times D \times U^0)$ and $U_\ell = C^0(D_\ell)$ imply inequality (3.1.3). Concerning the choice of V_ℓ and κ the examples of §2.2 still apply.

The assumptions (3.1.3) and (3.1.5) are relatively weak. E.g., the inequality (3.1.5) is to be fulfilled only by $K_\ell = K_\ell(u_\ell^*)$, not by $K_\ell(v_\ell)$ for $v_\ell \neq u_\ell^*$. Note that (3.1.3) does not imply the regularity assumption (2.2.6).

For the multigrid process it is necessary to consider the perturbed equations

$$u_\ell = \mathcal{K}_\ell(u_\ell) + f_\ell, \quad \ell = 0,1,\ldots, \tag{3.1.6a}$$

with sufficiently small perturbations f_ℓ. E.g., we may require

$$\|f_\ell\|_U \leq \varepsilon_\ell. \tag{3.1.6b}$$

For simplicity, one may set

$$U_\ell^0 := \{u_\ell\text{: solution of (3.1.6a) for } f_\ell \text{ with (3.1.6b)}\}.$$

Note 3.1.3 Let $u_\ell^* \in U_\ell^0$ be a solution of (3.1.2). By the inverse mapping theorem, the assumptions (3.1.3) and (2.2.5) establish the unique solvability of (3.1.6a, b) in a neighbourhood of u_ℓ^*, if $\varepsilon_\ell > 0$ from (3.1.6b) is sufficiently small. The inverse mapping $f_\ell \longmapsto u_\ell$ is denoted by Φ_ℓ: $u_\ell = \Phi_\ell(f_\ell)$ is the solution of Eq. (3.1.6a). The mapping Φ_ℓ is defined on $\{f_\ell \in U_\ell : \|f_\ell\|_U \leq \varepsilon_\ell\}$. Its derivative is

$(I - K_\ell(\Phi_\ell(f_\ell)))^{-1}$. In particular,

$$\Phi_\ell(0) = u_\ell^*, \quad \Phi'_\ell(0) = (I - K_\ell)^{-1} \tag{3.1.7}$$

holds.

3.2 Multigrid Method

3.2.1 Two-grid Iteration

Nonlinear two-grid iteration for solving $u_\ell = \mathcal{K}_\ell(u_\ell) + f_\ell$	(3.2.1)
$u'_\ell := \mathcal{K}_\ell(u_\ell^j) + f_\ell;$	(3.2.1a)
$d_{\ell-1} := r_*(u'_\ell - \mathcal{K}_\ell(u'_\ell) - f_\ell);$	(3.2.1b$_1$)
$\tilde{u}_{\ell-1} \in U^0_{\ell-1}$ must be given;	
possibly determined a posteriori by	
$\tilde{u}_{\ell-1} := r\, u'_\ell.$	(3.2.1b$_2$)
The defect of $\tilde{u}_{\ell-1}$ must be given, too :	
$\tilde{f}_{\ell-1} := \tilde{u}_{\ell-1} - \mathcal{K}_{\ell-1}(\tilde{u}_{\ell-1});$	(3.2.1b$_3$)
$\tilde{u}_\ell^{j+1} := u'_\ell + p[\Phi_{\ell-1}(\tilde{f}_{\ell-1} - s\, d_{\ell-1}) - \tilde{u}_{\ell-1}]/s$	(3.2.1b$_4$)

The nonlinear iteration requires a coarse-grid correction (3.2.1b$_1$), (3.2.1b$_4$) that is different from the linear version (2.3.1b). The real number s appearing in (3.2.1b$_4$) will be discussed in the sequel. Algorithm (3.2.1) is a genuine generalisation of the linear algorithm because of

<u>Note 3.2.1</u> If $\mathcal{K}_\ell$ and $\mathcal{K}_{\ell-1}$ are affine (i.e. $\mathcal{K}_k(u_k) = K_k u_k + g_k$), the algorithm (3.2.1) with arbitrary $\tilde{u}_{\ell-1} \in U_{\ell-1}$ and $s \in \mathbb{R}$ is algebraically equivalent to the linear iteration (2.3.1).

<u>Proof</u> Since $\Phi_k(f_k) = (I - K_k)^{-1}(g_k + f_k)$, the correction step (1b$_4$) equals

$$u_\ell^{j+1} = u'_\ell + p[(I - K_{\ell-1})^{-1}(g_{\ell-1} + \tilde{f}_{\ell-1} - s\, d_{\ell-1}) - \tilde{u}_{\ell-1}]/s$$

$$= u'_\ell + p[(I - K_{\ell-1})^{-1}(g_{\ell-1} + \{(I - K_{\ell-1})\tilde{u}_{\ell-1} - g_{\ell-1}\} - s\, d_{\ell-1}) - \tilde{u}_{\ell-1}]/s$$

$$= u'_\ell - p\,(I - K_{\ell-1})^{-1}\, d_{\ell-1}$$

for any $\tilde{u}_{\ell-1}$ and s. □

In the nonlinear case the Taylor expansion of the right hand side of $(3.2.1b_4)$ yields $u'_\ell - p(I - K_{\ell-1})^{-1}\, d_{\ell-1}$. This shows why $(3.2.1b_4)$ is the nonlinear counterpart of the linear correction step (2.3.1b).

The function $\Phi_{\ell-1}(\tilde{f}_{\ell-1} - s\, d_{\ell-1})$ is only defined if

$$\|\tilde{f}_{\ell-1} - s\, d_{\ell-1}\|_U \leq \varepsilon_{\ell-1}. \qquad (3.2.2)$$

<u>Note 3.2.2</u> Let $\tilde{u}_{\ell-1}$ be given a priori with, e.g.,

$$\|\tilde{f}_{\ell-1}\|_U \leq \varepsilon_{\ell-1}/2. \qquad (3.2.3)$$

Then the choice $s := \|\tilde{f}_{\ell-1}\|_U/(2\|d_{\ell-1}\|_U)$ ensures (3.2.2). The fixed choice of $\tilde{u}_{\ell-1}$ has the advantage that $\tilde{f}_{\ell-1}$ (cf. $(3.2.1b_3)$) has to be computed only once, whereas the definition of $\tilde{u}_{\ell-1}$ by $(3.2.1b_2)$ requires the computation of $\tilde{f}_{\ell-1}$ by $(3.2.1b_3)$ in every iteration.

<u>Note 3.2.3</u> Brandt's (1977) so-called 'FAS method' is obtained by $(3.2.1b_2)$ and $s = 1$; $\tilde{f}_{\ell-1}$ has to be recomputed in every iteration. The inequality (3.2.2) in question becomes

$$\|[ru'_\ell - \mathcal{K}_{\ell-1}(ru'_\ell)] - r[u'_\ell - \mathcal{K}_\ell(u'_\ell) - f_\ell]\|_U$$

$$= \|r\mathcal{K}_\ell(u'_\ell) - \mathcal{K}_{\ell-1}(ru'_\ell) + rf_\ell\|_U \leq \varepsilon_{\ell-1}.$$

The first term $\|r\mathcal{K}_\ell(u'_\ell) - \mathcal{K}_{\ell-1}(ru'_\ell)\|_U$ may be expected to be of $O(h_\ell^\kappa)$, provided that u'_ℓ is smooth. However, the smoothing step (3.2.1a) guarantees the smoothness of the error $u'_\ell - \Phi_\ell(f_\ell)$, though not necessarily of u'_ℓ itself.

3.2.2 Multigrid Iteration

The two-grid iteration (3.2.1) is hardly of practical interest, since each iteration requires the solution of $\Phi_{\ell-1}(\tilde{f}_{\ell-1} - s\, d_{\ell-1})$ of a nonlinear coarse-grid equation. In the multigrid algorithm nonlinear equations have to be solved only at the coarsest level $\ell = 0$. For that purpose we need some iterative method

$$(u_0^j, f_0) \longmapsto u_0^{j+1} = \tilde{\Phi}_0(u_0^j, f_0). \tag{3.2.4}$$

Note 3.2.4 If K_ℓ is contractive (i.e. $\|K_0(v_0) - K_0(w_0)\|_U < L\|v_0 - w_0\|_U$, $L < 1$), a certain number of Picard iterations may serve as $\tilde{\Phi}_0$. Another possibility is a Newton iteration $\tilde{\Phi}_0$ requiring an approximation of the Jacobi matrix $K_0(u_0^j)$.

The multigrid iteration needs auxiliary values

$$\tilde{u}_0 \in U_0^0,\ \tilde{u}_1 \in U_1^0, \ldots, \tilde{u}_{\ell-1} \in U_{\ell-1}^0 \tag{3.2.5a}$$

with defects

$$\tilde{f}_k = \tilde{u}_k - K_k(\tilde{u}_k), \quad 0 \leq k \leq \ell-1. \tag{3.2.5b}$$

These values may be given a priori (cf. (3.2.8b) or a posteriori analogously to (3.2.1b$_2$) by inserting statements

$$\tilde{u}_{\ell-1} := ru; \quad \tilde{f}_{\ell-1} := \tilde{u}_{\ell-1} - K_{\ell-1}(\tilde{u}_{\ell-1})$$

between (3.2.6b) and (3.2.6c$_1$). The first variant (2.3.2) becomes

Nonlinear multigrid iteration (first variant) for solving
$u_\ell = K_\ell(u_\ell) + f_\ell$ (3.2.6)

```
procedure NMGM(ℓ,u,f); integer ℓ; array u,f;
if ℓ = 0 then u := Φ̃_0(u,f) else                    (3.2.6a)
begin integer j; real s; array v,d;
      u := K_ℓ(u) + f;                                  (3.2.6b)
      d := r*(u - K_ℓ(u) - f);                          (3.2.6c_1)
      s := if d = 0 then 1 else σ_{ℓ-1}/‖d‖_U;          (3.2.6c_2)
      d := f̃_{ℓ-1} - s * d;                             (3.2.6c_3)
      v := ũ_{ℓ-1};                                     (3.2.6c_4)
      for j := 1,2 do NMGM (ℓ-1, v, d);                 (3.2.6c_5)
      u := u + p(v - ũ_{ℓ-1})/s                         (3.2.6c_6)
end;
```

According to Note 3.2.2, one should choose $\sigma_{\ell-1} \leqslant \varepsilon_{\ell-1}/2$. If $\tilde{f}_{\ell-1} \neq 0$ satisfies (3.2.3), the choice

$$\sigma_{\ell-1} = \|\tilde{f}_{\ell-1}\|_U$$

is permitted. The case of $d = 0$ in (3.2.6c_2) is uninteresting, since then the coarse-grid correction (3.2.6c) can be omitted on account of (3.2.6c_5) yielding $v = \tilde{u}_{\ell-1}$ for any $s \neq 0$.

Note 3.2.5 Iteration (3.2.6) requires no derivative. Only the evaluation of $K_\ell(u_\ell)$ is needed.

Also the second variant (2.4.2) can be generalised to nonlinear problems.

Nonlinear multigrid iteration (second variant) for solving
$u_\ell = K_\ell(u_\ell) + f_\ell$ (3.2.7)

<u>procedure</u> NMGM(ℓ,u,f); <u>integer</u> ℓ; <u>array</u> u,f;

<u>if</u> $\ell = 0$ <u>then</u> u := $\tilde{\Phi}_0$(u,f) <u>else</u> (3.2.7a)

<u>begin</u> <u>integer</u> j; <u>real</u> s; <u>array</u> v,u',d;

$u' := K_\ell(u) + f;$ (3.2.7b)

$d := K_{\ell-1}(ru') - K_{\ell-1}(ru) + r*(u - K_\ell(u') - f);$ (3.2.7c_1)

$u := u - pr(u - u');$ s := <u>if</u> d = 0 <u>then</u> 1 <u>else</u> $\sigma_{\ell-1}/\|d\|_U$;

(3.2.7c_2)

$d := \tilde{f}_{\ell-1} - s*d;$ (3.2.7c_3)

$v := \tilde{u}_{\ell-1};$ <u>for</u> j := 1,2 <u>do</u> NMGM(ℓ-1,v,d); (3.2.7c_4)

$u := u + p*(v - \tilde{u}_{\ell-1})/s$ (3.2.7c_5)

<u>end</u>;

In (3.2.7c_1) the difference $K_{\ell-1}(ru') - K_{\ell-1}(ru)$ may be replaced by $K_{\ell-1}(\tilde{u}_{\ell-1} + r(u' - u)) - \tilde{u}_{\ell-1} + \tilde{f}_{\ell-1}$.

The nonlinear formulation of the third multigrid variant (2.5.1) is obvious. Analogously to Note 3.2.1 one proves

<u>Note 3.2.6</u> If the mappings $K_k(u_k) = K_k u_k + g_k$, $0 \leq k \leq \ell$, are affine, the iterations (3.2.6) and (3.2.7) are algebraically equivalent to the respective linear multigrid iterations (2.3.2) and (2.4.2), independently of the choice of σ_k and $\tilde{u}_{k-1}$.

3.2.3 Nested Iteration

The nested iteration (3.2.8) is used not only to provide good starting guesses at the next level, but also to compute the auxiliary $\tilde{u}_k$ from (3.2.5a) and their defects (3.2.5b). When procedure NMGM is called in (3.2.8d) at level k, the values $\tilde{u}_0, \tilde{u}_1, \ldots, \tilde{u}_{k-1}$ and $\tilde{f}_0, \tilde{f}_1, \ldots, \tilde{f}_{k-1}$ are already computed. The statement (3.2.8b) can be omitted if (3.2.1$b_{2,3}$) is used inside (3.2.6) or (3.2.7).

Nested iteration for solving $u_k = K_k(u_k)$, $0 \le k \le \ell$	(3.2.8)
$\tilde{u}_0$: given approximation to u_0^*;	(3.2.8a)
<u>for</u> k := 1 <u>step</u> 1 <u>until</u> ℓ <u>do</u>	
<u>begin</u> $\tilde{f}_{k-1} := \tilde{u}_{k-1} - K_{k-1}(\tilde{u}_{k-1})$;	(3.2.8b)
$\tilde{u}_k := \tilde{p}\,\tilde{u}_{k-1}$;	(3.2.8c)
<u>for</u> j := 1 <u>step</u> 1 <u>until</u> i <u>do</u> NMGM(k,$\tilde{u}_k$,0)	(3.2.8d)
<u>end</u>;	

If K_ℓ is defined by Example 3.1.2, also the nonlinear counterpart of the nested iteration (2.7.6) with Nyström's interpolation can be applied.

3.2.4 Convergence

In §3.4.3 we shall analyse the convergence of the multigrid iteration and the accuracy of the results $\tilde{u}_k$ of the nested iteration. A necessary condition for convergence $u_\ell^j \to u_\ell^*$ is stated in

<u>Note 3.2.7</u> The exact solution u_ℓ^* is a fixed point of the multigrid iterations (3.2.6) and (3.2.7).

It will turn out that the nonlinear multigrid iteration converges asymptotically as fast as the linear iteration applied to the linearised problem $u_\ell = K_\ell u_\ell + f_\ell$ with K_ℓ from (3.1.4).

As with any nonlinear iteration, the multigrid iteration may fail if the starting guess is not close enough to the solution. The starting values $\tilde{u}_k$ from (3.2.8c) must be inside the 'convergence domain' $U_k^* := \{v_k \in U_k^0 : \text{iteration with } u_k^0 = v_k,\ f_k = 0 \text{ yields } u_k^i \to u_k^*\}$. However, $\tilde{p}\tilde{u}_{k-1} \in U_k^*$ cannot be guaranteed in general even if $\tilde{u}_{k-1}$ is very accurate, since $\tilde{p}u_{k-1}^* \notin U_k^*$ may hold. In that case, one should try to apply continuation techniques.

3.2.5 Computational Work

The work of vector additions and of multiplications by r, p or scalar factors can be neglected, since the dominating part of the computation is the evaluation of $K_\ell(v_\ell)$. Assume

$$v_\ell \longmapsto K_\ell(v_\ell) \text{ requires } \leq W_\ell \text{ operations,} \tag{3.2.9a}$$

$$(v_0, f_0) \longmapsto \tilde{\Phi}_0(v_0, f_0) \text{ requires } \leq W_\Phi \text{ operations.} \tag{3.2.9b}$$

Note 2.3.2 can be extended to the present case.

Note 3.2.8 The coarse-grid equations are solved in ($3.2.6c_5$) and ($3.2.7c_4$) by two iterations at level $\ell-1$ starting with $u^0_{\ell-1} = \tilde{u}_{\ell-1}$. The evaluation of $K_{\ell-1}(\tilde{u}_{\ell-1})$ can be omitted since $K_{\ell-1}(\tilde{u}_{\ell-1}) = \tilde{u}_{\ell-1} - \tilde{f}_{\ell-1}$.

Taking into account Note 3.2.8 one proves

Proposition 3.2.9 Assume $\tilde{u}_k$, $\tilde{f}_k$, $0 \leq k \leq \ell-1$, to be known. One multigrid iteration at level ℓ requires at most

$$W_\ell^{NMGM} = \begin{cases} 2W_\ell + 3\Sigma_{k=1}^{\ell-1} 2^{\ell-k-1} W_k + 2^\ell W_\Phi & \text{for 1st variant (3.2.6)} \\ \Sigma_{k=1}^{\ell} 2^{\ell-k+1} W_k + 2^{\ell-1} W_0 + 2^\ell W_\Phi & \text{for 2nd variant (3.2.7)} \\ 2W_\ell + \Sigma_{k-1}^{\ell-1} 2^{\ell-k} W_k + 2^\ell W_\Phi & \text{for 3rd variant} \end{cases} \tag{3.2.10}$$

operations. If $\tilde{u}_k$ and $\tilde{f}_k$ are computed in every iteration by means of ($3.2.1b_{2,3}$), the respective numbers are

$$W_\ell^{NMGM} = \begin{cases} \Sigma_{k=1}^{\ell} 2^{\ell-k+1} W_k + 2^{\ell-1} W_0 + 2^\ell W_\Phi & \text{for 1st variant.} \\ 2W_\ell + 5\Sigma_{k=1}^{\ell-1} 2^{\ell-k-1} W_k + 2^\ell (W_0 + W_\Phi) & \text{for 2nd variant,} \\ 2W_\ell + 3\Sigma_{k=1}^{\ell-1} 2^{\ell-k+1} W_k + 2^{\ell-1} W_0 + 2^\ell W_\Phi & \text{for 3rd variant.} \end{cases} \tag{3.2.11}$$

The nested iteration (3.2.8) requires at most

$$W_\ell^{\text{nested it.}} = \begin{Bmatrix} 2iW_\ell + \Sigma_{k=1}^{\ell-1}[1+i(3.2^{\ell-k}-1)]W_k + W_0 \\ 2iW_\ell + \Sigma_{k=1}^{\ell-1}[1+i(4.2^{\ell-k}-2)]W_k + [1+i(2^\ell-1)]W_0 \\ 2iW_\ell + \Sigma_{k=1}^{\ell-1}[1+i2^{\ell-k+1}]W_k + W_0 \end{Bmatrix} + i(2^{\ell+1}-2)W_\Phi \tag{3.2.12}$$

operations. If $\tilde{u}_k$, $\tilde{f}_k$ are computed in each iteration and if (3.2.8b) is omitted, the respective numbers for the nested iteration are

$$\Sigma_{k=1}^{\ell}(2^{\ell-k+2} - 2)W_k + (2^{\ell} - 1)(W_O + 2W_{\Phi}),$$

$$\Sigma_{k=1}^{\ell}(5.2^{\ell-k} - 3)W_k + 2(2^{\ell-1} - 1)(W_O + W_{\Phi}), \tag{3.2.13}$$

$$\Sigma_{k=1}^{\ell}(3.2^{\ell-k} - 1)W_k + (2^{\ell} - 1)(W_O + 2W_{\Phi}).$$

These formulae do not take into account that the evaluation of $r\, K_{\ell}(u_{\ell})$ may be cheaper than the evaluation of $K_{\ell}(u_{\ell})$.

3.3 Application to Nonlinear Elliptic Equations

Equation (3.1.1) of the second kind may be representative of equations other than integral equations. A prototype is studied in this section. Further applications to elliptic boundary value problems are mentioned in §3.5 and in Hackbusch (1980a).

3.3.1 Reformulation of the Continuous Boundary Value Problem

Consider the elliptic boundary value problem

$$-\Delta u(x,y) = g(x,y, u(x,y)) \text{ in } \Omega = (0,1) \times (0,1),$$

$$u = 0 \text{ on } \Gamma = \partial\Omega \tag{3.3.1}$$

with some (nonlinear) function $g \in C^0(\Omega \times \mathbb{R})$. Problem (3.3.1) is equivalent to

$$u = (-\Delta)^{-1}\, g(.,.,u), \tag{3.3.2}$$

where $w = (-\Delta)^{-1}v$ denotes the solution of $-\Delta w = v$ in Ω, $w = 0$ on Γ. Abbreviating the right side of (3.3.2) by

$$K(u) := (-\Delta)^{-1}\, g(.,.,u), \tag{3.3.3}$$

we obtain the equation (3.1.1), $u = K(u)$. The derivative of K is

$$K(v) := (-\Delta)^{-1}\, g_u(.,.,v),$$

if $g_u = \partial g/\partial u$ exists. The relation $w = (-\Delta)^{-1}v$ can be represented by means of Green's function $G(x,y;\xi,\eta)$: $w(x,y) = \int_\Omega G(x,y;\xi,\eta)v(\xi,\eta)d\xi d\eta$. Hence, formally, $K(u)$ is the weakly singular, nonlinear integral operator

$$K(u)\ (x,y) = \int_\Omega G(x,y;\xi,\eta)g(\xi,\eta,u(\xi,\eta))d\xi d\eta. \tag{3.3.4}$$

We have to check the regularity (2.1.4), $\|K\|_{V\leftarrow U} \leq C$, of $K = K(u^*)$, since otherwise there is no hope of satisfying (2.2.6). We may choose

$$U = C^0(\bar{\Omega}),\ V = C^{1+\alpha}(\bar{\Omega}) \text{ with fixed } \alpha \in [0,1). \tag{3.3.5}$$

The well-known estimate $\|(-\Delta)^{-1}w\|_{C^{1+\alpha}(\bar{\Omega})} \leq C_\Omega \|w\|_{C^0(\bar{\Omega})}$ yields

$$\|Kw\|_V = \|(-\Delta)^{-1}g_u(\cdot,\cdot,u^*)w\|_{C^{1+\alpha}(\bar{\Omega})} \leq C_\Omega \|g_u(\cdot,\cdot,u^*)\|_{C^0(\bar{\Omega})} \|w\|_{C^0(\bar{\Omega})},$$

hence

$$\|K\|_{V\leftarrow U} \leq C_\Omega \max_{(x,y)\in\bar{\Omega}} |g_u(x,y,u^*(x,y))|. \tag{3.3.6}$$

3.3.2 Discretisation

The boundary value problem (3.3.1) can be discretised by

$$L_\ell u_\ell = g_\ell(u_\ell), \tag{3.3.7}$$

where $g_\ell(u_\ell)$ is the grid function with values $g_\ell(u_\ell)(x,y) = g(x,y,u_\ell(x,y))$ on the grid

$$D_\ell = \{(x,y) = (\nu h_\ell, \mu h_\ell) \in \mathbb{R}^2:\ \nu,\mu \in \mathbb{Z},\ 0 < \nu,\mu < 1/h_\ell\},\ h_\ell = h_0/2^\ell. \tag{3.3.8}$$

L_ℓ may be the five-point scheme including the homogeneous boundary conditions. The reformulation (3.3.1) $\longmapsto$ (3.3.2) applied to (3.3.7) yields

$$u_\ell = K_\ell(u_\ell) \text{ with } K_\ell(u_\ell) := L_\ell^{-1} g_\ell(u_\ell). \tag{3.3.9}$$

Note that the discretisation of $u = K(u)$ by (3.3.9) is much easier than a discretisation of the integral operator (3.3.4). The performance of $u_\ell \longmapsto K_\ell(u_\ell)$ is relatively cheap:

Note 3.3.1 The evaluation of $K_\ell(u_\ell)$ requires the evaluation of $g_\ell(u_\ell)$ [$O(h_\ell^{-2})$ operations] and one call of a direct Poisson solver. E.g. Buneman's algorithm (cf. Meis and Marcowitz (1978)) can be used. The required work amounts to $O(h_\ell^{-2}|\log h_\ell|)$.

The derivative of $K_\ell(u_\ell)$ at v_ℓ equals

$$K_\ell(v_\ell) = L_\ell^{-1}g'_\ell(v_\ell), \text{ where } (g'_\ell(v_\ell)w_\ell)(x,y) = [\partial g(x,y,v_\ell)/\partial u]w_\ell(x,y).$$

One can prove

Lemma 3.3.2 Assume g, $g_u \in C^0(\bar{\Omega} \times \mathbb{R})$, and suppose that there are solutions u_ℓ^* of (3.3.7) bounded by $\|u_\ell^*\|_{C^0(D_\ell)} \leq \text{const}$ for $\ell = 0,1,2...$.

Then $K_\ell = K'_\ell(u_\ell^*)$ satisfies the regularity condition (2.2.6) for the choice $U_\ell = C^0(D_\ell)$, $V_\ell = C^{1+\alpha}(D_\ell)$, $\alpha < 1$. Also, (2.2.9) and (2.2.12) can be shown to hold with $\kappa = 1 + \alpha$.

Although Eq. (3.3.9) is not a discrete integral equation, the multigrid iterations (3.2.6) and (3.2.7) can be applied to Eq. (3.3.9), since according to Note 3.3.1 $K_\ell(u_\ell)$ can easily be evaluated. The convergence depends on the conditions (2.2.5), (2.2.6), (2.2.9), (2.2.11), (2.2.12), not on the fact that K_ℓ represents an integral equation.

Nevertheless, the abstract equation (3.3.9) of the second kind give rise to some other properties of the multigrid process. The reason is not the nonlinearity. The following comments hold even for a linear equation $u_\ell = K(u_\ell)$, as obtained for $g(.,.,u) = g_1 + g_2u$.

1) In the case of discrete integral equations (2.2.4), the entries $K_{\ell,ij}$ of the matrix K_ℓ are explicitly known and they are used for the matrix multiplication. In the case of a linear Eq. (3.3.9) the entries of K_ℓ are never computed and the mapping $u_\ell \longmapsto K_\ell u_\ell = K_\ell(u_\ell) - K_\ell(0)$ is not performed as a matrix multiplication.

2) As a consequence, the modifications mentioned in §2.8.2 cannot be generalised, since the matrix A_ℓ involves the knowledge of special matrix components.

3) Nyström's interpolation (cf. §2.7.3) cannot be extended to problem (3.3.9).

4) The computational work of the mapping $u_\ell \to K_\ell(u_\ell)$ is not necessarily proportional to $(\dim U_\ell)^2$. In fact, in the case of Eq. (3.3.9) the work is $O(n_\ell \log n_\ell)$, $n_\ell = \dim(U_\ell)$ (cf. Note 3.3.1).

5) In the case of Eq. (3.3.9) the evaluation of $r_{inj}K_\ell(v_\ell)$ is not cheaper than the evaluation of $K_\ell(v_\ell)$.

6) Often, $K_\ell(u_\ell)$ from (3.3.9) has to be approximated iteratively, as pointed out in §3.4.

3.3.3 Computational Work

According to Note 3.3.1, the work W_ℓ required in the evaluation of $K_\ell(u_\ell)$ satisfies

$$W_{\ell-1} \lesssim \frac{1}{4} W_\ell , \tag{3.3.10}$$

as for (one-dimensional) integral equations.

Note 3.3.3 Assuming (3.3.10) and neglecting W_Φ, one obtains the following operation counts for the nested iteration (3.2.8):

$$W_\ell^{\text{nested it.}} \lesssim W_\ell * \begin{cases} \frac{1}{3} + 4\frac{2}{3} \cdot i & \text{(1st variant)} \\ \frac{1}{3} + 5\frac{1}{3} \cdot i & \text{(2nd variant)} \\ \frac{1}{3} + 4 \cdot i & \text{(3rd variant).} \end{cases} \tag{3.3.11}$$

3.3.4 Numerical Example

Consider the problem (3.3.1) with $g(x,y,u) = e^u$:

$$-\Delta u = e^u \text{ in } \Omega = (0,1) \times (0,1),\ u = 0 \text{ on } \Gamma. \tag{3.3.12}$$

The five-point discretisation of (3.3.7) is chosen with $h_0 = 1/2$, $h_1 = 1/4,\dots,h_5 = 1/64$. At level $\ell = 0$ the discrete problem $u_0 = K_0(u_0) + f_0$ is a scalar equation ($\dim U_0 = 1$), which is solved iteratively by

$$v_0 \longmapsto \tilde{\Phi}_0(v_0, f_0) = 0.0668 - 1.07\, f_0.$$

The nested iteration (3.2.8) is stated with $\tilde{u}_0(\frac{1}{2},\frac{1}{2}) = 0.066819$. The number of iterations is $i = 1$. The same piecewise linear interpolation $\tilde{p}$ is used as inside the multigrid iteration NMGM (3.2.6). The resulting values $\tilde{u}_k$ at the midpoint $(\frac{1}{2},\frac{1}{2})$ are listed in Table 3.3.1. The iteration errors are much smaller than the discretisation errors. Richardson's extrapolation applied to the values $\tilde{u}_k(\frac{1}{2},\frac{1}{2})$, $k = 4,5,6$, given in Table 3.3.1 yields the number 0.078101022605, which coincides with the accurate value of $u(\frac{1}{2},\frac{1}{2})$ to nine digits. For more details, see Hackbusch (1979a).

Table 3.3.1

Results of the nested iteration applied to Eq. (3.3.12).

level k	0	1	2	3	4	5
grid size h_k	1/2	1/4	1/8	1/16	1/32	1/64
$\tilde{u}_k(\frac{1}{2},\frac{1}{2})$	.066819	.074579	.077190	.077872	.078043	.078087
observed rate	-	0.02	0.007	0.0026	0.001	0.0007

According to (3.3.11), the above computation requires a work $\leq 5W_5$, equivalent to five calls of a direct Poisson solver. Were the problem (3.3.7) solved by Newton's iteration, direct Poisson solvers could not have been applied, since the resulting linear equations $-\Delta w = g_u w$ have variable coefficients $g_u(x,y)$.

3.4 *Multigrid Methods with Iterative Computation of* $K_\ell(u_\ell)$

3.3.4 Notation

In the examples of §3.3 the computation of $K_\ell(u_\ell)$ involves the solution of a discrete Poisson equation. Applying a direct Poisson solver we are able to evaluate $K_\ell(u_\ell)$ explicitly. However, if Eq. (3.3.7) involves a more complicated operator L_ℓ, it cannot be inverted by a direct solver; one has to use an iterative method, e.g. a linear multigrid iteration. Thus, $K_\ell(u_\ell)$ will not be determined exactly. Instead, an approximation can be computed depending on the starting value and on the number of iterations. By

$$K_\ell(v_\ell;w_\ell,\mu)$$

we denote the result of μ iterations with starting iterate $w_\ell =: K_\ell(v_\ell;w_\ell, 0)$.

Example 3.4.1 Let $u_\ell^{j+1} = M_\ell u_\ell^j + N_\ell f_\ell$ be a (e.g. multigrid) iteration solving $L_\ell u_\ell = f_\ell$ with L_ℓ from (3.3.7). An iterative computation of $K_\ell(v_\ell) = L_\ell^{-1} g_\ell(v_\ell)$ from (3.3.9) is given by

$$w_\ell =: K_\ell(v_\ell;w_\ell,0) \longmapsto K_\ell(v_\ell;w_\ell,1) := M_\ell w_\ell + N_\ell g_\ell(v_\ell) \longmapsto \ldots$$

$$\longmapsto K_\ell(v_\ell;w_\ell,\mu) := M_\ell K_\ell(v_\ell;w_\ell,\mu-1) + N_\ell g_\ell(v_\ell) \longmapsto \ldots$$

converging to $K_\ell(v_\ell)$ if $\rho(M_\ell) < 1$. The explicit description of $K_\ell(v_\ell;w_\ell,\mu)$ is

$$K_\ell(v_\ell;w_\ell,\mu) = M_\ell^\mu w_\ell + \sum_{\chi=0}^{\mu-1} M_\ell^\chi N_\ell g_\ell(v_\ell), \tag{3.4.1a}$$

$$K_\ell(v_\ell;w_\ell,\mu) - K_\ell(v_\ell) = M_\ell^\mu(w_\ell - K_\ell(v_\ell)). \tag{3.4.1b}$$

According to (3.4.1b) we shall assume the error estimate

$$\| K_\ell(v_\ell;w_\ell,\mu) - K_\ell(v_\ell) \|_U \leq \varepsilon^\mu \| w_\ell - K_\ell(v_\ell) \|_U , \tag{3.4.2}$$

which is true for (elliptic) multigrid iterations.

The following algorithm has been described and analysed in Hackbusch (1981c):

3.4.2 Algorithm

The multigrid iteration (3.2.6) is to be reformulated with $K_\ell(\cdot)$ replaced by $K_\ell(\cdot;\cdot,\mu_\ell)$, μ_ℓ depending on ℓ. If some rough approximation $\tilde{u}_\ell$ of the solution $u_\ell = K_\ell(u_\ell) + f_\ell$ is known, the obvious starting guess of $K_\ell(\tilde{u}_\ell)$ is $\tilde{u}_\ell - f_\ell$. Hence, $K_\ell(\tilde{u}_\ell)$ will be substituted by $K_\ell(\tilde{u}_\ell;\ \tilde{u}_\ell - f_\ell,\ \mu_\ell)$.

Multigrid iteration (first variant) for solving $u_\ell = K_\ell(u_\ell) + f_\ell$ with interior iterations (3.4.3)

```
procedure  NMGM(ℓ,u,f);  integer  ℓ;  array  u,f;
if  ℓ = 0  then  u := Φ̃_0(u,f)  else                         (3.4.3a)
begin  real s;  integer j;  array v,d;
         u := K_ℓ(u ; u - f, μ_ℓ) + f;                        (3.4.3b)
         d := r*(u - K_ℓ(u; u - f, μ_ℓ) - f);                 (3.4.3c_1)
         if  d ≠ 0  then                                      (3.4.3c_2)
         begin  s := σ_{ℓ-1}/‖d‖_U;                           (3.4.3c_3)
                d := f̂_{ℓ-1} - s*d;  v := ũ_{ℓ-1};            (3.4.3c_4)
                for j := 1,2 do  NMGM (ℓ-1,v,d);              (3.4.3c_5)
                u := u + p*(v - ũ_{ℓ-1})/s                    (3.4.3c_6)
end     end;
```

The auxiliary values $\tilde{u}_k$, $\hat{f}_k$ $(k < \ell)$ are assumed to be related by

$$\hat{f}_k := \tilde{u}_k - K_k(\tilde{u}_k;\ \tilde{u}_k,\ \mu_k), \quad k = 0,1,\ldots,\ell-1. \qquad (3.4.4)$$

Under conditions (3.4.2) the mapping $v_\ell \mapsto K_\ell(v_\ell; v_\ell - f_\ell, \mu_\ell) + f_\ell$ has the fixed point $v_\ell = \Phi_\ell(f_\ell)$ (solution of $u_\ell = K_\ell(u_\ell) + f_\ell$). But a difficulty arises from the coarse-grid correction (3.4.3c).

Note 3.4.2 If the parameter s were fixed (in contrast to (3.4.3c_3)) as it is in the case of Note 3.2.3, the multigrid iteration (3.4.3) would not converge to the solution $u_\ell = K_\ell(u_\ell) + f_\ell$.

Proof Apply the iteration (3.4.3) to the exact solution $u_\ell = \Phi_\ell(f_\ell)$. As mentioned above, $d = 0$ results in (3.4.3c_1). Neglect (3.4.3$c_{2,3}$). The statements (3.4.3c_4) yield $v = \tilde{u}_{\ell-1}$ and $d = \hat{f}_{\ell-1}$. For the case $\hat{f}_k = \tilde{u}_k - K_k(\tilde{u}_k)$, $k < \ell$, one would obtain $v = \tilde{u}_{\ell-1}$ by (3.4.3c_5).

However, $\hat{f}_k$ from (3.4.4) does not necessarily equal $\tilde{u}_k - K_k(\tilde{u}_k)$; hence (3.4.3c$_6$) yields $u \neq u_\ell$. This proves that $u_\ell = \Phi_\ell(f_\ell)$ is not a fixed point if (3.4.3c$_2$) is omitted. Even with (3.4.3c$_2$) present, we can repeat the proof with u_ℓ^0 arbitrarily close to u_ℓ and obtain a next iterate u_ℓ^1 with $\|u_\ell^1 - u_\ell^0\| \geq C > 0$ for all u_ℓ^0 in a neighbourhood of u_ℓ. □

It turns out that the fixed point of iteration (3.4.3) with constant s is some $\hat{u}_\ell$ with

$$\|\hat{u}_\ell - u_\ell\|_U = O(\|\hat{f}_{\ell-1} - \tilde{u}_{\ell-1} + K_{\ell-1}(\tilde{u}_{\ell-1})\|_U/s).$$

By (3.4.2) the right side equals $O(\varepsilon^{\mu_{\ell-1}}\|\tilde{u}_{\ell-1} - K_{\ell-1}(\tilde{u}_{\ell-1})\|_U/s)$ $= O(\varepsilon^{\mu_{\ell-1}}\|\tilde{f}_{\ell-1}\|_U/s)$ with $\tilde{f}_{\ell-1}$ from (3.2.5b). The difference $\hat{u}_\ell - u_\ell$ is small if s is large. However, s must be small enough to ensure $\|\hat{f}_{\ell-1} - sd\|_U \leq \varepsilon_{\ell-1}$ (cf. (3.2.2)). Thanks to (3.4.3c$_3$), the value of s increases as u_ℓ^j tends to u_ℓ since $\|d\|_U$ decreases.

The nested iteration (3.2.8) becomes

Nested iteration for solving $u_k = K_k(u_k)$, $k = 0,1,\ldots,\ell$	(3.4.5)
$\tilde{u}_0$: given approximation to u_0^*;	(3.4.5a)
<u>for</u> k := 1 <u>step</u> 1 <u>until</u> ℓ <u>do</u>	
<u>begin</u> $\hat{f}_{k-1} := \tilde{u}_{k-1} - K_{k-1}(\tilde{u}_{k-1};\ u_{k-1},\ \mu_{k-1})$;	(3.4.5b)
$\tilde{u}_k := \tilde{p} * \tilde{u}_{k-1}$;	(3.4.5c)
<u>for</u> j := 1 <u>step</u> 1 <u>until</u> i <u>do</u> NMGM(k, $\tilde{u}_k$, 0);	(3.4.5d)
<u>comment</u> NMGM from (3.4.3)	
<u>end</u>;	

<u>Note 3.4.3</u> The algorithms (3.2.6) and (3.2.8) are special cases of (3.4.3) and (3.4.5), where $K_\ell(\cdot;\cdot,\cdot)$ is defined by $K_\ell(v_\ell;\ w_\ell,\ 0) = w_\ell$, $K_\ell(v_\ell;\ w_\ell,\ \mu) = K_\ell(v_\ell)$ for $\mu > 0$, and where $\mu_k = 1$. Then, formally, (3.4.2) holds with $\varepsilon = 0$.

3.4.3 Convergence Analysis

We recall that u_ℓ^* denotes the exact solution of (3.1.2).

Theorem 3.4.4 Assume

- (3.1.5): $\|M_\ell\|_{U\leftarrow U} \leqslant C_{TGM} h_\ell^\kappa, \ \kappa > 0$ (3.4.6a)

  (two-grid convergence of the linearised problem),

- (3.4.2): contractivity of $K_\ell(\cdot;\cdot,\cdot)$, (3.4.6b)

- $\|\tilde{\Phi}_0(u_0,f_0) - \Phi_0(f_0)\|_U \leqslant C_\Phi h_0^\kappa \|u_0 - \Phi_0(f_0)\|_U$ (3.4.6c)

  for all $\|f_0\|_U \leqslant \varepsilon_0$ (cf. (3.1.6b)) with $C_\Phi h_0^\kappa < 1$,

- h_0 sufficiently small, (3.4.6d)

- μ_k be such that

  $$\varepsilon^{\mu_k} \leqslant C_\varepsilon \, C_{TGM} h_k^\kappa, \quad k = 0,1,2,\ldots,\ell-1, \qquad (3.4.6e)$$

  for some C_ε,

- $\|\tilde{p}\|_{U\leftarrow U} \leqslant C_{\tilde{p}}$, (3.4.6f)

- $\|\tilde{u}_0 - u_0^*\|_U \leqslant C_{u,0} \, h_0^{(i+1)\kappa}$, (3.4.6g)

- $\|\tilde{p}\, u_{k-1}^* - \tilde{u}_k^*\|_U \leqslant C_C h_k^\kappa, \quad k = 1,2,\ldots,\ell,$ (3.4.6h)

- $h_{\ell-1}/h_\ell \leqslant C_H$ (3.4.6i)

- there be constants $\underline{C}$ and $\bar{C}$ such that σ_k satisfies

  $$\|\hat{f}_k - \tilde{f}_k\|_U \leqslant \underline{C}\, |\sigma_k|\, C_{TGM} h_k^\kappa, \quad 0 \leqslant k \leqslant \ell-1, \qquad (3.4.6j)$$

  $$|\sigma_k| \leqslant \varepsilon_k/2, \quad |\sigma_k| \leqslant \bar{C}\, C_{TGM} h_k^\kappa, \quad 0 \leqslant k \leqslant \ell-1, \qquad (3.4.6k)$$

  with $\tilde{f}_k$ from (3.2.1b$_3$),

- $\varepsilon_{k-1}/\varepsilon_k \leqslant (h_{k-1}/h_k)^\kappa$. (3.4.6l)

Then there are constants C_u, C_f and C_{NMGM} so that values $\tilde{u}_k$ and $\hat{f}_k$ produced by the nested iteration (3.4.5) satisfy

$$\|\tilde{u}_k - u_k^*\|_U \leq C_u h_k^\kappa (C_{NMGM} h_k^\kappa)^i, \quad 0 \leq k \leq \ell, \tag{3.4.7a}$$

$$\|\hat{f}_k\|_U \leq C_{\hat{f}} h_k^\kappa (C_{NMGM} h_k^\kappa)^i, \quad 0 \leq k \leq \ell-1, \tag{3.4.7b}$$

provided that $i \geq 1$. $C_{NMGM} h_k^\kappa$ is the contraction number of the nonlinear multigrid iteration (3.4.3). More precisely, if u_ℓ^0, f_ℓ, $\hat{f}_k$ satisfy

$$\|u_\ell^0 - \Phi_\ell(f_\ell)\|_U \leq C_u h_k^\kappa , \tag{3.4.8a}$$

$$\|f_\ell\|_U \leq \min(C_f h_\ell^\kappa, \varepsilon_\ell), \quad \|\hat{f}_k\|_U \leq \min(C_{\hat{f}} h_k^\kappa, \varepsilon_{k/2}), \quad 0 \leq k \leq \ell-1, \tag{3.4.8b}$$

then iteration (3.4.3) with $f = f_\ell$ yields u_ℓ^1 satisfying

$$\|u_\ell^1 - \Phi_\ell(f_\ell)\|_U \leq C_{NMGM} h_\ell^\kappa \|u_\ell^0 - \Phi_\ell(f_\ell)\|_U. \tag{3.4.9}$$

The assumptions (3.4.6j), (3.4.6k) seem complicated. We shall propose a simple choice of σ_k in Corallary 3.4.5. Inequality (3.4.6ℓ) excludes the case of rapidly decreasing ε_k. Usually, $\varepsilon_0 = \varepsilon_1 = \ldots$ will hold.

Proof 1) First we prove the convergence statement (3.4.9).

1a) Analysis of statement (3.4.3b). The error of u_ℓ^0 is $\delta_\ell^0 := u_\ell^0 - \Phi_\ell(f_\ell)$, $\Phi_\ell(f_\ell)$ solution of $u_\ell = K_\ell(u_\ell) + f_\ell$. Set

$$u_\ell' := K_\ell(u_\ell;\ u_\ell^0 - f_\ell,\ \mu_\ell) + f_\ell.$$

Then

$$\delta_\ell' := u_\ell' - u_\ell = K_\ell^0(u_\ell^0;\ u_\ell^0 - f_\ell,\ \mu_\ell) - K_\ell(u_\ell)$$

$$= K_\ell(u_\ell^0) - K_\ell(u_\ell) + K_\ell(u_\ell^0;\ u_\ell^0 - f_\ell,\ \mu_\ell) - K_\ell(u_\ell^0)$$

$$= K_\ell(u_\ell)\delta_\ell^0 + O(\|\delta_\ell^0\|_U^2) + \varepsilon^{\mu_\ell}\|(u_\ell^0 - f_\ell) - K_\ell(u_\ell^0)\|_U$$

by (3.4.2). Since $u^0_\ell - f_\ell - K_\ell(u^0_\ell) = \delta^0_\ell - (K_\ell(u^0_\ell) - K_\ell(u_\ell)) = O(\|\delta^0_\ell\|_U)$ and $K_\ell(u_\ell) = K_\ell + O(\|u_\ell - u^*_\ell\|_U) = K_\ell + O(\|f_\ell\|_U)$, we conclude that

$$\delta'_\ell = K_\ell\delta^0_\ell + O([\|\delta^0_\ell\|_U + \varepsilon^{\mu_\ell} + \|f_\ell\|_U]\|\delta^0_\ell\|_U). \tag{3.4.10}$$

1b) Analysis of statement (3.4.3c$_1$). Let

$$d_\ell := u'_\ell - K_\ell(u'_\ell;\ u'_\ell - f_\ell,\ \mu_\ell) - f_\ell.$$

This can be written as

$$d_\ell = u'_\ell - K_\ell(u'_\ell) - f_\ell + O(\varepsilon^{\mu_\ell}\|u'_\ell - K_\ell(u'_\ell) - f_\ell\|_U)$$

$$= [I - K_\ell(u_\ell)]\delta'_\ell + O(\|\delta'_\ell\|^2_U + \varepsilon^{\mu_\ell}\|\delta'_\ell\|_U)$$

$$= [I - K_\ell]\delta'_\ell + O([\|\delta'_\ell\|_U + \varepsilon^{\mu_\ell} + \|f_\ell\|_U]\|\delta'_\ell\|_U).$$

Eq. (3.4.10) implies

$$d_\ell = [I - K_\ell]K_\ell\delta^0_\ell + O([\|\delta^0_\ell\|_U + \varepsilon^{\mu_\ell} + \|f_\ell\|_U]\|\delta^0_\ell\|_U). \tag{3.4.11}$$

1c) Analysis of the coarse-grid equation. Relation $d_{\ell-1} := \hat{f}_{\ell-1} - srd_\ell$ fulfils

$$\|d_{\ell-1}\|_U \leqslant \|\hat{f}_{\ell-1}\|_U + |\sigma_{\ell-1}| \leqslant (C_{\hat{f}} + \bar{C}\, C_{TGM})h^\kappa_{\ell-1}$$

and $\|d_{\ell-1}\|_U \leqslant \varepsilon_{\ell-1}$. Choose C_f in (3.4.8b) large enough so that $C_f \geqslant C_{\hat{f}} + \bar{C}\, C_{TGM}$. Let $v_{\ell-1} = \Phi_{\ell-1}(d_{\ell-1})$ be the exact solution of the coarse-grid equation. The error of the starting value $v^0_{\ell-1} := \tilde{u}_{\ell-1}$ is

$$\|v^0_{\ell-1} - v_{\ell-1}\|_U = \|\Phi_{\ell-1}(\tilde{f}_{\ell-1}) - \Phi_{\ell-1}(d_{\ell-1})\| \leqslant C'\|\tilde{f}_{\ell-1} - d_{\ell-1}\|_U$$

$$= C'\|\tilde{f}_{\ell-1} - \hat{f}_{\ell-1} + srd_\ell\|_U \leqslant C'[\underline{C}\, C_{TGM}h^\kappa_{\ell-1} + 1]|\sigma_{\ell-1}|$$

$$\leqslant C'[\underline{C}\, C_{TGM}h^\kappa_{\ell-1} + 1]\,\bar{C}\, C_{TGM}h^\kappa_{\ell-1}$$

with $C' := \sup\{\|K_{\ell-1}(v_{\ell-1})\|_{U\leftarrow U} : v_{\ell-1} = \Phi_{\ell-1}(f_{\ell-1}),\ \|f_{\ell-1}\|_U \leq \varepsilon_{\ell-1}\}$. Choose C_u so large that $C'[\underline{C}\ C_{TGM}h^{\kappa}_{\ell-1} + 1]\bar{C}\ C_{TGM} \leq C_u$. Then, the data $d_{\ell-1}$ and $v^0_{\ell-1}$ satisfy (3.4.8a,b). By induction, two iterations at level ℓ-1 yield $v^2_{\ell-1}$ with

$$\|v^2_{\ell-1} - v_{\ell-1}\|_U \leq C^2_{NMGM}h^{2\kappa}_{\ell-1}\|v^0_{\ell-1} - v_{\ell-1}\|_U$$

$$\leq C^2_{NMGM}\ C'\ h^{2\kappa}_{\ell-1}(\|\delta f_{\ell-1}\|_U + |\sigma_{\ell-1}|),$$

where

$$\delta f_k := \tilde{f}_k - \hat{f}_k = \tilde{u}_k - K_k(\tilde{u}_k) - \hat{f}_k. \tag{3.4.12}$$

1d) Analysis of correction step (3.4.3c$_6$). The new error δ^1_ℓ is

$$\delta^1_\ell := u^1_\ell - u_\ell = u'_\ell - p(v^2_{\ell-1} - \tilde{u}_{\ell-1})/s - u_\ell$$

$$= u'_\ell + p[\Phi_{\ell-1}(\hat{f}_{\ell-1} - srd_\ell) - \Phi_{\ell-1}(\hat{f}_{\ell-1})]/s - u_\ell$$

$$+ p[(v^2_{\ell-1} - v_{\ell-1}) + \Phi_{\ell-1}(\hat{f}_{\ell-1}) - \Phi_{\ell-1}(\tilde{f}_{\ell-1})]/s$$

$$= \delta'_\ell - p[I - K_{\ell-1}(\Phi_{\ell-1}(\hat{f}_{\ell-1}))]^{-1}\ rd_\ell$$

$$+ O(|s|\|rd_\ell\|^2_U + C^2_{NMGM}h^{2\kappa}_\ell(\|\delta f_{\ell-1}\|_U + |\sigma_{\ell-1}|)/s + \|\delta f_{\ell-1}\|_U/s)$$

$$= K_\ell\delta^0_\ell - p[I - K_{\ell-1}]^{-1}r[I - K_\ell]K_\ell\delta^0_\ell$$

$$+ O(\|\hat{f}_{\ell-1}\|_U\|rd_\ell\|_U + |\sigma_{\ell-1}|\|rd_\ell\|_U$$

$$+ C^2_{NMGM}h^{2\kappa}_\ell(\|\delta f_{\ell-1}\|_U/|\sigma_{\ell-1}| + 1)\|rd_\ell\|_U$$

$$+ \|\delta f_{\ell-1}\|_U\ \|rd_\ell\|_U/|\sigma_{\ell-1}|)$$

since $s = \sigma_{\ell-1}/\|rd_\ell\|_U$. The first term equals $M_\ell \delta^0_\ell$. Thanks to (3.4.6a,e) and (3.4.11) one obtains

$$\delta^1_\ell = M_\ell \delta^0_\ell + O([\|\hat{f}_{\ell-1}\|_U + |\sigma_{\ell-1}| + \|\delta f_{\ell-1}\|_U/|\sigma_{\ell-1}| + C^2_{NMGM} h^{2\kappa}_\ell$$

$$+ \|\delta^0_\ell\|_U + \varepsilon^{\mu_\ell} + \|f_\ell\|_U]\|\delta^0_\ell\|_U), \tag{3.4.13}$$

$$\|\delta^1_\ell\|_U \leq C_{TGM} h^\kappa_\ell \{1 + C[\bar{C} + \underline{C} + C_\varepsilon + (C_{\hat{f}} + 2C_c + C_f + C^2_{NMGM} h^\kappa_\ell)$$

$$/C_{TGM}]\}\|\delta^0_\ell\|_U.$$

Make the ansatz $C_{NMGM} = \alpha\, C_{TGM}$. The expression in braces becomes $\{1 + C[\ldots + \alpha^2\, C_{TGM} h^\kappa_\ell]\}$. By (3.4.6d), α can be chosen such that $\{\ldots\} \leq \alpha$ yielding

$$\|\delta^1_\ell\|_U \leq \alpha\, C_{TGM} h^\kappa_\ell \|\delta^0_\ell\|_U = C_{NMGM} h^\kappa_\ell \|\delta^0_\ell\|_U$$

and proving (3.4.9).

1e) Start of the induction proof. In step 1c) we used induction. At the lowest level $\ell = 0$ the claimed inequality (3.4.9) follows from (3.4.6c) if we choose $C_{NMGM} \geq C_\Phi$.

2) Next, we prove (3.4.7a,b) by induction.

2a) $k = 0$. Eq. (3.4.6g) implies (3.4.7a), if C_u is chosen such that $C_{u,0} \leq C_u\, C^i_{NMGM}$. Eq. (3.4.7b) follows from

2b) (3.4.7a) implies (3.4.7b). There is a constant C with

$$\|\tilde{f}_k\|_U = \|\tilde{u}_k - K_k(\tilde{u}_k)\|_U = \|\tilde{u}_k - K_k(\tilde{u}_k) - (u^*_k - K_k(u^*_k))\|_U$$

$$\leq C\|\tilde{u}_k - u^*_k\|_U$$

where $u^*_k = \Phi_k(0)$ is the solution of $u_k = K_k(u_k)$. Eq. (3.4.2) shows

$$\|\delta f_k\|_U = \|K_k(\tilde{u}_k;\, \tilde{u}_k,\, \mu_k) - K_k(\tilde{u}_k)\|_U$$

$$\leq \varepsilon^{\mu_\ell}\|\tilde{u}_k - K_k(\tilde{u}_k)\|_U = \varepsilon^{\mu_\ell}\|\tilde{f}_k\|_U \tag{3.4.14}$$

(cf. (3.4.12)). The combination yields

$$\|\hat{f}_k\|_U \leq \|\tilde{f}_k\|_U + \|\delta f_k\|_U \leq (1 + \varepsilon^{\mu_\ell})\|\tilde{f}_k\|_U \leq C^*\|\tilde{u}_k - u_k^*\|_U.$$

Set $C_{\hat{f}} = C_u C^*$; then (3.4.7b) is valid.

2c) It remains to prove (3.4.7a) for $k = \ell$, assuming that (3.4.7) holds for $k < \ell$. The error of the starting guess $u_\ell^0 := \tilde{p}\,\tilde{u}_{\ell-1}$ (cf. (3.4.5c)) is $\delta_\ell = u_\ell^0 - u_\ell^*$. It is bounded by

$$\|\delta_\ell^0\|_U = \|\tilde{p}\,\tilde{u}_{\ell-1} - u_\ell^*\|_U = \|\tilde{p}(\tilde{u}_{\ell-1} - u_{\ell-1}^*) + \tilde{p}\,\tilde{u}_{\ell-1} - u_\ell^*\|_U$$

$$\leq C_c h_\ell^\kappa + C_{\tilde{p}}\, C_u h_{\ell-1}^\kappa\, (C_{NMGM} h_{\ell-1}^\kappa)^i.$$

Choose C_u such that

$$C_u \geq C_c/[1 - C_{\tilde{p}} C_H^\kappa\, (C_{NMGM} h_{\ell-1}^\kappa)^i],$$

which is possible by (3.4.6d), and $i > 0$. One obtains $\|\delta_\ell^0\|_U \leq C_u h_\ell^\kappa$. Since (3.4.8a,b) are satified, (3.4.9) implies (3.4.7a). □

The choice of $\sigma_{\ell-1}$ is still to be discussed. According to (3.4.13), the optimal value of $\sigma_{\ell-1}$ is $O(\|\delta f_{\ell-1}\|_U^{1/2})$. However, the magnitude $\|\delta f_{\ell-1}\|_U^{1/2}$ is not known in practice. A very simple and practical choice is proposed in

<u>Corollary 3.4.5</u> Set $\sigma_{\ell-1} := \|\hat{f}_{\ell-1}\|_U$ with $\hat{f}_{\ell-1}$ from (3.4.5b). This value vanishes if and only if $\tilde{u}_{\ell-1} = u_{\ell-1}^*$ (i.e. $\tilde{f}_{\ell-1} = \hat{f}_{\ell-1} = 0$). The inequalities (3.4.6j,k) are fulfilled with $\underline{C} = C_\varepsilon/(1 - C_\varepsilon C_{TGM} h_0^\kappa)$.

<u>Proof</u> From $\|\delta f_k\|_U \leq \varepsilon^{\mu_k}\|\tilde{f}_k\|_U$ (cf. (3.4.14)) and $\|\tilde{f}_k\|_U \leq \|\hat{f}_k\|_U + \|\delta f_k\|_U$ one concludes $\|\tilde{f}_k\|_U \leq \|\hat{f}_k\|_U/(1 - \varepsilon^{\mu_k})$. Hence, $\sigma_k = 0$ implies $\hat{f}_k = \tilde{f}_k = 0$ and $\tilde{u}_k = u_k^*$. The term $\|\delta f_k\|_U/|\sigma_k|$ is bounded by $\varepsilon^{\mu_k}/(1 - \varepsilon^{\mu_k}) \leq C_\varepsilon\, C_{TGM} h_k^\kappa/(1 - C_\varepsilon\, C_{TGM} h_k^\kappa)$, proving the definition of $\underline{C}$.

□

To avoid $\sigma_{\ell-1} = 0$ and the effects of rounding error one should modify the choice σ_k by

$$\sigma_k := \|\hat{f}_k\|_U + O(\sqrt{\|\tilde{u}_k\|_U \cdot \text{eps}}), \qquad (3.4.15)$$

where eps is the relative machine precision.

Corollary 3.4.6 If $K_\ell(u_\ell)$ is evaluated exactly (i.e. $\tilde{K}_\ell(u_\ell;\cdot,\cdot) = K_\ell(u_\ell)$), the conditions (3.4.6b,e,j) can be omitted. Then $\hat{f}_k = \tilde{f}_k$.

There are several modifications of the nested iteration (3.4.5). Instead of computing $\hat{f}_{k-1}$ in (3.4.5b) after the multigrid iterations at level k-1, one may define $\hat{f}_{k-1}$ together with the defect (3.4.3c_1) during the last multigrid iteration at level k-1. Then $\|\hat{f}_{k-1}\|_U$ and $\|\delta f_{k-1}\|_U$ equal $O(\|\delta^{i-1}_{k-1}\|_U)$. Possible choices of σ_{k-1} are $\|\hat{f}_{k-1}\|_U^{1/2}$ and $\|\hat{f}_{k-1}\|_U$, or (3.4.15).

3.5 Applications to Elliptic Boundary Value and Eigenvalue Problems

3.5.1 Elliptic Boundary Value Problems

In §3.3 we mentioned an application of the multigrid iteration of the second kind to equations of the form $-\Delta u = g(u)$. The technique can be extended to

$$L(u) := L^I(u) - L^{II}(u) = 0,$$

which can be reformulated as the fixed point equation

$$u = \Phi^I(L^{II}(u)) \qquad (\Phi^I : \text{inverse of } L^I),$$

provided that L^I contains the principal part of the (nonlinear) elliptic operator, whereas L^{II} is of lower order. For examples, see Hackbusch (1979a).

Interesting applications arise from systems of weakly coupled elliptic equations. A prototype of this form is

$$-\Delta v = g^I(v,w), \quad -\Delta w = g^{II}(v,w) \text{ in } \Omega, \quad v = w = 0 \text{ on } \Gamma,$$

which yields the system

$$v = (-\Delta)^{-1} g^I(v,w), \quad w = (-\Delta)^{-1} g^{II}(v,w)$$

of two equations of the second kind. Set $u := (v,w)^T$. The evaluation of

$$K(u) := ((-\Delta)^{-1}g^I(u),\quad (-\Delta)^{-1}g^{II}(u))^T$$

requires only the solution of two scalar Poisson equations. In the discrete case of $u_\ell = K_\ell(u_\ell)$ one can make use of Poisson solvers.

Storage can be saved if the components of the elliptic system are coupled only by (linear or nonlinear) boundary conditions. Such systems arise, for instance, from optimal control problems (Hackbusch (1980b)). We give a particular example.

Let $\Omega = (0,\pi) \times (0,\pi)$, $\Gamma = \partial\Omega$, $\Gamma_1 = \{(x_1,x_2) \in \Gamma : x_1 = \pi\} = \{\pi\} \times (0,\pi)$, $\Gamma_2 = \Gamma \setminus \Gamma_1$. For any $u \in L^2(\Gamma_1)$, $y(u)$ is the solution of

$$\{-(1 + x_1^2)^{-1}(\partial^2/\partial x_1^2) - (\partial^2/\partial x_2^2)\}\, y(u) = 0 \text{ in } \Omega,$$
$$(1 + \pi^2)^{-1}(\partial y(u)/\partial x_1) = u \text{ on } \Gamma_1,\quad y(u) = 0 \text{ on } \Gamma_2. \tag{3.5.1}$$

We seek the control u minimising a given functional

$$J(v) = \|y(v)|_{\Gamma_1} - \sin x_2\|^2_{L^2(\Gamma_1)} + (1 + \pi^2)^{-1}\|v\|^2_{L^2(\Gamma_1)}.$$

The solution u (optimal control) is determined by

$$u = -(1 + \pi^2)p(u)|_{\Gamma_1}, \tag{3.5.2}$$

where $p(u)$ is the solution of the adjoint problem

$$-(\partial^2/\partial x_1^2)[p(u)/(1 + x_1^2)] - (\partial^2/\partial x_2^2)p(u) = 0 \text{ in } \Omega,$$
$$(\partial/\partial x_1)[p(u)/(1 + x_1^2)] = y(u)|_{\Gamma_1} - \sin x_2 \text{ on } \Gamma_1, \tag{3.5.3}$$
$$p(u) = 0 \text{ on } \Gamma_2.$$

Eq. (3.5.2) can be used to eliminate Eq. (3.5.1). Then we obtain a system of two elliptic equations for y and p, which are coupled by means of the boundary conditions. On the other hand, Eqs. (3.5.1), (3.5.3), (3.5.2) describe mappings $u \mapsto y(u) \mapsto p(u) \mapsto u$, that give rise to an equation $u = Ku + f$ of the second kind. It can be shown that K satisfies the regularity assumption (2.1.4) for $U = L^2(\Gamma_1)$, $V = H^2(\Gamma_1) \cap H_0^1(\Gamma_1)$.

Replacing the differential equations by difference schemes, we obtain $u_\ell = K_\ell u_\ell + f_\ell$ with K_ℓ fulfilling the regularity assumption (2.2.6) for the discrete analogues U_ℓ, V_ℓ. Conditions (2.2.9) and (2.2.12) hold with $\kappa = 2$. The evaluation of K_ℓ requires the solution of the difference counterparts of Eqs. (3.5.1), (3.5.3). The comparison of the rates of the multigrid convergence shows that the rate is decreasing as $O(h_\ell^2)$, confirming $\kappa = 2$. The nested iteration (3.4.5) with i = 1 yields the following values $u_\ell(\pi/2)$. For details see Hackbusch (1980b).

Table 3.5.1

Results of nested iteration (3.4.5) applied to (3.5.1-3)

step size	$h_0=\pi/2$	$h_1=\pi/4$	$h_2=\pi/8$	$h_3=\pi/16$	$h_4=\pi/32$	$h_5=\pi/64$
$\tilde{u}(\pi/2)$	0.0126	0.0919	0.1286	0.1456	0.1505	0.15128
iteration error $(\tilde{u}_\ell - u_\ell)(\pi/2)$	0.0	9.2-4	7.8-3	2.5-3	1.9-4	8.2-6
discretisation error $(u_\ell - u)(\pi/2)$	1.4-1	5.9-2	1.5-2	3.3-3	7.8-4	1.9-4

3.5.2 Eigenvalue Problems

Consider the eigenvalue problem

$$u = \lambda\, ku, \qquad u \neq 0, \tag{3.5.4}$$

where k may represent a linear integral operator (Chatelin (1983)). Let

$$u_\ell = \lambda_\ell k_\ell u_\ell, \qquad u_\ell \neq 0, \tag{3.5.5}$$

be the discrete analogues. The condition $u_\ell \neq 0$ may be substituted by $\phi_\ell(u_\ell) \neq 0$, where ϕ_ℓ is some linear or nonlinear functional. Examples are

$$\phi_\ell(u_\ell) = \|u_\ell\|^2 \qquad \text{(Euclidean norm)},$$

$$\phi_\ell(u_\ell) = u_\ell(\xi) \qquad (\xi \in D_\ell \text{ fixed}).$$

The second choice makes sense only if the eigenvector u_ℓ does not vanish at the grid point ξ. A possible system of equations for λ_ℓ and u_ℓ is

$$u_\ell = \lambda_\ell k_\ell u_\ell, \qquad \phi_\ell(u_\ell) = 1.$$

It is even more convenient to normalise u_ℓ by $\phi_\ell(u_\ell) = \lambda_\ell$ and to eliminate λ_ℓ:

$$u_\ell = \phi_\ell(u_\ell)k_\ell u_\ell =: \mathcal{K}_\ell(u_\ell). \tag{3.5.6}$$

This is an equation of the form (3.1.2). Eq. (3.5.6) can be solved by iteration (3.2.6). Eq. (3.5.6) and the continuous analogue

$$u = \phi(u)ku =: \mathcal{K}(u) \tag{3.5.7}$$

make sense in view of the following lemmata:

<u>Lemma 3.5.1</u> Let $u^* \neq 0$ and λ^* satisfy the original problem (3.5.4). Assume $\lambda^* \in R(u^*)$, where

$$R(u) := \{\phi(\alpha u) : \alpha \in \mathbb{C}\} \subset \mathbb{C} \qquad (u \in U).$$

Then $u = \alpha u^*$ with suitable $\alpha \in \mathbb{C}$ is a solution of Eq. (3.5.7). On the other hand, if $u \neq 0$ is a solution of Eq. (3.5.7), then u is also a solution of the eigenvalue problem (3.5.4) with $\lambda = \phi(u)$.

<u>Lemma 3.5.2</u> Let $0 \neq u \in U$ be a solution of Eq. (3.5.7), $u = \mathcal{K}(u)$. Assume

$$\|k\|_{V \leftarrow U} \leq C, \qquad V \text{ compactly embedded in } U,$$

(cf. (2.1.4), Note 2.1.4). Suppose that $\lambda = \phi(u)$ be a single eigenvalue of k, i.e.

$$(I - \lambda k)w = \rho u \Rightarrow \rho = 0, \quad w \in \text{span}(u).$$

Further, let the derivative ϕ' of ϕ satisfy

$$\phi'(u)u \neq 0, \quad \phi'(u) : U \to \mathbb{C} \text{ bounded}.$$

Then the derivative $K = \mathcal{K}'(u)$ is compact and satisfies (2.1.4) and

$$\|(I - K)^{-1}\|_{U \leftarrow U} \leq C.$$

Thus, the solution u of Eq. (3.5.7) is isolated.

Proof Since $Kv = \phi(u)kv + (\phi'(u)v)ku$, the estimate (2.1.4) follows and implies compactness of K. Therefore, it suffices to show that $v = Kv$ implies $v = 0$. Now $v = Kv$ is equivalent to

$$(I - \lambda k)v = \rho u, \text{ where } \lambda = \phi(u),\ \rho = \phi'(u)v/\phi(u).$$

Since λ is a single eigenvalue, one concludes $v = \alpha u$, $\rho = 0$. Hence, $\rho = \alpha\phi'(u)u/\phi(u) = 0$, implying $\alpha = 0$ and $v = 0$. □

Similarly, one can prove the regularity assumption (2.2.6). The consistency condition (2.2.9) on K_ℓ follows from the corresponding condition on k_ℓ and simple assumptions on ϕ_ℓ (Hackbusch (1980d)).

Remark 3.5.3 The above reformulation of (3.5.4) as (3.5.6) can also be applied if the eigenvalue appears nonlinearly: $u = k(\lambda)u$. In that case, $\mathcal{K}_\ell$ becomes $\mathcal{K}_\ell(u_\ell) := k_\ell(\phi_\ell(u_\ell))u_\ell$.

An example of $u = k(\lambda)u$ is given in §3.5.4. The operator k may arise from an elliptic boundary value problem. In § 3.5.3 we show that even non-standard problems such as the Steklov eigenvalue problem can easily be solved.

3.5.3 First Example: Steklov Eigenvalue Problem

Consider the boundary value problem

$$-\Delta y = 0 \text{ in } \Omega = (0,1) \times (0,1), \tag{3.5.8a}$$

$$\partial y/\partial n = \lambda y \text{ on } \Gamma = \partial\Omega. \tag{3.5.8b}$$

This can be reformulated as (3.5.1), $u = \lambda ku$, with

$$k : U \subset L^2(\Gamma) \to V \subset H^1(\Gamma)$$

defined as follows. Let $U = \{v \in L^2(\Gamma) : \int_\Gamma v\,d\Gamma = 0\}$. For $u \in U$ the Neumann problem

$$-\Delta y = 0 \text{ in } \Omega, \quad \partial y/\partial n = u \text{ on } \Gamma,$$

has solutions y determined up to a constant. We make y unique if we require $y/_\Gamma \in U$. Set $ku := y/_\Gamma$. It lies in $V = U \cap H^1(\Gamma)$. The eigenvalue problem $u = \lambda ku$ is equivalent to (3.5.8). According to §3.5.2, (3.5.8) is equivalent to $u = \mathcal{K}(u)$. Analogously, one can reformulate the discrete counterpart of (3.5.7), $u_\ell = \mathcal{K}_\ell(u_\ell)$, where u_ℓ is defined on the boundary Γ_ℓ.

In the following, the functional ϕ_ℓ is defined by $\phi_\ell(u_\ell) = u_\ell(0,0)$. In order to compute the fifth eigenvalue, one may start with the grid size $h_0 = 1/4$. At level $\ell = 0$ the fifth eigenvalue and corresponding eigenvector can be computed to be

$$\tilde{\lambda}_0 = \tilde{u}_0(0,0) = 4.628, \; \tilde{u}_0(\tfrac{1}{4}, 0) = -0.726, \; \tilde{u}_0(\tfrac{1}{2}, 0) = -3.125.$$

The remaining values of $\tilde{u}_0$ on Γ_0 are defined by symmetries

$$u_0(x,y) = u_0(1 - x,y) = u_0(x,1 - y). \tag{3.5.9}$$

Using this starting value $\tilde{u}_0$, the nested iteration (3.2.8) with $i = 1$ yields the numbers of Table 3.5.2.

Table 3.5.2

Fifth eigenvalue of the Steklov problem (3.5.8)

step size	h_0=1/4	h_1=1/8	h_2=1/16	h_3=1/32	h_4=1/64
$\tilde{\lambda} = \tilde{u}_\ell(0,0)$	4.6277	4.6918	4.6562	4.651034	4.647942
exact λ_ℓ	4.6277	4.6690	4.6552	4.649417	4.647822

Although the fifth eigenvalue is a double one, no extra computational difficulties arise since the symmetry (3.5.9) determines the eigenfunctions uniquely. For further details, see Hackbusch (1980d).

3.5.4 Second Example: Nonlinear Eigenvalue Problem

Consider the integro-differential equation

$$-u''(x) = \int_0^1 e^{\lambda|x-y|} u(y)dy; \quad u(0) = u(1) = 0; \quad u \not\equiv 0. \tag{3.5.10}$$

This nonlinear eigenvalue problem can be written as

$$u = k(\lambda)\, u, \qquad u \neq 0,$$

where $k(\lambda)v$ is the solution w of

$$-w''(x) = \int_0^1 e^{\lambda|x-y|}v(y)\,dy, \quad w(0) = w(1) = 0.$$

According to Remark 3.5.3, one obtains the nonlinear equation $u = K(u)$ with $K(u) := k(\phi(u))u$. Here, we use $\phi(v) = v(1/2)$.

The discrete nonlinear equation $u_\ell = K_\ell(u_\ell)$ results by replacing $-u''$ with second differences and the integral with the trapezoidal rule. Starting with

$$\tilde{u}_0(0) = \tilde{u}_0(1) = 0,\ \tilde{u}_0(1/4) = \tilde{u}_0(3/4) = 6.925668,\ \tilde{u}_0(1/2) = 7.734301$$

for step size $h_0 = 1/4$, we calculate the eigenvalues without difficulty. In Hackbusch (1980d) we propose to modify the statement (3.2.8c) of the nested iteration (3.2.8) by $\tilde{u}_k := s*\tilde{p}*\tilde{u}_{k-1}$, where $s := (5\tilde{\lambda}_{k-1} - \tilde{\lambda}_{k-2})/(4\phi_k(\tilde{p}\tilde{u}_{k-1}))$. Then the eigenvalue approximation $\phi_k(\tilde{u}_k)$ is the value extrapolated from $\tilde{\lambda}_{k-1}$ and $\tilde{\lambda}_{k-2}$.

Table 3.5.3

Results of problem (3.5.10)

h_ℓ	1/4	1/8	1/16	1/32	1/64
λ_ℓ	7.734	6.535	6.3024	6.2975	6.2339

3.5.5 Elliptic Boundary Value Problems (Revisited)

In §2.6.2 we formulated an elliptic problem as a Fredholm integral equation of the second kind. In this section we discuss integral equations of the first kind. Again, we consider Laplace's equation

$$\Delta u(\zeta) = 0 \quad \text{in} \quad \mathbb{R}^2 \setminus \Gamma, \tag{3.5.11a}$$

$$u(\zeta) = F(\zeta) \quad \text{on} \quad \zeta \in \Gamma, \tag{3.5.11b}$$

with $u(\zeta) = B\,\ell n(\zeta) + O(1)$ as $|\zeta| \to \infty$. We represent $u(\zeta)$ by $\int_\Gamma V(z)\,\ell n|\zeta - z|\,d\Gamma_z$ with V satisfying the integral equation

$$G(\zeta) = \int_\Gamma V(z)\ \ell n\ |\zeta - z|\, d\,\Gamma_z \tag{3.5.12}$$

of the first kind. Let $z(t)$, $0 \leqslant t \leqslant 1$, be a parametrisation of Γ. Eq. (3.5.12) becomes

$$g(t) = \int_0^1 v(s)\ \ell n|z(t)-z(s)|ds, \quad 0 \leqslant t \leqslant 1 \tag{3.5.13}$$

where $g(t) := G(z(t))$ and $v(s) := V(z(s))$. To avoid nontrivial solutions of the homogeneous problem, one should consider the system

$$g(t) = \int_0^1 v(s)\ \ell n|z(t)-z(s)|ds + \omega, \quad \int_0^1 v(s)ds = B, \tag{3.5.14}$$

with given g, B for the unknowns v, ω. However, to simplify the situation, we shall discuss Eq. (3.5.13) instead of (3.5.14).

The multigrid iteration (2.3.2) applies to integral equations of the second kind only. Eq. (3.5.13) can be rewritten as

$$g(t) = \int_0^1 v(s)\ \ell n\delta(s-t)ds + \int_0^1 v(s)\ \ell n\ \frac{|z(s)-z(t)|}{\delta(s-t)}ds, \tag{3.5.15}$$

where $\delta(s-t) = \min\{|s-t|, |s-t-1|, |s-t+1|\}$ is the distance modulo 1. Eq. (3.5.15) reads as

$$g = Av + kv, \tag{3.5.16}$$

where the first integral operator A is a convolution, whereas the second one has a smooth (nonsingular) kernel, provided that Γ (hence $z(.)$) is smooth enough. It can be shown that A^{-1} is a pseudo-differential operator of first order. If $\Gamma \in C^r$ with $r > 2$, $k: L^2(0,1) \to H^{r-1}(0,1)$ holds and $A^{-1} : H^{r-1}(0,1) \to H^{r-2}(0,1)$ implies $K := -A^{-1}k: L^2(0,1) \to H^{r-2}(0,1)$. Therefore,

$$v = Kv + f \text{ with } K = -A^{-1}k,\ f = A^{-1}g, \tag{3.5.17}$$

is an equation of the second kind, to which the multigrid process applies.

The discretisation (e.g. by the Galerkin method) leads us to the system

$$A_\ell v_\ell + k_\ell v_\ell = g_\ell, \tag{3.5.18}$$

where $A_\ell = (a_{jk}^{(\ell)})$ is a $n_\ell \times n_\ell$ Toeplitz matrix, i.e. $a_{jk}^{(\ell)} = a_{j'k'}^{(\ell)}$ for $j - k \equiv (j' - k') \mathrm{mod}\, n_\ell$. The Galerkin discretisation (3.5.18) and its discretisation error are analysed by Hsiao, Kopp and Wendland (1980).

According to Eq. (3.5.17), the discrete equation can be written as

$$v_\ell = K_\ell v_\ell + f_\ell, \tag{3.5.19}$$

with $K_\ell = -A_\ell^{-1} k_\ell$, $f_\ell = A_\ell^{-1} g_\ell$. Let T_ℓ be the discrete Fourier transform $v_\ell \longmapsto \hat{v}_\ell = T_\ell v_\ell$. Since for any Toeplitz matrix A_ℓ the transformed matrix $D_\ell := T_\ell^{-1} A_\ell T_\ell$ is diagonal, A_ℓ^{-1} equals $T_\ell D_\ell^{-1} T_\ell^{-1}$. The multiplication $K_\ell v_\ell$ requires the usual multiplication $k_\ell v_\ell$ ($O(n_\ell^2)$ operations), while T_ℓ^{-1} and T_ℓ can be performed by fast Fourier transforms ($O(n_\ell \log n_\ell)$ operations). One concludes that the computational work of the multigrid iteration applied to (3.5.19) is almost the same as for a usual integral equation (2.2.4) of the second kind.

We conclude with numerical results for the system (3.5.14). Let Γ be an ellipse with axes of length 2 and 1. The discrete equations (3.5.18) are obtained by a Galerkin approach with piecewise constant finite elements. At the lowest level $\ell=0$ the dimension of the finite element space is $n_0 = 4$. The observed multigrid convergence rates are as follows.

n_ℓ	16	32	64
multigrid rate	0.79	0.37	0.19

Obviously, the rates behave as $O(h_\ell^\kappa)$ with $\kappa = 1$.

3.6 Application to Parabolic Problems

We begin with the time-periodic problem, since this can more easily be formulated. The more interesting parabolic boundary control problem is discussed in §3.6.2.

3.6.1 Time-Periodic Parabolic Problems

Let Ω be a domain in $\mathbb{R}^d$ with boundary Γ. $(0,T)$ is a fixed time interval. Set $Q = \Omega \times (0,T)$ and $\Sigma = \Gamma \times (0,T)$. We consider the parabolic equation

$$y_t + L_y = g_1 \text{ in } Q, \tag{3.6.1a}$$

subject to some boundary condition

$$By = g_2 \text{ in } \Sigma. \tag{3.6.1b}$$

The solution has to be T-periodic:

$$y(\cdot,T) = y(\cdot,0) \text{ in } \Omega. \tag{3.6.1c}$$

The usual initial value problem consists of (3.6.1a,b) and $y(\cdot,0) = u$. Denote its solution by $y(x,t;u)$ and define $K(u) := y(\cdot,T;u)$. Therefore, Eq. (3.6.1a-c) is equivalent to $u = K(u)$. Here, K is affine: $K(u) = Ku + f$ with f depending on g_1 and g_2.

Discretising (3.6.1a,b) e.g. by a totally implicit difference scheme, we obtain a discrete analogue $K_\ell(u_\ell)$. It involves the solution of one discrete initial-boundary value problem. The regularity assumption (2.2.6) is discussed in Hackbusch (1981d). Therein, numerical examples are reported. In the case of $\Omega \subset \mathbb{R}^1$ and discretisation parameters $\Delta t_\ell = O(\Delta x_\ell^2)$, the nested iteration (2.7.1) requires less than 4.6 W_ℓ operations, where W_ℓ is the work required by one discrete initial-boundary value problem at level ℓ.

The method described above has been applied by Schippers (1980a, 1982a and 1982c) to oscillating disk flow (rotating flow due to an infinite disk performing torsional oscillations). The governing nonlinear equations for the unknowns f, g, h are

$$\frac{\omega}{\Omega} f_t = \frac{\Omega}{2\omega} f_{zz} + 2h\, f_z - f^2 + g^2 - k,$$

$$\frac{\omega}{\Omega} g_t = \frac{\Omega}{2\omega} g_{zz} + 2h\, g_z - 2fg,$$

$$h_z = f.$$

3.6.2 Optimal Control Problems for Parabolic Equations

Let Ω, Q, Σ, $(0,T)$ be defined as in §3.6.1, and $\Sigma_0 \subset \Sigma$. For any $u \in L^2(\Sigma_0)$ the state $y(u)$ is defined as the solution of

$$y_t(u) + A\, y(u) = g_1 \quad \text{in } Q, \tag{3.6.2a}$$

$$By(u) = u \text{ on } \Sigma_0, \quad By(u) = g_2 \text{ on } \Sigma \setminus \Sigma_0, \tag{3.6.2b}$$

$$y(\cdot,0;u) = y_0 \quad \text{on } \Omega, \tag{3.6.2c}$$

where A is an elliptic differential operator and B a boundary operator. We seek u minimising a cost function, e.g.

$$J(v) = \|y(\cdot,T;v) - z\|^2_{L^2(\Omega)} + \|v\|^2_{L^2(\Sigma_0)}$$

for some given z. The optimal control u is characterised by

$$u = -p(u)|_{\Sigma_0}, \tag{3.6.3}$$

where p(u) is the solution of a certain adjoint parabolic equation

$$-p_t(u) + A^* p(u) = 0 \text{ in } Q, \tag{3.6.4a}$$

$$Cp(u) = 0 \qquad \text{on } \Sigma, \tag{3.6.4b}$$

$$p(\cdot,T;u) = y(\cdot,T;u) - z \text{ on } \Omega \tag{3.6.4c}$$

(Hackbusch (1979b)). Eliminating u in (3.6.2b) by (3.6.3), we obtain a coupled system of two parabolic equations, where the second one (Eq. (3.6.4)) has reversed time direction. The discretised problem would be a large sparse system.

Instead, we consider another representation of the problem. Eq. (3.6.2) describes a mapping $u \to y(u)$, while Eq. (3.6.4) defines a mapping $y(u) \to p(u)$. Set $K(u) := -p(u)|_{\Sigma_0}$. This is affine : $K(u) = Ku + f$ with f depending on g_1, g_2, y_0, z. Eq. (3.6.3) is equivalent to $u = K(u)$.

Replacing Eqs. (3.6.2) and (3.6.4) by difference schemes, we can define analogous equations $u_\ell = K_\ell(u_\ell)$. The evaluation of $K_\ell(u_\ell)$ requires the solution of two discrete initial-boundary value problems, namely $u_\ell \longmapsto y_\ell(u_\ell)$ and $y_\ell(\cdot,T;u_\ell) \longmapsto p_\ell(u_\ell)$.

Consider the particular case of $\Omega = (0,1)$, $T = 1$, $\Sigma_0 = \{0\} \times (0,T)$, $A = A^* = -\partial^2/\partial x^2$, $B = C = \partial/\partial n$. Then the equations for y(u), p(u) become

$$y_t = y_{xx} + g_1, \; -p_t = p_{xx}, \; y(x,0) = y_0, \; p(x,T) = y(x,T) - z,$$

$$-y_x(0,t) = u(t), \; y_x(1,t) = g_2, \; p_x(0,t) = p_x(1,t) = 0.$$

It can be shown that K is a bounded mapping from $U = L^2(\Sigma_0)$ into $V = H^{\frac{1}{2}}(\Sigma_0)$, while $K_\ell = K'_\ell$ satisfies the analogous discrete estimates. The inequalities (2.2.9) and (2.2.12) hold with $\kappa = 1/2$ and $h_\ell = \Delta t_\ell = O(\Delta x_\ell^2)$.

The following numerical results are obtained for

$$g_1 = -\frac{\pi}{2}\cos(\frac{\pi}{2}(1-x)),\ g_2 = 0,\ y_0 = -\frac{\pi}{2}\cos(\frac{\pi}{2}(1-x)),\ z = 1 - \frac{\pi}{2}\cos(\frac{\pi}{2}(1-x)).$$

Table 3.6.1 shows grid sizes Δx_ℓ, Δt_ℓ and the convergence rates ρ_ℓ of the multigrid iteration (2.3.2). The rates decrease as $O(\Delta t_\ell^{\frac{1}{2}})$ confirming the above value of $\kappa = 1/2$.

Table 3.6.1

Multigrid convergence rates

ℓ	0	1	2	3	4	5
Δt_ℓ	1/2	1/4	1/8	1/16	1/32	1/64
Δx_ℓ	1/4	1/8	1/8	1/16	1/16	1/32
ρ_ℓ	-	0.086	0.063	0.044	0.030	0.019

Extensions to control problems with bounded controls are discussed in Hackbusch (1981a).

REFERENCES

Abramov, A.A. (1962) On the separation of the principal part of some algebraic problems, *Ž. Vyčisl. Mat.* 2,1, pp. 141-145.

Anselone, Ph., M. (1971) Collectively compact operator approximation theory and applications to integral equations, Prentice-Hall, Englewood Cliffs.

Atkinson, K. (1973) Iterative variants of the Nyström method for the numerical solution of integral equations, *Numer. Math.*, **22**, pp. 17-31.

Atkinson, K. (1976) An automatic program for linear Fredholm integral equations of the second kind, *ACM Trans. Math. Software*, **2**, pp. 154-171.

Baker, C.T.H. (1977) The numerical treatment of integral equations, Oxford University Press, London.

Brakhage, H. (1960) Über die numerische Behandlung von Integralgleichungen nach der Quadraturformelmethode, *Numer. Math.*, **2**, pp. 183-196.

Brandt, A. (1977) Multi-level adaptive solutions to boundary-value problems, *Math. Comp.*, **31**, pp. 333-390.

Chatelin, F. (1983) Spectral approximation of linear operators, Academic Press, New York.

Hackbusch, W. (1979a) On the fast solution of nonlinear elliptic equations, *Numer. Math.*, **32**, pp. 83-95.

Hackbusch, W. (1979b) On the fast solving of parabolic boundary control problems, *SIAM J. Control Optim.*, **17**, pp. 231-244.

Hackbusch, W. (1980a) The fast numerical solution of very large elliptic difference schemes, *J. Inst. Math. Appl.*, **26**, pp. 119-132.

Hackbusch, W. (1980b) On the fast solving of elliptic control problems, *J. Optim. Theory Appl.*, **31**, pp. 565-581.

Hackbusch, W. (1980c) Numerical solution of nonlinear equations by a multigrid iteration of the second kind. In "Numerical methods for nonlinear problems", Vol. 1, Proceedings, Swansea, Sept. 1980, C. Taylor, E. Hinton and D.R.V. Owen (eds.), Pineridge Press, Swansea, pp. 1041-1050.

Hackbusch, W. (1980d) Multigrid solutions to linear and nonlinear eigenvalue problems for integral and differential equations, Report 80-3, Universität zu Köln, to appear in Rostock Math. Colloqu.

Hackbusch, W. (1981a) Numerical solution of linear and nonlinear parabolic control problems. In "Optimization and optimal control", Proceedings, Oberwolfach, March 1980, A. Auslender, W. Oettli and J. Stoer (eds.). Lecture notes in Control and Information Sci., **30**, Springer-Verlag, Berlin, pp. 179-185.

Hackbusch, W. (1981b) Die schnelle Auflösung der Fredholmschen Integralgleichung zweiter Art, Beiträge, *Numer. Math.*, **9**, pp. 47-62.

Hackbusch, W. (1981c) Error analysis of the nonlinear multigrid method of the second kind, *Appl. Math.*, **26**, pp. 18-29.

Hackbusch, W. (1981d) Fast numerical solution of time-periodic parabolic problems by a multigrid method, *SIAM J. Sci. Statist. Comp.*, **2**, pp. 198-206.

Hackbusch, W. (1982) Multigrid convergence theory, In "Multigrid methods", W. Hackbusch and U. Trottenberg (eds.), Proceedings, Köln-Porz, Nov. 1981, Lecture Notes in Mathematics, 960, Springer-Verlag, Berlin, pp. 177-219.

Hackbusch, W. and Trottenberg, U. (1982) Multigrid methods, Proceedings, Köln-Porz, Nov. 1981. Lecture Notes in Mathematics, 960, Springer-Verlag, Berlin.

Hemker, P.W. and Schippers, H. (1981) Multiple grid methods for the solution of Fredholm integral equations of the second kind, *Math. Comp.*, **36**, pp. 215-232.

Hsiao, G.C., Kopp, P. and Wendland, W.L. (1980) A Galerkin collocation method for some integral equations of the first kind, *Computing*, **25**, pp. 89-130.

Lučka, A. (1980) Proekcionno-iterativnye metody rešenija differencial'nych i integral'nich uravnenij, Naukova Dumka, Kiev.

Mandel, J. (1983) On multilevel iterative methods for integral equations of the second kind and related problems, unpublished.

McCormick, S. (1980) Mesh refinement methods for integral equations. In "Numerical treatment of integral equations", J. Albrecht and L. Collatz (eds.), ISNM 53, Birkhäuser, Basel, pp. 183-190.

Meis, Th. and Marcowitz, U. (1978) Numerische Behandlung partieller Differentialgleichungen, Springer-Verlag, Berlin. English Translation (1981) Numerical solution of partial differential equations, Springer-Verlag, New York.

Nowak, Z.P. (1982) Use of the multigrid method for Laplacian problems in three dimensions. In. "Multigrid methods", Proceedings, Köln-Porz, Nov. 1981, W. Hackbusch and U. Trottenberg (eds.). Lecture Notes in Mathematics, 960, Springer-Verlag, Berlin, pp. 576-598.

Nyström, E.J. (1928) Uber die praktische Auflosung von linearen Integralgleichungen mit Anwendungen auf Randwertaufgaben der Potentialtheorie, *Soc. Sci. Fenn Comment Phys.-Math.*, **4**, 15 pp. 1-52.

Oskam, B. and Fray, J.M.J. (1982) General relaxation schemes in multigrid algorithms, *J. Comp. Phys.*, **48**, pp. 423-440.

Schippers, H. (1979) Multigrid techniques for the solution of Fredholm integral equations of the second kind. In "Colloquium; numerical treatment of integral equations", H.J.J. Te Riele (ed.), Proceedings, Amsterdam, 1979, MG Syllabus 41, Mathematisch Centrum, Amsterdam, pp. 29-46.

Schippers, H. (1980a) Multiple grid methods for oscillating disc flow. In "Boundary and interior layers - computation and asymptotic methods", J.J.H. Miller (ed.), Proceedings, Dublin, June 1980, Boole Press, Dublin, pp. 410-414.

Schippers, H. (1980b) The automatic solution of Fredholm equations of the second kind, Report NW 99/80, Mathematisch Centrum, Amsterdam.

Schippers, H. (1982a) Application of multigrid methods for integral equations to two problems from fluid dynamics, *J. Comp. Phys.*, **48**, pp. 441-461.

Schippers, H. (1982b) On the regularity of the principal value of the double-layer potential, *J. Eng. Math.*, **16**, pp. 59-76.

Schippers, H. (1982c) Multiple grid methods for equations of the second kind with application in fluid mechanics, Doctoral thesis, Mathematisch Centrum, Amsterdam.

Schippers, H. (1982d) Multigrid methods for boundary integral equations, Report NLR, Mathematisch Centrum, Amsterdam.

Schneider, C. (1977) Beiträge zur numerischen Behandlung schwachsingulärer Fredholmscher Integralgleichungen zweiter Art, Thesis, Mainz.

Šišov, V.S. (1962) The determination of eigenvalues and eigenfunctions of a linear integral operator with a symmetrical kernel by means of group elimination of unknowns, *Ž. vyčisl. mat.* 2,3, pp. 398-410.

Wolff, H. (1979) Multigrid techniek voor het oplossen van Fredholm - integraalvergelijkingen van de tweede soort, Report NN 19/79, Mathematisch Centrum, Amsterdam.

SOME IMPLEMENTATIONS OF MULTIGRID LINEAR SYSTEM SOLVERS

P.W. Hemker and P.M. de Zeeuw

(Centre for Mathematics and Computer Science, The Netherlands)

ABSTRACT

In this paper portable and efficient FORTRAN implementations for the solution of linear systems by multigrid are described. They are based on ILU- or ILLU- relaxation. Scalar and vector versions are compared. Also a complete formal description of a more general multigrid algorithm is given in ALGOL 68.

1. INTRODUCTION

At the moment several implementations of multigrid methods are known for the solution of linear systems that arise from the discretization of more or less general elliptic partial differential equations (Dendy, (1982), Foerster and Witsch (1982), Hemker, Kettler, Wesseling and de Zeeuw (1983)). Also some experiences for computations on vector machines such as the CRAY 1 or the CYBER 205 have been reported (Barkai and Brandt (1983), Dendy (1983), Hemker, Wesseling and de Zeeuw (1983)). It appears that really efficient programs are now available. E.g. for the Poisson equation a code has been developed (Barkai and Brandt (1983)) for the CYBER 205, that solves the problem "up to truncation error" in 0.36 μsec per meshpoint. It will be clear that -even with the present day computer technology- such a high speed can be obtained only when the computer code is specially tuned for the one particular problem and for the one particular machine.

In this paper we discuss the implementation of multigrid methods, not for a particular machine or problem, but for general elliptic 7-point difference equations and in a machine independent programming language. We describe two FORTRAN codes of which the purpose is to provide the user with a program that efficiently solves a large class of difference equations. A first code of this type was introduced by Wesseling (1982a). The codes are autonomous, i.e. they solve the linear systems of equations just like any standard subroutine for the solution of linear systems. The user has to specify only the matrix and the right hand side. Two versions of the codes are available -both in portable FORTRAN- one for use on scalar- the other for vector- (=pipeline) computers.

In section 2 of this paper we describe the problems to be solved. In section 3 we give an outline of the MG-algorithms used. The structure of the FORTRAN implementation is given in section 4 and in section 5 some remarks are made about computing times. In the first appendix, we present an ALGOL 68 program that gives a complete formal description of the flexible algorithm as mentioned in section 3. In a second appendix we give the user interfaces of the FORTRAN codes.

2. THE DIFFERENCE PROBLEM

We consider the scalar linear second order elliptic PDE in two dimensions

$$a_{11}u_{xx} + 2\,a_{12}u_{xy} + a_{22}u_{yy} + a_1u_x + a_2u_y + a_0 = f, \tag{2.1a}$$

on a rectangle $\Omega \subset \mathbb{R}^2$, with variable coefficients a_{ij}, a_i and with boundary conditions

$$\begin{cases} u_n + \alpha\, u_s + \beta\, u = \gamma & \text{on } \Gamma_N, \\ u = g & \text{on } \Gamma_D, \end{cases} \tag{2.1b}$$

where $\Gamma_N \cup \Gamma_D = \delta\Omega$. The subscripts n and s denote the derivates normal to and along the boundary. If the equation (2.1) is discretized on a regular triangulation of the rectangle as given in Fig. 1, then the discretization obtained by a simple finite element method (with piecewise linear trial- and test-functions on the triangulation) will be a linear system

$$A_h\, u_h = f_h, \tag{2.2}$$

with a regular 7-diagonal structure. We consider codes for the solution of these linear systems. The 7-point discretization is the simplest one in which also cross-derivatives u_{xy} can be represented. It does not seem worthwhile to consider more complex difference molecules because the solution of higher order discretizations can be performed by means of defect correction iteration in which only systems of the above mentioned form have to be solved.

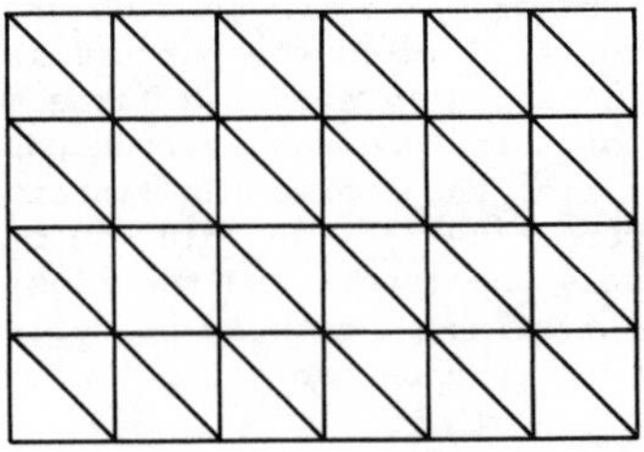

Fig. 1

On the rectangle Ω equidistant computational grids Ω^k, $k = 0,1,2,\ldots,\ell$, are defined by

$$\Omega^k = \{(x_1,x_2) \mid x_i = m_i 2^{-k},\ m_i = 0,1,\ldots,N_i 2^k\}. \tag{2.3}$$

To obtain a solution u_h on Ω^h, for the codes we consider, the user has to define the matrix A_h and the right hand side vector f_h only for the discretization on the finest grid $\Omega^\ell := \Omega^h$.

The regular structure of the domain and the regular 7-point structure of the difference equations allows a simple structure of the data that are to be transferred to and from the programs. The solution and the right-hand-side can be stored in the most straightforward way in a 1- or 2- dimensional array. The coefficient matrix is stored similarly, by its diagonals.

There are many possible ways to solve the system (2.2) by multigrid. Based on previous work (Hemker, (1982), Hemker (1984), Hemker, Kettler, Wesseling and de Zeeuw (1983), Hemker, Wesseling and de Zeeuw (1983), Kettler (1982), Wesseling (1982a), Wesseling (1982b)), in this paper we select two particularly efficient strategies for which FORTRAN codes have been made available and we give the description of a more general multigrid algorithm. A detailed ALGOL 68 program which implements this more general algorithm is included in appendix 1. It can be used to experiment with the different possibilities.

3. THE MULTIGRID CYCLING ALGORITHM

The general multigrid algorithm for the solution of (2.2) is an iterative cycling procedure in which discretizations of (2.1) on all grids Ω^k, $k = 0,1,\ldots,\ell$, are used. We denote these discretizations by $A_k u_k = f_k$, $k = 0,1,\ldots,\ell$; k denotes the "level of discretization" and we take $A_k := A_h$ and $f_k := f_h$.

One multigrid iteration cycle on level k is defined by the subsequent execution of

(1) p relaxation sweeps applied to the system $A_k u_k = f_k$,
(2) the application of a "coarse grid correction", and
(3) again q relaxation sweeps for $A_k u_k = f_k$.

The coarse grid correction consists of: (1) the computation of

$$f_{k-1} := R_{k-1,k} (f_k - A_k \tilde{u}_k), \tag{3.1}$$

where $\tilde{u}_k$ is the current approximation to the solution and $R_{k-1,k}$ is a restriction operator which represents the current residual on the next coarser level; (2) the computation of $\tilde{u}_{k-1}$, an approximation to the solution of the correction equation

$$A_{k-1} u_{k-1} = f_{k-1}. \tag{3.2}$$

This approximation is obtained by application of s multigrid iteration cycles on level k-1, with a zero starting approximation; and (3) updating the current solution $\tilde{u}_k$ by

$$\tilde{u}_k := \tilde{u}_k + P_{k,k-1}\, \tilde{u}_{k-1}, \tag{3.3}$$

where the prolongation operator $P_{k,k-1}$ denotes the interpolation from level k-1 to k.

On the coarsest level another method (at choice) can be used for the computation of $\tilde{u}_0$.

In principle, the parameters p, q and s and the operators $R_{k-1,k}$, A_{k-1}, $P_{k,k-1}$ are free to be chosen. Obvious restrictions are p+q>=1 and 1<=s<=3. A natural choice for combination with the finite element discretization (2.2) is the use of a piecewise linear interpolation over triangles in Ω^{k-1} for $P_{k,k-1}$. The corresponding restriction is the transposed operation $R_{k-1,k} = P^T_{k,k-1}$. This prolongation and restriction are exactly the 7-point prolongation and restriction as described in Wesseling (1982b). With these $P_{k,k-1}$ and $R_{k-1,k}$ the finite element discrete operators on coarser grids are easily derived from the fine grid finite element discretization by

$$A_{k-1} = R_{k-1,k}\, A_k\, P_{k,k-1}, \quad k = \ell, \ell-1, \ell-2, \ldots, 1. \tag{3.4}$$

Thus, the coarser grid discretizations are obtained by algebraic manipulation only.

An ALGOL 68 program, based on these choices for the operators is presented as a worked-out illustration in appendix 1. The multigrid cycling procedure is given in <u>proc</u> MG. It is imbedded in a complete solution procedure <u>proc</u> MGM, which also checks the consistency of the input data, which generates the coarse grid operators by (3.4) and which constructs an initial estimate by "full multigrid", i.e. first it finds an approximate solution on the coarser grid and interpolates this to the next finer ones. The parameters p, q, s, the relaxation procedure and the stopping strategy are still to be chosen. For a set of default parameters (that can be changed by the user) an autonomous procedure is given in <u>proc</u> SOLVE SYS. This procedure requires as data only the matrix A_h, the right hand side vector f_h and the number of levels ℓ. It delivers the solution u_h without further interference by the user.

In the procedure MGM the user can select his own multigrid strategy (p,q,s) and he may select from different relaxation procedures: Point Gauss Seidel, Line Gauss Seidel or Incomplete Line LU-decomposition relaxation. V-cycles are obtained by s=1, W-cycles by s=2.

4. THE STRUCTURE OF THE FORTRAN IMPLEMENTATIONS

Less flexible but more efficient implementations have been written in FORTRAN. Here we consider two versions of the general MG-algorithm. Both use p= 0, s= q= 1 as the strategy. The first version (MGD1) uses Incomplete LU-decomposition (ILU-) relaxation as the relaxation procedure (Wesseling (1982a)), the other (MGD5) uses Incomplete Line LU-decomposition (ILLU-) relaxation (Kettler (1982)).

MGD1 is particularly efficient because of the smoothing properties of the ILU-relaxation (Hemker (1982), Kettler (1982)) and the efficient residual computation. In this version on each level the 7-diagonal matrix A_k is decomposed as

$$A_k = L_k U_k - C_k,$$

where L_k is a lower-triangular matrix (with unity on the main diagonal) and U_k is an upper-triangular matrix. The requirement that L_k and U_k have non-zero diagonals only where A_k has, determines L_k and U_k. The remainder matrix C_k has only two non-zero diagonals of which the elements are easily derived from L_k and U_k.

One relaxation sweep of ILU-relaxation corresponds to the solution of the system

$$L_k U_k u_k^{(i+1)} = f_k + C_k u_k^{(i)}.$$

After such a relaxation sweep the residual is efficiently computed by

$$r_k^{(i+1)} := f_k - A_k u_k^{(i+1)} = C_k (u_k^{(i+1)} - u_k^{(i)}).$$

The other relaxation method, ILLU-relaxation, which is due to J.A. Meyerink, is described in Kettler (1982) and in more detail in Wesseling (these proceedings). A complete description in ALGOL 68 is found in the ALGOL 68 program in the appendix 1.

The global structure of both MGD1 and MGD5 is the same. First, in a preparational phase, the sequence of coarse grid operators is constructed by a subroutine RAP, according to (3.4). Then the decomposition is performed (in DECOMP). Finally, in the cycling phase, at most MAXIT iterations of the cycling process are performed. On the basis of intermediate results -the detection of a small residual norm- the iteration can be stopped earlier. This necessitates the computation of this norm (in VL2NOR) in each cycle.

The following is an outline in quasi FORTRAN of the multigrid cycling process in MGD1. At all computational levels $k = 1,2,\ldots,\ell$, the matrix decomposition $A_k = L_k U_k - C_k$ is available. At the beginning (or end) of each MG-iteration cycle, u_ℓ contains the current solution and r_ℓ the corresponding residual. If no initial estimate is available we take $u_\ell \equiv 0$ and $r_\ell \equiv f$.

`C       THE MGD1 ITERATION PROCESS`	
`        DO 100 N=1, MAXIT`	
`        CALL RESTRI(F,R,L-1)`	$f_{\ell-1} = R_{\ell-1,\ell}\, r_\ell,$
`        DO 10 K=L-2, 1, -1`	
`        CALL RESTRI(F,F,K)`	$f_k = R_{k,k+1}\, f_{k+1},$
`     10 CONTINUE`	
`        CALL SOLVE(U,F,1)`	$u_1 = (L_1\, U_1)^{-1}\, f_1,$
`        DO 20 K=2,L-1`	
`        CALL PROLON(U,U,K)`	$u_k = P_{k,k-1}\, u_{k-1},$
`        CALL CTUPF(V,U,F,K)`	$v_k = C_k\, u_k + f_k,$
`        CALL SOLVE(U,V,K)`	$u_k = (L_k\, U_k)^{-1}\, v_k,$
`     20 CONTINUE`	
`        CALL PROLON(R,U,L)`	$r_\ell = P_{\ell,\ell-1}\, u_{\ell-1},$
`        DO 30 J=1,NF`	
`        R(J)=R(J)+U(J)`	$r_\ell = r_\ell + u_\ell,$
`     30 CONTINUE`	
`        CALL CTUPF(V,R,F,L)`	$v_\ell = C_\ell\, r_\ell + f_\ell,$
`        CALL SOLVE(U,V,L)`	$u_\ell = (L_\ell\, U_\ell)^{-1}\, v_\ell,$
`        CALL CTUMV(U,R)`	$r_\ell = C_\ell\, (u_\ell - r_\ell),$
`        RES = VL2NOR(R)`	$\Vert r_\ell \Vert_2$
`        IF(RES .LT. TOL) GOTO 200`	
`    100 CONTINUE`	
`    200 CONTINUE`	

In the actual implementation of MGD1, the matrix A_k is not kept in storage, but it is overwritten by L_k and U_k. At minimal costs, the remainder matrix C_k is recomputed each time from L_k and U_k (in the subroutines CTUMV and CTUPF).

The other program, MGD5, with ILLU-relaxation, is less efficient for problems like the Poisson equation, but it is more suitable for problems such as the convection-diffusion or the anisotropic diffusion equation, in which a small parameter multiplies the highest derivatives (Hemker (1984), Kettler (1982)).

The cycling process in MGD5 is similar to the one in MGD1. In this case, however, the matrices A_k are not overwritten and the residual is computed in a straightforward way.

```
C       THE MGD5 ITERATION PROCESS

        DO 100 N=1,MAXIT
```

`      CALL RESTRI(F,R,L-1)`	$f_{\ell-1} = R_{\ell-1,\ell}\, r_\ell,$
`      DO 10 K=L-2, 2, -1`	
`      CALL RESTRI(F,F,K)`	$f_k = R_{k,k+1}\, f_{k+1},$
`   10 CONTINUE`	
`      CALL RESTRI(U,F,1)`	$u_1 = R_{1,2}\, f_2,$
`      CALL SMOOTH(U,F,1)`	relax on level 1,
`      DO 20 K=2,L-1`	
`      CALL PROLON(U,U,K)`	$u_k = P_{k,k-1}\, u_{k-1},$
`      CALL SMOOTH(U,F,K)`	relax on level k,
`   20 CONTINUE`	
`      CALL PROLON(R,U,L)`	$r_\ell = P_{\ell,\ell-1}\, u_{\ell-1},$
`      DO 30 J=1,NF`	
`      U(J)=U(J)+R(J)`	$u_\ell = u_\ell + r_\ell,$
`   30 CONTINUE`	
`      CALL SMOOTH(U,F,L)`	relax on level ℓ,
`      CALL RESIDU(R,F,U)`	$r_\ell = f_\ell - A_\ell\, u_\ell,$
`      RES = VL2NOR(R)`	$\Vert r_\ell \Vert_2$
`      IF(RES .LT. TOL) GOTO 200`	
`  100 CONTINUE`	
`  200 CONTINUE`	

All subroutines in the iteration processes in MGD1 or MGD5 have their own particular features that make them more or less feasible for vectorization. This will be shown in section 5.

5. THE EFFICIENCY OF THE FORTRAN IMPLEMENTATIONS

Both algorithms MGD1 and MGD5 have been coded in portable ANSI-FORTRAN. The codes pass the PFORT verifier, except that more complex subscript expressions appear than (I*M+N). (These expressions, where I is variable and M and N are constants, are the only ones that are allowed for subscripting by PFORT.) In this portable FORTRAN, optimized versions for scalar- and vector- architecture have been constructed. The corresponding codes are called MGD1S, MGD1V, MGD5S and MGD5V. They are all in the form of a FORTRAN subroutine. Their user-interface is given in appendix 2. The different versions run on several machines among which are the CYBER 205 and the CRAY 1.

If run on scalar architecture, after the preparational phase, the computing time for the programs is proportional to the number of iteration steps and to the number of points in the finest grid. The preparational work to generate the coarse grid operators and to form their decompositions is roughly equivalent to 3 iteration sweeps. The computing times for the scalar optimized versions on the CYBER 170 and the CYBER 205 (using scalar architecture) are given in table 5.1.

Table 5.1

Computing times for MGD1 and MGD5 in scalar mode, in μsec/(meshpoint.cycle).

	MGD1S	MGD5S
CYBER 170	15.4	24.9
CYBER 205	8.1	11.1

The relative time spent in the different subroutines (as defined in the previous section) is slightly different for the different machines (compilers). These times are given in table 5.2. We notice that the time to compute the prolongations, the restrictions and the norms is small compared to the relaxation or the residual computations. Further we see e.g. that the time spent in CTUMV is 3/4 of the time spent in CTUPF, as is expected (CTUPF runs over all points, whereas CTUMV only works on points on the finest grid).

Table 5.2

The time spent in the different subroutines in scalar mode, expressed in the time spent in a complete iteration cycle.

code machine	MGD1S CY 170	MGD1S CY 205	MGD5S CY 170	MGD5S CY 205
RAP	2.32	1.50	1.40	1.10
DECOMP	0.86	1.40	0.76	1.90
PROLON	0.072	0.063	0.05	0.046
RESTRI	0.089	0.040	0.06	0.030
VL2NOR	0.040	0.044	0.025	0.032
SOLVE	0.33	0.30		
CTUMV	0.15	0.22		
CTUPF	0.22	0.29		
RESIDU			0.16	0.14
SMOOTH			0.65	0.72

To run portable FORTRAN programs on a vector architecture we have to rely on the auto-vectorization capabilities of the available compilers. Both on the CRAY 1 and on the CYBER 205 we found it possible to vectorize all nonrecursive inner loops in this way. The length of the vectors in the experiments was $(2^{k+1} + 1)^j$ with j=1 or j=2 and $k=1,\dots,\ell$, where ℓ denotes the finest level of discretization. Most loops run over lines in the grid (j=1), but in a number of cases loops run over the entire net (j=2).

Some comparisons of the CRAY 1 and the CYBER 205 have been given in Hemker, Wesseling and de Zeeuw (1983) There it was shown that the essential difference between both machines in these computations is the fact that the CYBER 205 is not very effective for loops with a stride unequal to 1. This is particularly important in the restriction and the prolongation, where frequently strides 2 occur. For the restriction the improvement of vector- over scalar- computing time was a factor 4.2-5.6 (ℓ=5,6) for the CRAY 1 and 1.2-2.2 (ℓ=5,6,7) for the CYBER 205.

Nevertheless, it was also shown that -although an essential part of the computation contains recursive loops- a reasonable gain of efficiency was obtained for MGD1 using the CRAY 1 or CYBER 205 vector architecture.

Since the experiences reported in Hemker, Wesseling and de Zeeuw (1983), a new compiler for the CYBER 205 became available (FORTRAN 2.0). With this compiler it was possible to obtain in portable language a more efficient implementation of some recursive loops, whereas with the previous compiler reference had to be made to special "stacklib" routines.

With the portable FORTRAN program on the CYBER 205, an acceleration factor 3.3-4.6 is obtained for MGD1 (acceleration of MGD1V in vector mode on a two-pipe CYBER 205 over MGD1S in scalar mode on the same CYBER). The program MGD5 is less amenable to vectorization. Its acceleration factor is only 2.1-2.3. Details of the performance of the different subroutines under vector-mode computation are given in table 5.3. In this table we see the CP-times that are spent in the different subroutines of MGD1 and MGD5, when the vector version is run for one iteration cycle on the CYBER 205.

Table 5.3

The time (in m.sec.) for the different subroutines in the vector implementations MGD1V and MGD5V on the CYBER 205 (two pipes, FORTRAN 2.0 compiler). Between brackets the acceleration factor (compared with the scalar versions in scalar mode).

grid	65*65		129*129		257*257	
RAP	20	(2.8)	49	(4.2)	143	(5.6)
DECOMP(MGD1)	12	(4.0)	43	(4.4)	161	(4.6)
DECOMP(MGD5)	29	(3.1)	96	(3.7)	352	(4.0)
CYCLE(MGD1)	1.1	(3.3)	3.3	(4.1)	11.6	(4.6)
CYCLE(MGD5)	2.3	(2.1)	8.2	(2.3)	32.0	(2.3)
PROLON	0.9	(2.4)	2.1	(4.1)	5.9	(5.7)
RESTRI	1.2	(1.3)	3.0	(1.8)	9.5	(2.2)
VL2NOR	0.1	(15)	0.4	(14.8)	1.6	(15.6)
SOLVE	6.8	(1.6)	22.5	(1.8)	82.5	(1.9)
CTUMV	0.3	(25)	1.3	(22.8)	5.8	(20.4)
CTUPF	0.5	(20)	1.8	(21.6)	8.0	(19.4)
RESIDU	0.7	(9)	3.1	(8.0)	13.2	(7.9)
SMOOTH	19.3	(1.8)	72.3	(1.8)	287.5	(1.8)

In table 5.4 we show the megaflop rates for the different subroutines. These rates are defined as the number of floating point operations per second divided by 1.OE+6. One can consider these numbers as a measure of how well the subroutines are suited for the hardware. For different sizes of the finest grid, the rates for the vector- and scalar-version are given for the CYBER-205 (two pipes, with autovectorization via the FORTRAN 2.0 compiler). For the 65*65 grid also the rate for the CYBER 170-750 (with FORTRAN IV) is shown.

The CP-times used for the computation of the megaflop rate is the time spent in the subroutines on the finest and on all coarser grids. As can be expected for the vectormachine, the numbers are dependent on the vectorlengths (i.e. the number of points in the x-direction or the total number of gridpoints) and whether or not strides greater than one occur. If we compare the first column for the rates of the 129*129 grid with the first column for the rates of the 257*257 grid, we see both increases and decreases. The increases are explained by vectorlengths increasing from 129 to 257, the decreases are explained by vectorlengths increasing from 129*129 to 257*257 = 66049 which makes splitting of the long vectors necessary because of the restricted number of vectoraddresses (namely 65535) on the CYBER-205.

Table 5.4

Megaflop rates for the different subroutines. For each grid the rates for the efficient vector implementation (1st column) and the efficient scalar version (2nd column) on a two-pipe CYBER-205 (FORTRAN 2.0) are given. For the 65*65-grid also the rate for the CYBER 170-750 (FORTRAN IV) is shown (3rd column).

finest grid		65*65			129*129		257*257	
RAP	(MGD1,MGD5)	13.7	4.9	1.8	21.4	5.1	28.7	5.1
DECOMP	(MGD1)	8.6	2.1	1.8	9.4	2.1	9.9	2.1
DECOMP	(MGD5)	7.1	2.3	2.6	8.4	2.3	9.0	2.3
CYCLE	(MGD1)	15.5	4.7	2.6	20.3	4.9	23.0	5.0
CYCLE	(MGD5)	12.1	5.8	2.6	13.3	5.9	13.6	5.9
PROLON	(MGD1,MGD5)	11.5	4.7	2.2	19.0	4.7	26.5	4.6
RESTRI	(MGD1,MGD5)	8.7	6.9	1.6	13.1	7.4	16.3	7.5
VL2NOR	(MGD1,MGD5)	84.5	5.6	3.2	83.2	5.6	82.6	5.6
SOLVE	(MGD1)	11.8	7.5	3.7	13.9	7.7	15.0	7.7
CTUMV	(MGD1)	84.5	3.4	2.6	76.8	3.4	68.3	3.4
CTUPF	(MGD1)	68.5	3.4	2.4	74.5	3.4	66.3	3.4
RESIDU	(MGD5)	84.5	9.4	3.6	75.2	9.4	70.1	8.9
SMOOTH	(MGD5)	9.8	5.5	2.8	10.2	5.5	10.1	5.5

6. APPENDICES

6.1 Appendix 1

In this appendix the text is given of an ALGOL 68 program which implements a general multigrid algorithm. The solutions and the right hand sides are represented in nets, i.e. two-dimensional arrays corresponding to the grid Ω^k. The matrices in netmats, i.e. three-dimensional arrays; here the first 2 indices denote the equation (corresponding to a gridpoint), the 3rd index denotes the diagonal (for details, see the comments on page 98).

bristol algol68 text PWH/15/12/83 1

```
 begin   # solution of a linear system by multigrid    #
         # a complete description                      #
         # not an optimal efficient implementation     #

# mode declarations                            #

    mode   net      =  ref [, ] real ;
    mode   netmat   =  ref [,,] real ;

# elementary  operators                        #

    op   zero   = ( ref [] real   a )  ref [] real :
                  ( for  i  from    lwb a  to    upb a
                    do  a[i]:= 0.0  od ; a );
    op   zero   = ( net   a )  net :
                  ( for  i  from  1 lwb a  to  1 upb a
                    do   zero  a[i,]  od ; a );
    op   zero   = ( netmat  a )  netmat :
                  ( for  i  from  1 lwb a  to  1 upb a
                    do   zero  a[i,,]  od ; a );
    op   +:=    = ( net   aa,bb ) net :
                  (  int   l1 = 1 lwb aa, l2 = 2 lwb aa,
                           u1 = 1 upb aa, u2 = 2 upb aa;
                     for  i  from  l1  to  u1  do
                     for  j  from  l2  to  u2  do
                              aa[i,j]+:=  bb[i,j]  od  od ; aa );

# prolongation: linear interpolation           #

    proc  lin int pol = ( net  net ) net :
    begin   int  l1 = 1 lwb net,  l2 = 2 lwb net,
                 b1 = 1 upb net,  b2 = 2 upb net;
            heap [2*l1:2*b1,2*l2:2*b2] real  fine;
            int  jj;  real  u2,u3,u4;
                 ref [] real  uip= net[l1,@l2],
                              upp= fine[2*l1,@2*l2];
                 jj:= 2*l2; upp[jj]:= u4:= uip[l2];
                  for  jp  from  l2+1  to  b2
                  do  u3:= u4; u4:= uip[jp];
                      upp[jj+:=1]:= (u3+u4)/2;
                      upp[jj+:=1]:=      u4
                  od ;
            for  ip  from  l1+1  to  b1
            do   ref [] real  ui   = net [ip-1  ,@  l2],
                              uip = net [ip    ,@  l2],
                              umm = fine[2*ip-1,@2*l2],
                              upp = fine[2*ip  ,@2*l2];
                jj:= 2*l2; u2:= ui[l2]; u4:= uip[l2];
                umm[jj]:= (u2+u4)/2;  upp[jj]:= u4;
                 for  jp  from  l2+1  to  b2
                 do  jj+:= 1;  u2:= ui [jp];
                     u3:= u4;  u4:= uip[jp];
                     umm[jj] := (u2+u3)/2;
                     upp[jj] := (u3+u4)/2;
                     jj+:= 1;
                     umm[jj] := (u2+u4)/2;
                     upp[jj] :=     u4
                 od
            od ; fine
    end ;
```

bristol algol68 text PWH/15/12/83 2

```
# interpolation: quadratic on finer grids      #

    proc  sqr int pol = ( net  net ) net :
    if    int  l1 = 1 lwb net,  l2 = 2 lwb net,
               b1 = 1 upb net,  b2 = 2 upb net;
          odd (b1-l1)  or   odd (b2-l2)
    then  lin int pol (net)
    else  int  ll1 = 2*l1, ll2 = 2*l2;
          heap [ll1:2*b1,ll2:2*b2] real  fine;

       int  jj, jp;
       real  x1, x2, x3, y1, y2, y3, z1, z2, z3, yy2, yy3, zz2, zz3;
       ref [] real  ui= net[ l1,@l2], fi= fine[ll1,];
       fi[ll2]:= x1:= ui[l2]; jj:= ll2+1;
       for  j  from  l2+1  by  2  to  b2-1
       do   x2:= ui[j]; x3:= ui[j+1];
            fi[jj:jj+3] :=(  (  3*(x1 + 2*x2) -  x3 )/8, x2,
                             (  -x1 + 3*(2*x2 + x3 ))/8, x3  );
            jj +:= 4; x1:= x3
       od ;
       for  ii  from  l1+1  by  2  to  b1-1
       do   ref [ ] real  uim= net[ii-1,@l2], uii= net[ii  ,@l2],
                          uip= net[ii+1,@l2];
            ref [,] real  finei = fine[2*ii-1:2*ii+2,@ll2];

            x3:=          uim[l2]  /8;
            y3:= ( yy3:= uii[l2] )/4;
            z3:= ( zz3:= uip[l2] )/8;
            finei[,ll2]:= ( 3*(x3+y3) - z3, yy3, 3*(y3+z3) - x3, zz3 );

            for  jj  from  l2+1  by  2  to  b2-1
            do   jp:= jj+1;        x1:= x3; y1:= y3; z1:= z3;
                 x2:=          uim[jj]  /4; x3:=          uim[jp]  /8;
                 y2:= ( yy2:= uii[jj] )/4; y3:= ( yy3:= uii[jp] )/4;
                 z2:= ( zz2:= uip[jj] )/4; z3:= ( zz3:= uip[jp] )/8;

                 finei[,2*jj-1:2*jj+2]:=
                 ((2*(x2+y1)-z1+y2-x3,
                             2*(x2+y2)-x1+y1-z1,
                                        3*(x3+y2)-z1,
                                                   3*(x3+y3)-z3   ),
                  (2*(y1+y2)-x1+x2-x3,   yy2,
                                        2*(y2+y3)-z1+z2-z3,   yy3 ),
                  (3*(z1+y2)-x3,
                             2*(y2+z2)-x3+y3-z3,
                                        2*(z2+y3)-x3+y2-z1,
                                                   3*(z3+y3)-x3   ),
                  (3*(z1+z2)-z3,  zz2,  3*(z3+z2)-z1,          zz3 ))
       od   od ;
      fine
    fi ;
```

bristol algol68 text PWH/15/12/83 3

```
 # restriction: transposed linear interpolation #

     proc  lin  weight = ( net  ffi) net :
     begin  int  l1 = (1 lwb ffi) over 2, u1 = (1 upb ffi) over 2,
                 l2 = (2 lwb ffi) over 2, u2 = (2 upb ffi) over 2;
         heap  [l1:u1,l2:u2] real  fco;
         int  ti,tk,tkp;
         real  ffb,ffd,ffe;

              zero  fco[l1,];
         for  i  from  l1  to  u1-1
         do  ti:= i+i; fco[i+1,l2]:= 0;

         for  k  from  l2  to  u2-1
         do  tk:= k+k; tkp:= tk+2; ffe:= ffi[ti+1,tk+1];
               fco[i  ,k+1]+:= ffe+( ffb:= ffi[ti  ,tk+1] );
               fco[i+1,k  ] := ffe+( ffd:= ffi[ti+1,tk  ] );
            ((fco[i  ,k  ]+:= ffd+ffb)*:=0.5)+:= ffi[  ti,  tk]
         od ;
               fco[i+1,u2]  := ffd:= ffi[ti+1,tkp ];
            ((fco[i  ,u2 ]+:= ffd      )*:=0.5)+:= ffi[  ti,2*u2]
         od ;

         for  k  from  l2  to  u2-1
         do  tk:= k+k; tkp:= tk+2;
              fco[u1,k+1]+:=        ( ffb:= ffi[2*u1,tk+1] );
            ((fco[u1 ,k ]+:=  ffb     )*:=0.5)+:= ffi[2*u1,  tk]
         od ;
             (fco[u1 ,u2]                *:=0.5)+:= ffi[2*u1,2*u2];
              fco
     end ;

 # residual evaluation                                 #

     proc  residual = ( netmat  m,  net  u,f ) net :
     begin  int  l1= 1 lwb u, l2= 2 lwb u,
                 u1= 1 upb u, u2= 2 upb u;
        heap [l1:u1,l2:u2] real  s;

        ref [ ] real  uim:= u[l1,@l2], ui, uip:= u[l1,@l2];
        for  i  from  l1  to  u1
        do  ( ui:= uip; i = u1 !  skip  ! uip:= u[i+1,@l2] );
            # where the matrix does not define the netmat m, #
            # m should contain zeroes !                      #
             ref  [] real  si = s[i,@l2], fi = f[i,@l2];
             ref [,] real  mi = m[i,@l2,@-3];

             int  jm:= l2, jj, jp:= l2;
             for  j  from  l2  to  u2
             do  ( jj:= jp; j=u2 !  skip  ! jp+:= 1 );

                  ref [] real  mij = mi[jj,@-3];
                 si[jj]:= fi[jj] - (mij[-3]*uim[jj] + mij[-2]*uim[jp] +
                 mij[-1]*ui [jm]  + mij[ 0]*ui [jj] + mij[ 1]*ui [jp] +
                 mij[ 2]*uip[jm]  + mij[ 3]*uip[jj]);

                  jm := jj
             od ; uim:= ui
        od ;s
    end ;
```

```
# coarse grid operator construction                    #

    proc  rap = ( netmat  afi) netmat :
    begin  int  l1 = (1 lwb afi) over 2, u1 = (1 upb afi) over 2,
                l2 = (2 lwb afi) over 2, u2 = (2 upb afi) over 2;
        heap  [l1:u1,l2:u2,-3:3] real  aco;
        real  q= 0.25;
        int  ti,tip,tk,tkp;

       [1:3,1:3,-3:3] real  fine;
        ref [] real
               a = fine[1,1,@-3], b = fine[1,2,@-3], c = fine[1,3,@-3],
               d = fine[2,1,@-3], e = fine[2,2,@-3], f = fine[2,3,@-3],
               g = fine[3,1,@-3], h = fine[3,2,@-3], j = fine[3,3,@-3];
        ref  real
               aa =a[ 0], ab =a[ 1], ad =a[ 3],
               ba =b[-1], bb =b[ 0], bc =b[ 1], bd =b[ 2], be =b[ 3],
               cb =c[-1], cc =c[ 0], ce =c[ 2], cf =c[ 3],
               da =d[-3], db =d[-2], dd =d[ 0], de =d[ 1], dg =d[ 3],
               eb =e[-3], ec =e[-2], ed =e[-1], ee =e[ 0],
                                     ef =e[ 1], eg =e[ 2], eh =e[ 3],
               fc =f[-3], fe =f[-1], ff =f[ 0], fh =f[ 2], fj =f[ 3],
               gd =g[-3], ge =g[-2], gg =g[ 0], gh =g[ 1],
               he =h[-3], hf =h[-2], hg =h[-1], hh =h[ 0], hj =h[ 1],
               jf =j[-3], jh =j[-1], jj =j[ 0];


# orientation:


      aco = coarse          k-1         k
                  --------------------------->  y
       !
       !          fine      1     2     3
       !
       !  i-1       1       a --  b --  c
       !                    !  /  !  /  !
       !            2       d --  e --  f
       !                    !  /  !  /  !
       !  i         3       g --  h --  j
   x   v


  the slice [i,j,] corresponds to the coefficients in equation (i,j);
  the slice [ ,,k] corresponds to matrix diagonals as follows:


     [,,-3] :  n                                 the difference star:
     [,,-2] :  n-e
     [,,-1] :  w                                         -3    -2
     [,, 0] :  p     (the main diagonal)                  !   /
     [,, 1] :  e                                   -1  -  0  -  1
     [,, 2] :  s-w                                      /  !
     [,, 3] :  s                                       2    3

 #
```

```
         zero  aco[ l1,  ,];
     for  i  from  l1  to  u1-1
     do  ti:= i+i; tip:= ti+2;

         zero  aco[i+1,l2,];
     for  k  from  l2  to  u2-1
     do  tk:= k+k; tkp:= tk+2;
        fine[1:3,1:3,]:= afi[ti:tip,tk:tkp,];
         ref [] real  a = aco[i   ,k   ,@-3],
                      c = aco[i   ,k+1,@-3],
                      g = aco[i+1,k   ,@-3],
                      j = aco[i+1,k+1,@-3];

       #aa#((a[ 0]+:= (ab+ba+ad+da)*2+ bb+dd+bd+db )*:=q)+:=aa;
       #cc#  c[ 0]+:= (ce+ec+cb+bc)*2+ ee+bb+be+eb+ef+fe;
       #gg#  g[ 0]+:= (ge+eg+gd+dg)*2+ ee+dd+de+ed+eh+he;
       #jj#  j[ 0] := fh+hf;
       #ac#( a[ 1]+:= (ab+bc)*2 + bb+be+db+de)*:=q;
       #ca#( c[-1]+:= (ba+cb)*2 + bb+eb+bd+ed)*:=q;
       #ag#( a[ 3]+:= (ad+dg)*2 + dd+bd+de+be)*:=q;
       #ga#( g[-3]+:= (da+gd)*2 + dd+db+ed+eb)*:=q;
       #gc#( g[-2] := (ge+ec)*2 + ee+he+de+hf+db+ef+eb)*:=q;
       #cg#( c[ 2] := (eg+ce)*2 + ee+eh+ed+fh+bd+fe+be)*:=q;
       #gj#  g[ 1] := eh+hf+ef;
       #jg#  j[-1] := he+fh+fe;
       #cj#  c[ 3] := eh+ef+fh;
       #jc#  j[-3] := he+fe+hf
     od ;
       fine[1:3,1,]:= afi[ti:tip,tkp,];
        ref [] real  a = aco[i   ,u2,@-3],
                     g = aco[i+1,u2,@-3];
       #aa#((a[ 0]+:= (ad+da)*2 + dd)*:= q)+:=aa;
       #gg#  g[ 0]+:= (gd+dg)*2 + dd;
       #ga#( g[-3]+:= (gd+da)*2 + dd)*:=q;
       #ag#( a[ 3]+:= (ad+dg)*2 + dd)*:=q;
             g[-2] := g[ 1]:= 0.0
     od ;

     for  k  from  l2  to  u2-1
     do  tk:= k+k; tkp:= tk+2;
        fine[1,1:3,]:= afi[tip,tk:tkp,];
        ref [] real  a = aco[u1,k   ,@-3],
                     c = aco[u1,k+1,@-3];
       #aa#((a[ 0]+:= (ab+ba)*2 + bb)*:= q)+:=aa;
       #cc#  c[ 0]+:= (cb+bc)*2 + bb;
       #ca#( c[-1]+:= (cb+ba)*2 + bb)*:=q;
       #ac#( a[ 1]+:= (ab+bc)*2 + bb)*:=q;
             c[ 2] := c[ 3]:= 0.0
     od ;
       #aa#(aco[u1,u2,0]*:=q)+:=afi[2*u1,2*u2,0];
       aco
 end ;
```

```
# point relaxation procedure                              #

    proc  pgs  relax = ( ref  netmat  dec, netmat  m,  net  u,f) void :
    begin  # point gauss seidel (pgs) #
       int  l1:= 1 lwb u, u1:= 1 upb u, start1, step1, stop1,
            l2:= 2 lwb u, u2:= 2 upb u, start2, step2, stop2;

       to  ( symmetric ! 2 ! 1)
       do  ( backward ! start1:= u1; step1:= -1; stop1:= l1
                      ! start1:= l1; step1:=  1; stop1:= u1 );
           ( reverse  ! start2:= u2; step2:= -1; stop2:= l2
                      ! start2:= l2; step2:=  1; stop2:= u2 );

            for  i  from  start1  by  step1  to  stop1
            do   ref []  real   fi= f[i,@l2], uim= u[(i>l1!i-1!i),@l2],
                                ui= u[i,@l2], uip= u[(i<u1!i+1!i),@l2];
                 ref [,] real   mi= m[i,@l2,@-3];
                 for  j  from  start2  by  step2  to  stop2
                 do  int   jm= (j>l2!j-1!j), jp= (j<u2!j+1!j);
                     ref [] real  mij = mi[j,@-3];

                     ui[j]:=       ( mij[-3]*uim[j]+mij[-2]*uim[jp]+
                     mij[-1]*ui [jm]    -     fi[j]+mij[ 1]*ui [jp]+
                     mij[ 2]*uip[jm]+mij[ 3]*uip[j] )/   -mij[ 0]

                 od
            od   ;
           ( symmetric! reverse:= not reverse; backward:= not backward)
       od
    end ;


# line  relaxation procedure                              #

    proc  lgs  relax = ( ref  netmat  dec, netmat  m,  net  u,f) void :
    begin  # line gauss seidel (lgs)  #

       int  st = ( zebra ! 2 ! 1 );
       int  l1:= 1 lwb u, u1:= 1 upb u, start, step, stop;

       proc  line relax = (  ref [ ] real  um,u,up,f,
                             ref [,] real  m ) void :
       begin   ref [] real  b= m[, 1], n = m[,-3], ne= m[,-2],
                            a= m[, 0], s = m[, 3], sw= m[, 2],
                            c= m[,-1];
             #not existing matrix elements: c[l]= b[k]= 0 !!#

            int  l= lwb f, k= upb f; [l:k] real  aa;
            int  i:=l;  real  g:= 0, p; aa[l]:= 1.0;

            for  j  from  l  to  k
            do  aa[j]:= a[j] - b[i]*    (  p:= c[j]/aa[i] );
                g     := f[j]        - n[j]*um[j] -
                       sw[j]*up[i] - s[j]*up[j] - g*p;
                ( j<k ! g  -:=                 ne[j]*um[j+1] );
                u[ j]:= g;  i:= j
            od ;
            for  j  from  k  by  -1  to  l
            do  u [j]:= g := ( u[j] - b[j]*g )/aa[j]  od
       end ;
```

```
    for  k  to  ( symmetric  or  zebra ! 2 ! 1)
    do  ( backward ! start := u1; step := -st; stop := l1
                   ! start := l1; step :=  st; stop := u1 );
        (   zebra
        ! ( symmetric /=  odd (k+start) ! start+:=  sign step )
        # ( symmetric ! even-odd ! odd-even ) half step #);

         for  i  from  start   by  step   to  stop
         do  line relax ( u[ (i>l1!i-1!i),], u[i,],
                          u[ (i<u1!i+1!i),], f[i,], m[i,,@-3] )
         od ;
        (symmetric ! backward:=  not  backward )
    od
  end ;

# illu  relaxation procedure                          #

  proc  illu relax = ( ref  netmat  dec, netmat jac, net  u,f) void :
  begin  int  l1= 1 lwb u , u1= 1 upb u, l2= 2 lwb u, u2= 2 upb u;
       (  netmat (dec) :=:  netmat ( nil ) ! illudec (jac,dec) );
       [l1:u1,l2:u2] real  du,rh;

        proc  soll = ( int  i,  net  r) void :
       ( ref [] real  l = dec[i,,-1],  d = dec[i,,0],
                      u = dec[i, , 1], z = r  [i,  ];
         for  j  from  l2+1  to  u2  do  z[j]+:= l[j]*z[j-1]  od ;
         for  j  from  l2    to  u2  do  z[j]*:= d[j]         od ;
         for  j  from  u2-1  by  -1  to  l2
                                     do  z[j]+:= u[j]*z[j+1]  od
       );

       rh:= residual(jac,u,f);
       soll(l1,rh);
        for  i  from  l1+1  to  u1
        do  for  j  from  l2  to  u2
            do  rh[i,j]-:= jac[i,j,-3]*rh[i-1,j  ] +
                     ( j<u2 ! jac[i,j,-2]*rh[i-1,j+1] ! 0.0 )
            od ;
            soll(i,rh)
        od ;
       du[u1,]:=rh[u1,];
        for  i  from  u1-1  by  -1  to  l1
        do  for  j  from  l2  to  u2
            do  du[i,j] := jac[i,j, 3]*du[i+1,j  ] +
                     ( j>l2 ! jac[i,j, 2]*du[i+1,j-1] ! 0.0 )
            od ;
            soll(i,du);
            for  j  from  l2  to  u2
            do  du[i,j] := rh[i,j] - du[i,j]  od
        od ;

        for  i  from  l1  to  u1  do
        for  j  from  l2  to  u2  do
             u[i,j]+:= du[i,j]
        od   od
   end ;
```

```
bristol algol68 text                                    PWH/15/12/83      8


 # illu decomposition procedure                        #

    proc  illudec = ( netmat  jac,  ref  netmat  decomp ) void :
    begin  int  l1= 1 lwb jac, u1= 1 upb jac,
                l2= 2 lwb jac, u2= 2 upb jac;
           int  ip;
           real  dd,ll,ii,l dinv u;
          [l2:u2,-1:+1] real  d;
          [l2:u2,-2:+2] real  dinv;
          [l2:u2,-1:+2] real  l dinv;
           heap [l1:u1,l2:u2,-1:+1] real  dec;

          d[l2:u2,-1:+1]:= jac[l1,l2:u2,-1:+1];
          dd:= dec[l1,l2,0]:= 1.0/d[l2,0];
           for  j  from  l2  to  u2-1
           do  dec[l1,j  ,+1]:=     -d[j  ,+1]*dd;
               dec[l1,j+1,-1]:= ll:=-d[j+1,-1]*dd;
               dec[l1,j+1, 0]:= dd:= 1.0/( d[j+1, 0] + d[j,1]*ll )
           od ;

           for  i  from  l1  to  u1-1
           do   ip:= i+1;
                     dinv[u2,0]:= ii:= dec[i,u2,0];
                for  j  from  u2-1  by  -1  to  l2
                do  dinv[ j,0]:= ii:= dec[i, j,0] +
                                  ii * dec[i,j,1]*dec[i,j+1,-1]
                od ;

                for  k  to  2  do
                for  j  from  u2  by -1  to  l2+k  do
                     dinv[j  ,-k]:= dinv[j    ,1-k]*dec[i,j-k+1,-1];
                     dinv[j-k, k]:= dinv[j-k+1,k-1]*dec[i,j-k  ,+1]
                od   od ;

                for  k  from              -1  to          2   do
                for  j  from  l2+(k=-1!1!0)  to  u2-(k=2!2!1)
                do   l dinv[j ,k]:= jac[ip,j  ,-3]*dinv[j  ,k  ] +
                                  jac[ip,j  ,-2]*dinv[j+1,k-1]
                od ;
                                                          ( k<1 !
                     l dinv[u2,k]:= jac[ip,u2,-3]*dinv[u2 ,k  ]     )
                od ;

                for  k  from              -1  to           1   do
                for  j  from  l2+(k=-1!1!0)  to  u2-(k=1!1!0)
                do   l dinv u := l dinv[j,k  ]*jac[i,j+k   ,3];
                                                         (j+k<u2 !
                     l dinv u+:= l dinv[j,k+1]*jac[i,j+k+1,2]        );
                     d[j,k]    :=     jac[ip,j,k] - l dinv u
                od   od ;

              dd:= dec[ip,l2,0]:= 1.0/d[l2,0];
               for  j  from  l2  to  u2-1
               do  dec[ip,j  ,+1]:=     -d[j  ,+1]*dd;
                   dec[ip,j+1,-1]:= ll:=-d[j+1,-1]*dd;
                   dec[ip,j+1, 0]:= dd:= 1.0/( d[j+1, 0] + d[j,1]*ll )
           od   od ;
          decomp:= dec
     end ;
```

```
# linear algebra solution procedure          #

    proc  mgm = ( ref [] netmat  lh,  ref [] net  uh,fh,
                int  itmax,p,q,s,t,
                proc ( ref  netmat , netmat , net , net ) void  relax,
                ref [] netmat decomp, ref  int  itused,
                proc ( int , netmat , net , net ) bool  goon mgm,
                proc ( int , string ) void  fail) void :
    begin  int  l= upb uh, r = s;
           ref [] netmat  lhdec =
                      ( decomp :=:  ref [] netmat ( nil )
                      !  loc [0:l] netmat  ! decomp );

         proc  mg = ( int l) void :
         # one multigrid cycle on level l #
          if    l = 0
          then  relax(lhdec[0],lh[0],uh[0],fh[0])
          else  # pre-relaxation #
                 to  p  do  relax(lhdec[l],lh[l],uh[l],fh[l]) od ;

                # coarse grid correction #
                fh[l-1]:= lin weight( residual (lh[l],uh[l],fh[l]) );
                 zero  uh[l-1];
                 to  (l=1!t!s)  do  mg (l-1)  od ;
                uh[l] +:= lin int pol ( uh[l-1]);

                # post-relaxation #
                 to  q  do  relax(lhdec[l],lh[l],uh[l],fh[l]) od
          fi ;

          int  err =            # check consistency data        #
         (   lwb  uh /= 0  or   lwb fh /= 0  or   lwb lh /= 0
                           or   upb fh /= 1  or   upb lh /= 1 ! 1
         !:  netmat  ll = lh[l];
             3 lwb ll /=-3  or  3 upb ll /= 3                   ! 2
         !:  net  ff = fh[l];
             int  l1 := 1 lwb ff,     u1 := 1 upb ff,
                  l2 := 2 lwb ff,     u2 := 2 upb ff;
                  l1 /= 1 lwb ll  or  u1 /= 1 upb ll  or
                  l2 /= 2 lwb ll  or  u2 /= 2 upb ll            ! 3
         !:  int  tpl = 2**l;
            l1 mod tpl /=0  or  u1 mod tpl/=0  or
            l2 mod tpl /=0  or  u2 mod tpl/=0                   ! 4
         !: l1:= l1 over tpl; u1:= u1 over tpl;
            l2:= l2 over tpl; u2:= u2 over tpl;
            ( itused <= 0
            ! uh[0]:=  zero   heap [l1:u1,l2:u2] real
            );         s <= 0  or  s > 3   or  t <= 0           ! 5
         !: itmax<0  or  p<0  or  q<0                           ! 6
         !:  lwb lhdec /= 0  or   upb lhdec /=1                 ! 7
         !  0  );
         ( err>0 ! fail ( err," mgm "));

          if  itused < 0       # no coarse operators available #
          then  # create galerkin approximations #
                for  i  from  l  by  -1  to  1
                do  lh[i-1]:= rap(lh[i]);
                    fh[i-1]:= lin weight(fh[i])
                od ; itused:= 0
          fi ;
```

```
            if  itused = 0        # no initial estimate available #
            then   for  i  from  0  to  1
                   do  lhdec[i]:=  nil   od ;

                  # apply full multigrid #
                   to  t  do  mg(0)  od ;
                   for  k  to  l-1
                   do   uh[k]:= sqr int pol (uh[k-1]);
                        to  r   do   mg (k)  od
                   od ; uh[l]:= sqr int pol (uh[l-1]);
                  goon mgm (itused,lh[l],uh[l],fh[l])
            fi ;

            to  itmax             # multigrid iteration             #
            while   mg (l); itused +:= 1;
                    goon mgm (itused, lh[l], uh[l], fh[l])
            do   skip   od
    end ;


# black box solution procedure                  #

    proc  solve sys =( int  l, ref  netmat  lh, ref  net  uh,fh) void :
          # solves the linear system lh*uh = fh                  #
          ([0:l] netmat  matrix; [0:l] net  rhs,solution;
           matrix[l]:= lh; rhs[l]:= fh;
           mgm(matrix,solution,rhs,mgitmax,mgp,mgq,mgs,mgt,mgrelax,
                nil , loc  int  := -1, mgm goon, fail);
           uh:= solution[l] );

# default global parameters     #

    bool  symmetric:=  false , backward:=  false ,
          reverse   :=  false , zebra    :=  false ;
    int   mgitmax   :=  8,
          mgp:= 0  ,  mgq:= 1,
          mgs:= 1  ,  mgt:= 1;
    proc  ( ref  netmat , netmat , net , net ) void
          mgrelax := illu relax;
    proc  mgm goon:= ( int  itnum, netmat lh, net uh,fh) bool :
                      true ;
    proc  fail    := ( int  n,[] char  text) void :
                     ( print((newline,text,n,newline)); stop);


#example program #

    int  l:= 4;

    netmat  matrix          :=  loc [0:2**l,0:2**l,-3:3] real ;
    net     solution, rhs :=  loc [0:2**l,0:2**l      ] real ;

   read((matrix,rhs));
   solve sys (l,matrix,solution,rhs);
   print(solution)

 end
```

6.2 Appendix 2

In this second appendix we give the user interfaces of the FORTRAN subroutines MGD1V (or MGD1S) and MGD5V (or MGD5S). We include also examples of a calling program. A tape with the complete programs can be obtained from the authors.

```
BRISTOL FORTRAN COMMENTS                              PWH/19/12/83      1

      SUBROUTINE MGD1V(A,U,RHS,UB,US,TEMP,LEVELS,NXC,NYC,NXF,NYF,NF,NM,
     .ISTART,MAXIT,TOL,IOUT,RESNO)
      COMMON /POI/ NGP(12),NGRIDX(12),NGRIDY(12)
      COMMON /CPU/ CP(9)
      DIMENSION A(NM,7),U(NM),UB(NF),RHS(NM),US(NM),TEMP(NXF),IOUT(5)
C-----------------------------------------------------------------------
C
C     PURPOSE
C     -------
C
C     THIS PROGRAM SOLVES A USER PROVIDED 7-POINT DIFFERENCE
C     EQUATION ON A RECTANGULAR GRID.
C
C     MATHEMATICAL METHOD
C     -------------------
C
C     SAWTOOTH MULTIGRID CYCLING
C     (I.E. ONE SMOOTHING-SWEEP AFTER EACH COARSE GRID CORRECTION)
C     WITH SMOOTHING BY INCOMPLETE CROUT-DECOMPOSITION,
C          7-POINT PROLONGATION AND RESTRICTION,
C          GALERKIN APPROXIMATION OF COARSE GRID MATRICES.
C
C***********************************************************************
C
C                    ****   PARAMETERS   ****
C
C***********************************************************************
C ---
C                    (INPUT DATA - SIZE OF PROBLEM)
C     LEVELS         NUMBER OF LEVELS IN MULTIGRID METHOD
C                    SHOULD BE .GE.2 AND .LE.12
C     NXC,NYC        NUMBER OF VERTICAL, HORIZONTAL GRID-LINES
C                    ON COARSEST GRID
C     NXF,NYF        NUMBER OF VERTICAL, HORIZONTAL GRID-LINES
C                    ON FINEST GRID
C     NF             NUMBER OF GRID-POINTS OF FINEST GRID
C     NM             NUMBER OF GRID-POINTS ON ALL GRIDS TOGETHER
C
C                    NOTE THAT THE FOLLOWING RELATIONS SHOULD HOLD,
C                    ---------------------------------------------
C                      NF=NXF*NYF
C                      NXF=(NXC-1)*(2**(LEVELS-1))+1
C                      NYF=(NYC-1)*(2**(LEVELS-1))+1
C
C                    THE PROGRAM CHECKS THE CONSISTENCY OF THESE DATA
C
C                    EXAMPLES
C                    --------
C
C                    LEVELS =    2    3     4     5      6      7
C                    NXC    =    3    3     3     3      3      3
C                    NYC    =    3    3     3     3      3      3
C                    NXF    =    5    9    17    33     65    129
C                    NYF    =    5    9    17    33     65    129
C                    NF     =   25   81   289  1089   4225  16641
C                    NM     =   34  115   404  1493   5718  22359
C
C                    LEVELS =    2    3     4     5      6      7
C                    NXC    =    5    5     5     5      5      5
C                    NYC    =    5    5     5     5      5      5
C                    NXF    =    9   17    33    65    129    257
C                    NYF    =    9   17    33    65    129    257
C                    NF     =   81  289  1089  4225  16641  66049
C                    NM     =  106  395  1484  5709  22350  88399
C
```

```
C ---
C     ISTART    (INPUT)
C               =1 IF THE USER PROVIDES AN INITIAL ESTIMATE
C                  OF THE SOLUTION IN UB
C               =Ø IF NO INITIAL ESTIMATE IS PROVIDED IN UB
C ---
C     MAXIT     (INPUT)
C               MAXIMUM NUMBER OF MULTIGRID ITERATIONS
C ---
C     TOL       (INPUT)
C               TOLERANCE DESIRED BY THE USER, TOL IS A BOUND OF THE
C               L2-NORM OF THE RESIDUAL
C               REMARK  IF EITHER MAXIT ITERATIONS OR THE TOLERANCE HAVE
C               ------  BEEN ACHIEVED,THEN MULTIGRID CYCLING IS STOPPED.
C ---
C     IOUT      (INPUT)
C               INTEGER ARRAY DIMENSIONED AS IOUT(5) THAT CONTROLS
C               THE AMOUNT OF OUTPUT DESIRED BY THE USER.
C               SMALLER IOUT-VALUES MEAN LESS OUTPUT,
C               POSSIBLE VALUES ARE ,
C               IOUT(1)=1 CONFIRMATION OF INPUT DATA
C                       Ø NONE
C               IOUT(2)=2 MATRICES AND RIGHT-HAND SIDES ON ALL LEVELS
C                       1 MATRIX AND RIGHT-HAND SIDE ON HIGHEST LEVEL
C                       Ø NONE
C               IOUT(3)=2 MATRIX-DECOMPOSITIONS ON ALL LEVELS
C                       1 MATRIX-DECOMPOSITION ON HIGHEST LEVEL
C                       Ø NONE
C               IOUT(4)=3 NORMS OF RESIDUALS, REDUCTION FACTORS,
C                         FINAL RESIDUAL, FINAL SOLUTION
C                       2 NORMS OF RESIDUALS, REDUCTION FACTORS,
C                         FINAL RESIDUAL
C                       1 NORMS OF RESIDUALS, REDUCTION FACTORS
C                       Ø NONE
C               IOUT(5)=1 THE TIME SPENT IN VARIOUS SUBROUTINES
C                       Ø NONE
C                         REMARK  CLOCK ROUTINES ARE NOT STANDARD
C                         ------  FORTRAN. TO OBTAIN TIMINGS THE USER
C                                 SHOULD ADAPT THE SUBROUTINE TIMING,
C                         IT SHOULD DELIVER THE CPU-TIME ELAPSED.
C ---
C     A         (INPUT)
C               REAL ARRAY DIMENSIONED AS A(NM,7)
C               THE USER HAS TO INITIALIZE A( 1,1),..,A( 1,7)
C                                          .          .
C                                          A( K,1)    A( K,7)
C                                          ..         .
C                                          A(NF,1),..,A(NF,7)
C               WITH THE MATRIX CORRESPONDING TO THE FINEST GRID.
C               THE ORDERING OF THE POINTS IN THE GRID IS AS FOLLOWS
C               THE SUBSCRIPT K=(J-1)*NXF+I CORRESPONDS TO THE POINT
C
C               (X,Y) = ( I*H , J*H )
C                            X     Y
C                                      I=1,...,NXF  J=1,...,NYF
```

```
C               THE 7-POINT DIFFERENCE MOLECULE AT THE POINT WITH
C               SUBSCRIPT K=(J-1)*NXF+I IS POSITIONED IN THE X,Y-PLANE
C               AS FOLLOWS
C
C
C                       Y,J
C                        +
C                        +
C                        +   A(K,6)   A(K,7)
C                        +      .        .
C                        +   A(K,3)   A(K,4)   A(K,5)
C                        +      .        .        .
C                        +            A(K,1)   A(K,2)
C                        +               .        .
C                        +
C                        O+ + + + + + + + + + + + + + + X, I
C
C       IMPORTANT THE USER HAS TO PROVIDE THE MATRIX A ONLY ON THE FINEST
C       --------- GRID.
C       IMPORTANT THE USER HAS TO TAKE CARE THAT PARTS OF THE MOLECULES
C       --------- OUTSIDE THE DOMAIN ARE INITIALIZED TO ZERO, OTHERWISE
C                                                       ----
C               WRONG RESULTS ARE PRODUCED.
C       IMPORTANT THE COEFFICIENT MATRIX A IS OVERWRITTEN BY THE PROGRAM.
C       --------- AFTER A CALL OF MGD1V (DECOMP),A CONTAINS THE INCOMPLETE
C               CROUT DECOMPOSITIONS.
C ---
C       RHS     (INPUT)
C               REAL ARRAY DIMENSIONED AS RHS(NM)
C               THE USER HAS TO INITIALIZE RHS(1),....,RHS(NF) WITH
C               THE RIGHT-HAND SIDE OF THE EQUATION.
C               THE ORDERING IS THE SAME AS INDICATED FOR ARRAY A.
C       IMPORTANT THE USER HAS TO PROVIDE THE RIGHT-HAND SIDE OF THE
C       --------- DISCRETIZED EQUATION ONLY ON THE FINEST GRID
C ---
C       U       (OUTPUT)
C               REAL ARRAY DIMENSIONED AS U(NM)
C               CONTAINS THE (APPROXIMATE) NUMERICAL SOLUTION AFTER A
C               CALL OF MGD1V.
C ---
C       UB      (WORKSPACE/INPUT)
C               REAL ARRAY DIMENSIONED AS UB(NF)
C               IS USED AS A SCRATCH ARRAY.  IF ISTART=1 THEN UB(1),...
C               ..,UB(NF) SHOULD CONTAIN AN INITIAL ESTIMATE OF THE
C               SOLUTION PROVIDED BY THE USER.
C               AFTER A CALL OF MGD1V, UB CONTAINS THE RESIDUAL OF THE
C               THE NUMERICAL SOLUTION.
C ---
C       US      (WORKSPACE)
C               REAL ARRAY DIMENSIONED AS US(NM)
C               IS USED AS A SCRATCH ARRAY
C ---
C       TEMP    (WORKSPACE)
C               REAL ARRAY DIMENSIONED AS TEMP(NXF)
C               IS USED AS A (SMALL) SCRATCH ARRAY.
C               IF THE SCALAR VERSION OF SUBROUTINE SOLVE (DENOTED BY
C               COMMENT CARDS BEGINNING WITH CSC) IS USED THEN IT IS
C               SUFFICIENT TO DIMENSION TEMP AS TEMP(1).
C ---
C       RESNO   (OUTPUT)
C               THIS VARIABLE CONTAINS THE L2-NORM OF THE RESIDUAL AT
C               THE END OF EXECUTION OF MGD1V.
C
C-----------------------------------------------------------------------
```

BRISTOL FORTRAN COMMENTS PWH/19/12/83 4

```
C-----------------------------------------------------------------------
C     THIS IS AN EXAMPLE OF A MAIN PROGRAM USING MGD1V
C-----------------------------------------------------------------------
C
C     ACTUAL USER PROVIDED DIMENSION STATEMENTS,
C
      DIMENSION A(88399,7),RHS(88399),U(88399),US(88399),UB(66Ø49),
     .TEMP(257),IOUT(5)
C
C     USER DATA STATEMENTS,
C
      DATA NXC,NYC,NXF,NYF/5,5,257,257/
      DATA LEVELS,NM,NF/7,88399,66Ø49/
      DATA MAXIT,ISTART/1Ø,Ø/
      DATA IOUT(1),IOUT(2),IOUT(3),IOUT(4),IOUT(5)/1,Ø,Ø,1,1/
C
C     PROBLEM SET UP
C
      CALL MATRHS(A,RHS,NM,NXF,NYF)
C***********************************************************************
C    MATRHS IS A SUBROUTINE WHICH FILLS THE MATRIX AND THE RIGHT-HAND
C    SIDE, IT DOES NOT BELONG TO THE PACKAGE AND IS ONLY AN EXAMPLE.
C***********************************************************************
C
C     SOLUTION OF THE LINEAR SYSTEM
C
      CALL MGD1V(A,U,RHS,UB,US,TEMP,LEVELS,NXC,NYC,NXF,NYF,NF,NM,
     .ISTART,MAXIT,Ø.Ø,IOUT,RESNO)
C
C     POSSIBLE REFINEMENT OF THE SOLUTION, 5 MORE ITERATIONS
C
C     CALL CYCLES(A,U,RHS,UB,US,TEMP,LEVELS,NXF,NF,NM,1,5,Ø.Ø,IOUT,
C    .RESNO)
C
C     POSSIBLE REFINEMENT UNTIL RESIDUAL NORM .LT. 1.ØE-12
C
C     CALL CYCLES(A,U,RHS,UB,US,TEMP,LEVELS,NXF,NF,NM,1,3Ø,1.ØE-12,IOUT,
C    .RESNO)
C
      STOP
      END
```

```
      SUBROUTINE MGD5V(A,V,RHS,VB,LDU,WORK,LEVELS,NXC,NYC,NXF,NYF,
     .                 NF,NM,ISTART,MAXIT,TOL,IOUT,RESNO)
      COMMON /POI/ NGP(12),NGRIDX(12),NGRIDY(12)
      COMMON /CPU/ CP(10)
      REAL LDU
      DIMENSION A(NM,7),V(NM),VB(NM),RHS(NM),LDU(NM,3),
     .          WORK(NXF,9),IOUT(5)
C-----------------------------------------------------------------------
C
C     PURPOSE
C     -------
C
C     THIS PROGRAM SOLVES A USER PROVIDED 7-POINT DIFFERENCE
C     EQUATION ON A RECTANGULAR GRID.
C
C     MATHEMATICAL METHOD
C     -------------------
C
C     SAWTOOTH MULTIGRID CYCLING
C     (I.E. ONE SMOOTHING-SWEEP AFTER EACH COARSE GRID CORRECTION)
C     WITH SMOOTHING BY INCOMPLETE LINE LU-DECOMPOSITION,
C          7-POINT PROLONGATION AND RESTRICTION,
C          GALERKIN APPROXIMATION OF COARSE GRID MATRICES.
C
C***********************************************************************
C
C                    ****   PARAMETERS    ****
C
C***********************************************************************
C ---
C                    (INPUT DATA - SIZE OF PROBLEM)
C     LEVELS         NUMBER OF LEVELS IN MULTIGRID METHOD
C                    SHOULD BE .GE.3 AND .LE.12
C     NXC,NYC        NUMBER OF VERTICAL, HORIZONTAL GRID-LINES
C                    ON COARSEST GRID, NXC SHOULD BE .GE.5
C                                 AND NYC SHOULD BE .GE.3
C     NXF,NYF        NUMBER OF VERTICAL, HORIZONTAL GRID-LINES
C                    ON FINEST GRID
C     NF             NUMBER OF GRID-POINTS OF FINEST GRID
C     NM             NUMBER OF GRID-POINTS ON ALL GRIDS TOGETHER
C
C                    SEE COMMENTS IN MGD1V FOR FURTHER DETAILS.
C                    -----------------------------------------

C     ISTART      (INPUT)
C     MAXIT       (INPUT)
C     TOL         (INPUT)
C     IOUT        (INPUT)
C     A           (INPUT)
C     RHS         (INPUT)
C
C                 THESE INPUT PARAMETERS HAVE THE SAME MEANING AS IN MGD1V
C                 --------------------------------------------------------
C                 THE ONLY DIFFERENCE IS THAT THE ARRAY A WILL NEVER BE
C                 OVERWRITTEN BY MGD5V.
C ---
C     LDU         (OUTPUT)
C                 REAL ARRAY DIMENSIONED AS LDU(NM,3)                  -
C                 LDU CONTAINS DECOMPOSITIONS OF ALL TRIDIAGONAL BLOCKS D
C                                                                      J
```

```
BRISTOL FORTRAN COMMENTS                              PWH/19/12/83     6

C ---
C      V            (INPUT/OUTPUT)
C                   REAL ARRAY DIMENSIONED AS V(NM)
C                   IF ISTART=1 THEN V(1),...,V(NF) SHOULD CONTAIN AN
C                   INITIAL ESTIMATE OF THE SOLUTION PROVIDED BY THE USER.
C                   IF ISTART=Ø THEN V IS INITIALIZED TO ZERO.(SUBR. PREPAR)
C                   AFTER A CALL OF MGD5V, V CONTAINS THE (APPROXIMATE)
C                   NUMERICAL SOLUTION.
C ---
C      VB           (WORKSPACE/OUTPUT)
C                   REAL ARRAY DIMENSIONED AS VB(NF)
C                   AFTER A CALL OF MGD5V, VB CONTAINS THE RESIDUAL OF THE
C                   NUMERICAL SOLUTION V.
C ---
C      WORK         (WORKSPACE)
C                   REAL ARRAY DIMENSIONED AS WORK(NXF,9)
C                   IS USED AS A (SMALL) SCRATCH ARRAY
C ---
C      RESNO        (OUTPUT)
C                   THIS VARIABLE CONTAINS THE L2-NORM OF THE RESIDUAL AT
C                   THE END OF EXECUTION OF MGD5V.
C
C-----------------------------------------------------------------------------


C-----------------------------------------------------------------------------
C      THIS IS AN EXAMPLE OF A MAIN PROGRAM USING MGD5V
C-----------------------------------------------------------------------------
C      ACTUAL USER PROVIDED DIMENSION STATEMENTS,
C
       REAL LDU
       DIMENSION A(88399,7),RHS(88399),V(88399),VB(88399),
      .LDU(88399,3),WORK(257,9),IOUT(5)
C
C      USER DATA STATEMENTS,
C
       DATA NXC,NYC,NXF,NYF/5,5,257,257/
       DATA LEVELS,NM,NF/7,88399,66Ø49/
       DATA MAXIT,ISTART/1Ø,Ø/
       DATA IOUT(1),IOUT(2),IOUT(3),IOUT(4),IOUT(5)/1,Ø,Ø,1,1/
C
C      PROBLEM SET UP
C
       CALL MATRHS(A,RHS,NM,NXF,NYF)
C*****************************************************************************
C     MATRHS IS A SUBROUTINE WHICH FILLS THE MATRIX AND THE RIGHT-HAND
C     SIDE, IT DOES NOT BELONG TO THE PACKAGE AND IS ONLY AN EXAMPLE.
C*****************************************************************************
C
C      SOLUTION OF THE LINEAR SYSTEM
C
       CALL MGD5V(A,V,RHS,VB,LDU,WORK,LEVELS,
      .          NXC,NYC,NXF,NYF,NF,NM,ISTART,MAXIT,Ø.Ø,IOUT,RESNO)
C
C      POSSIBLE REFINEMENT OF THE SOLUTION, 5 MORE ITERATIONS
C
C      CALL CYCLES(A,V,RHS,VB,LDU,WORK,LEVELS,NXF,NF,NM,
C     .            1,5,Ø.Ø,IOUT,RESNO)
C      POSSIBLE REFINEMENT UNTIL RESIDUAL NORM .LT. 1.ØE-12
C
C      CALL CYCLES(A,V,RHS,VB,LDU,WORK,LEVELS,NXF,NF,NM,
C     .             1,3Ø,1.ØE-12,IOUT,RESNO)
C
       STOP
       END
```

6.3 Appendix 3

In this appendix we give a full description in FORTRAN of our implementation of the ILLU-decomposition. First we give a brief description of that decomposition and the corresponding relaxation sweep. Let the seven diagonal matrix A correspond with the following molecule:

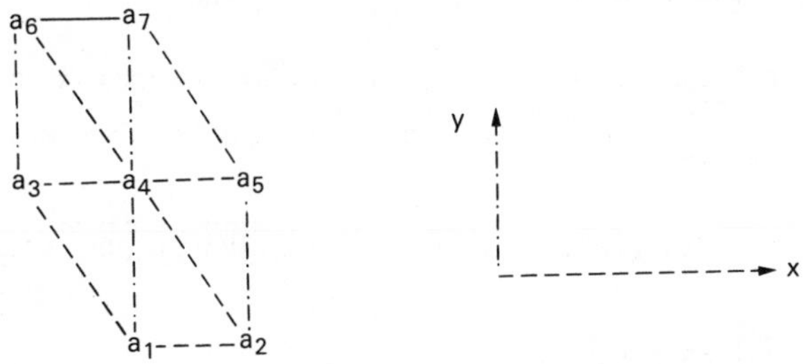

Let the matrix A be decomposed in block tridiagonal form:

$$A = L + D + U = \begin{pmatrix} D_1 & U_1 & & & \\ L_2 & D_2 & U_2 & & \\ & L_3 & D_3 & U_3 & \\ & \cdot & \cdot & \cdot & \\ & & L_i & D_i & U_i \\ & & & L_n & D_n \end{pmatrix}$$

L_i $i = 2\ (1)\ n$ corresponds with a_1 and a_2,

D_i $i = 1\ (1)\ n$ corresponds with a_3, a_4 and a_5,

U_i $i = 1\ (1)\ n-1$ corresponds with a_6 and a_7.

Then the ILLU-decomposition is defined by L, $\bar{D}$, U, with

$$\bar{D}_1 = D_1,$$
$$\bar{D}_j = D_j - \text{tridiag}\,(L_j\ \bar{D}_{j-1}^{-1}\ U_{j-1}),$$
$$\text{for } j = 2\ (1)\ n.$$

The tridiagonal matrix $\bar{D}$ is stored by means of its exact decomposition $\mathcal{L}$, $\mathcal{D}$, $\mathcal{U}$. ($\mathcal{L}$ and $\mathcal{U}$ are bidiagonal, $\mathcal{D}$ is a main diagonal, the main diagonals of $\mathcal{L}$ and $\mathcal{U}$ are equal to one.)

Let $u^{(i)}$ be an approximate solution of $Au = f$, then an ILLU-relaxation sweep reads:

Step 1: compute $r := f - A\,u^{(i)}$;

Step 2: solve $(L+\bar{D})\,\bar{D}^{-1}\,(\bar{D}+U)\ v = r$;

Step 3: $u^{(i+1)} := u^{(i)} + v$.

```
      SUBROUTINE DECOMP(A1,A2,A3,A4,A5,A6,A7,N,M,NM)
C-----------------------------------------------------------------------
C     INCOMPLETE CROUT-DECOMPOSITION (ILU-DECOMPOSITION) OF THE SEVENDIA
C     GONAL MATRIX A REPRESENTED BY A1,A2,A3,A4,A5,A6,A7.
C     A IS OVERWRITTEN BY ITS DECOMPOSITION.
C     THE MAIN DIAGONAL OF L IS ONE EVERYWHERE, THE OTHER DIAGONALS OF L
C     ARE STORED IN A1, A2, A3.
C     THE DIAGONALS OF U ARE STORED IN  A4, A5, A6, A7.
C     M IS THE NUMBER OF GRIDPOINTS IN THE X-DIRECTION,
C     N IS THE NUMBER OF GRIDPOINTS IN THE Y-DIRECTION,
C     NM=N*M.
C
C     NOTE  THE LOOPS 6, 10, 20, 30, 40, 50, 60, 400 ARE AUTOMATICALLY
C     ----  VECTORIZED.
C           THE LOOPS 5 AND 55 ARE RECURSIVE AND WILL THEREFORE NOT BE
C           VECTORIZED.
C
C-----------------------------------------------------------------------
      DIMENSION A1(NM),A2(NM),A3(NM),A4(NM),A5(NM),A6(NM),A7(NM)
      A4J=A4(1)
      DO 5 J=2,M
      A3(J)=A3(J)/A4J
      A4(J)=A4(J)-A3(J)*A5(J-1)
      A4J=A4(J)
    5 CONTINUE
      DO 6 J=2,M
      A6(J)=A6(J)-A3(J)*A7(J-1)
    6 CONTINUE
      M1=M-1
      JB=1
      JE=M
      DO 100 K=2,N
      JB=JB+M
      JE=JE+M
      DO 10 J=JB,JE
      A1(J)=A1(J)/A4(J-M)
   10 CONTINUE
      DO 20 J=JB,JE
      A2(J)=(A2(J)-A1(J)*A5(J-M))/A4(J-M1)
   20 CONTINUE
      DO 30 J=JB,JE
      A3(J)=A3(J)-A1(J)*A6(J-M)
   30 CONTINUE
      DO 40 J=JB,JE
      A4(J)=A4(J)-A2(J)*A6(J-M1)-A1(J)*A7(J-M)
   40 CONTINUE
      DO 50 J=JB,JE
      A5(J)=A5(J)-A2(J)*A7(J-M1)
   50 CONTINUE
      A4J=A4(JB-1)
      DO 55 J=JB,JE
      A3(J)=A3(J)/A4J
      A4(J)=A4(J)-A3(J)*A5(J-1)
      A4J=A4(J)
   55 CONTINUE
      DO 60 J=JB,JE
      A6(J)=A6(J)-A3(J)*A7(J-1)
   60 CONTINUE
  100 CONTINUE
C-----------------------------------------------------------------------
C     FOR ILU-RELAXATION THE RECIPROCAL OF A4 IS NEEDED, NOT A4 ITSELF.
C-----------------------------------------------------------------------
      DO 400 JJ=1,NM,65535
      JJE=(JJ-1)+MIN0(65535,NM-(JJ-1))
      DO 400 J=JJ,JJE
      A4(J)=1.0/A4(J)
  400 CONTINUE
      RETURN
      END
```

```
      SUBROUTINE ILLUDC(A,DIMA,L,D,U,NX,NY,NXY,WORK)
C-----------------------------------------------------------------------
C     INCOMPLETE LINE LU (ILLU-DECOMPOSITION) OF THE SEVENDIAGONAL
C     MATRIX A.  A REMAINS INTACT, L D AND U ARE FILLED IN WITH THE
C     DECOMPOSITIONS OF
C                               _
C                               D      J = 1(1)NY
C                                J
C
C     NX IS THE NUMBER OF GRIDPOINTS IN THE X-DIRECTION,
C     NY IS THE NUMBER OF GRIDPOINTS IN THE Y-DIRECTION,
C     NXY=NX*NY
C
C-----------------------------------------------------------------------
C
      INTEGER DIMA
      REAL L
      DIMENSION A(DIMA,7),L(NXY),D(NXY),U(NXY),WORK(NX,9)
      CALL TRIDEC(A(1,3),A(1,4),A(1,5),L,D,U,NX)
      NPOLD=1
      DO 100 J=2,NY
      NPNEW=NPOLD+NX
      CALL BLOCKS(A(NPOLD,1),A(NPNEW,1),DIMA,
     .            L(NPOLD),D(NPOLD),U(NPOLD),
     .            L(NPNEW),D(NPNEW),U(NPNEW),NX,
     .            WORK(1,1),WORK(1,2),WORK(1,3),WORK(1,4),WORK(1,5),
     .            WORK(1,6))
      NPOLD=NPNEW
  100 CONTINUE
      RETURN
      END
      SUBROUTINE TRIDEC(DM,DZ,DP,LJ,DJ,UJ,NX)
C-----------------------------------------------------------------------
C                                                      _
C     PERFORMS DECOMPOSITION OF A TRIDIAGONAL MATRIX REPRESENTED BY DM,
C     DZ, DP.
C     THE DECOMPOSITION CONSISTS OF A LOWER TRIANGULAR BIDIAGONAL MATRIX
C     LJ, AN UPPER TRIANGULAR BIDIAGONAL MATRIX UJ AND AN ONE DIAGONAL
C     MATRIX DJ, THE MAIN DIAGONALS OF LJ AND UJ EQUAL ONE.
C     NX IS THE NUMBER OF POINTS IN THE X-DIRECTION.
C
C     NOTE  LOOP 20 IS AUTOMATICALLY  VECTORIZED.
C     ----  LOOP 10 IS RECURSIVE AND WILL THEREFORE NOT BE VECTORIZED.
C
C-----------------------------------------------------------------------
      REAL LJ
      DIMENSION DM(NX),DZ(NX),DP(NX),LJ(NX),DJ(NX),UJ(NX)
      DJ(1)=1.0/DZ(1)
      DJIM1=DJ(1)
      DO 10 I=2,NX
      LJ(I)=-DM(I)*DJIM1
      DJ(I)=1.0/(DZ(I)+LJ(I)*DP(I-1))
      DJIM1=DJ(I)
   10 CONTINUE
      NX1=NX-1
      DO 20 I=1,NX1
      UJ(I)=-DP(I)*DJ(I)
   20 CONTINUE
      RETURN
      END
```

```
      SUBROUTINE BLOCKS(AJM1,AJ,DIMA, LJM1,DJM1,UJM1, LJ,DJ,UJ,NX,
     .                  QM2,QM1,QZE,QP1,QP2, LD)
C-----------------------------------------------------------------------
C     INCOMPLETE LINE LU DECOMPOSITION (ILLU-DECOMPOSITION) OF J-TH ROW
C     OF BLOCKS OF THE SEVENDIAGONAL MATRIX A.
C     AJ   IS  J    TH ROW OF BLOCKS OF A,
C     AJM1 IS (J-1) TH ROW OF BLOCKS OF A,
C     LJM1, DJM1, UJM1 ARE (J-1) TH ROWS OF L, D, U WHICH REPRESENT
C     BIDIAGONAL MATRICES (MAIN DIAGONALS EQUAL ONE) WHICH PRODUCT IS
C         -
C         D
C          (J-1)
C     LJ, DJ, UJ BECOME THE J TH ROWS OF L, D, U AFTER A CALL OF BLOCKS.
C     NX IS THE NUMBER OF GRIDPOINTS IN THE X-DIRECTION.
C     QM2,QM1,QZE,QP1,QP2,LD ARE WORK ARRAYS.
C
C     NOTE  THE LOOPS 10, 30, 40, 51, 52, 53, 54, 60, 70, 80 ARE AUTOMA-
C     ----  TICALLY VECTORIZED.
C           LOOP 20 IS RECURSIVE AND WILL THEREFORE NOT BE VECTORIZED.
C
C-----------------------------------------------------------------------
      INTEGER DIMA
      REAL LJM1,LJ,LD
      DIMENSION AJM1(DIMA,7),AJ(DIMA,7),LJM1(NX),DJM1(NX),UJM1(NX),
     .          LJ(NX),DJ(NX),UJ(NX),
     .          QM2(NX),QM1(NX),QZE(NX),QP1(NX),QP2(NX),
     .          LD(NX,4)
C-----------------------------------------------------------------------
C                                          - -1
C     FIRST STEP - COMPUTATION OF 5-DIAG( D     ),
C                                          J-1
C                  RESULTING DIAGONALS ARE QM2, QM1, QZE, QP1, QP2
C-----------------------------------------------------------------------
      NX1=NX-1
      NX2=NX-2
      DO 10 I=1,NX1
      QZE(I)=UJM1(I)*LJM1(I+1)
   10 CONTINUE
      QZE(NX)=DJM1(NX)
      QZEIP1=QZE(NX)
      DO 20 II=1,NX1
      I=NX-II
      QZE(I)=DJM1(I)+QZE(I)*QZEIP1
      QZEIP1=QZE(I)
   20 CONTINUE
      DO 30 I=2,NX1
      QM1(I)=LJM1(I)*QZE(I)
      QP1(I)=UJM1(I)*QZE(I+1)
   30 CONTINUE
      QP1(1)=UJM1(1)*QZE(2)
      QM1(NX)=LJM1(NX)*QZE(NX)
      DO 40 I=3,NX2
      QM2(I)=LJM1(I-1)*QM1(I)
      QP2(I)=UJM1(I)*QP1(I+1)
   40 CONTINUE
      QP2(1)=UJM1(1)*QP1(2)
      QP2(2)=UJM1(2)*QP1(3)
      QM2(NX1)=LJM1(NX2)*QM1(NX1)
      QM2(NX)=LJM1(NX1)*QM1(NX)
```

```
C-----------------------------------------------------------------------
C                                                        - -1
C     SECOND STEP - COMPUTATION OF  4 DIAGONALS OF  L D
C                                                     J J-1
C-----------------------------------------------------------------------
      QM1(1)=0.0
      QM2(2)=0.0
      QP2(NX1)=0.0
      QP1(NX)=0.0
      DO 51 I=1,NX1
      LD(I,1)=AJ(I,1)*QM1(I)+AJ(I,2)*QM2(I+1)
   51 CONTINUE
      DO 52 I=1,NX1
      LD(I,2)=AJ(I,1)*QZE(I)+AJ(I,2)*QM1(I+1)
   52 CONTINUE
      DO 53 I=1,NX1
      LD(I,3)=AJ(I,1)*QP1(I)+AJ(I,2)*QZE(I+1)
   53 CONTINUE
      DO 54 I=1,NX1
      LD(I,4)=AJ(I,1)*QP2(I)+AJ(I,2)*QP1(I+1)
   54 CONTINUE
      LD(NX,1)=AJ(NX,1)*QM1(NX)
      LD(NX,2)=AJ(NX,1)*QZE(NX)
C-----------------------------------------------------------------------
C                                         -                    - -1
C     THIRD AND FOURTH STEP - COMPUTATION OF D = D - 3-DIAG( L D    U  )
C                                           J   J           J J-1  J-1
C     -
C     D  IS REPRESENTED BY QM1, QZE, QP1
C      J
C-----------------------------------------------------------------------
      DO 60 I=2,NX
      QM1(I)=AJ(I,3)-LD(I,1)*AJM1(I-1,7)-LD(I,2)*AJM1(I   ,6)
   60 CONTINUE
      DO 70 I=1,NX1
      QZE(I)=AJ(I,4)-LD(I,2)*AJM1(I   ,7)-LD(I,3)*AJM1(I+1,6)
   70 CONTINUE
      DO 80 I=1,NX2
      QP1(I)=AJ(I,5)-LD(I,3)*AJM1(I+1,7)-LD(I,4)*AJM1(I+2,6)
   80 CONTINUE
      QZE( NX)=AJ( NX,4)-LD( NX,2)*AJM1( NX,7)
      QP1(NX1)=AJ(NX1,5)-LD(NX1,3)*AJM1( NX,7)
C-----------------------------------------------------------------------
C                                                                -
C     FIFTH STEP -  COMPUTATION OF DECOMPOSITION L ,D ,U   OF  D
C                                                 J  J  J       J
C-----------------------------------------------------------------------
      CALL TRIDEC(QM1,QZE,QP1,LJ,DJ,UJ,NX)
      RETURN
      END
```

7. REFERENCES

Barkai, D. and Brandt, A. (1983) Vectorized Multigrid Poisson Solver for the CDC CYBER 205, *Appl. Math. Comp.*, **13**, pp. 215-227.

Dendy (Jr.), J.E. (1982) Black Box multigrid, *J. Comp. Phys.*, **48**, pp. 366-386.

Dendy (Jr.), J.E. (1983) Black Box Multigrid for Non-symmetric Problems, *Appl. Math. Comp.*, **13**, pp. 261-283.

Foerster, H. and Witsch, K. (1982) Multigrid software for the solution of elliptic problems on rectangular domains MGOO (Release 1), in Multigrid Methods, (W. Hackbusch and U. Trottenberg, eds.), Lecture Notes in Mathematics, 960, Springer Verlag, Berlin.

Hemker, P.W. (1982) On the comparison of Line-Gauss Seidel and ILU relaxation in multigrid algorithms, in Computational and asymptotic methods for boundary and interior layers (J.J.H. Miller, ed.), Boole Press, Dublin, pp. 269-277.

Hemker, P.W. (1984) Multigrid methods for problems with a small parameter, in Numerical Analysis (D.F. Griffiths, ed.), Proceedings of Dundee Conference 1983, Springer Lecture Series in Mathematics, Springer Verlag, Berlin.

Hemker, P.W., Kettler, R., Wesseling, P. and de Zeeuw, P.M. (1983) Multigrid Methods: Development of Fast Solvers, *Appl. Math. Comp.*, **13**, pp. 311-326.

Hemker, P.W., Wesseling, P. and de Zeeuw, P.M. (1983) A portable vector code for autonomous multigrid modules, in PDE Software: Modules, Interfaces and Systems (B. Engquist, ed.), Proceedings IFIP WG 2.5 Working Conference, (North Holland).

Kettler, R. (1982) Analysis and comparison of relaxation schemes in robust multigrid and preconditioned conjugate gradient methods, in Multigrid Methods (W. Hackbusch and U. Trottenberg, eds.), Springer Lecture Series in Mathematics, 960, Springer Verlag, Berlin, pp. 502-534.

Wesseling, P. (1982a) A robust and efficient multigrid method, in Multigrid Methods (W. Hackbusch and U. Trottenberg, eds.), Springer Lecture Series in Mathematics, 960, Springer Verlag, Berlin, pp. 614-630.

Wesseling, P. (1982b) Theoretical and practical aspects of multigrid method, *SIAM J. Sci. Stat. Comp.*, **3**, pp. 387-407.

MULTIGRID AND CONJUGATE GRADIENT METHODS AS CONVERGENCE ACCELERATION TECHNIQUES

P. Sonneveld and P. Wesseling

(Delft University of Technology)

and

P.M. de Zeeuw

(Centre for Mathematics and Computer Science, Amsterdam)

ABSTRACT

Multigrid and conjugate gradient type techniques for the acceleration of iterative methods are discussed. A detailed discussion is given of incomplete factorizations. The theoretical background of the classical conjugate gradient method and preconditioning is briefly reviewed. A conjugate gradient type method for non-symmetric-positive-definite systems is presented. Multigrid methods are discussed, and two portable, autonomous computer codes are introduced. Multigrid treatment of convection-diffusion entails special difficulties, and ways to overcome these are outlined. Numerical experiments on a set of test problems are reported. Efficiency and robustness of several conjugate gradient and multigrid methods are compared and discussed.

1. INTRODUCTION

Finite difference and element discretizations of partial differential equations give rise to large sparse systems of equations, which in this paper will be assumed to have been linearized. In practice, the number of unknowns can be quite large, and solution methods must exploit the sparsity and the structure of the system. This can be done with direct methods, using sparse matrix techniques, or by iterative methods, which will be considered here.

Classical iterative methods, a review of which is given by Young (1971), are relatively simple to implement, but converge slowly for large problems. In recent years conjugate gradient (CG) and multigrid (MG) methods have been drawing increased attention as powerful and rather general techniques to accelerate the convergence of iterative methods. Our aim is to discuss recent progress that has been made with these techniques. A new CG method for systems that are not symmetric positive definite will be presented.

Both CG and MG have a wider use and significance than just being acceleration techniques. But the present viewpoint makes it possible to grasp the main principles in a simple manner, and furthermore, it brings out the main similarities and differences of the two methods.

The authors beg forgiveness for focussing on developments with which they are especially familiar.

2. ITERATIVE METHODS AND INCOMPLETE FACTORIZATIONS

The problem to be solved is a linear algebraic system denoted as

$$Ay = b. \tag{2.1}$$

Stationary iterative methods for the solution of (2.1) can usually be written as

$$y^{n+1} = y^n + B(b-Ay^n). \tag{2.2}$$

For example, for the Gauss-Seidel method (GS) we have $B = (D+L)^{-1}$, and for the successive overrelaxation method (SOR) we have $B = \omega(D+\omega L)^{-1}$, with L, D and U defined by

$$A = L + D + U, \tag{2.3}$$

where L and U are the lower and upper triangular parts of A, and $D = \text{diag}(A)$.

For the error $e^n = y^n - y$ we find from (2.2)

$$e^n = (I-BA)^n e^0. \tag{2.4}$$

For special cases the spectral radius $\rho(I-BA)$ is known. For example, if A is the familiar 5-point finite difference discretization of Laplace's equation on the unit square with Dirichlet boundary conditions and mesh-size h in both directions, we have

$$\text{GS: } \rho(I-BA) = \cos^2\pi h = 1 - \pi^2h^2 + O(h^4), \tag{2.5}$$

$$\text{SOR: } \rho(I-BA) = \frac{1-\sin\pi h}{1+\sin\pi h} = 1 - 2\pi h + O(h^2). \tag{2.6}$$

Let us define the computational cost W of an algorithm to be the number of operations from the set $\{+,-,*,/\}$. In practice computer time will depend on W, but also on the programming language, the skill of the programmer, the frequency of the occurrence of indirect addressing, type of machine etc., but still, W defined above is a convenient yardstick for measuring computational complexity.

Let N ($=h^{-2}$ in two dimensions) be the number of unknowns. Then the cost of one GS or SOR iteration is $W = O(N)$. The required number of iterations n for a desired residual or error reduction ε follows from

$$(\rho(I-BA))^n \leqslant \varepsilon. \tag{2.7}$$

From (2.5) and (2.6) we find, if ε is fixed (independent of h) that $n = O(N)$, $O(N^{\frac{1}{2}})$, for GS and SOR respectively, resulting in the following estimates for the total computational cost:

$$\text{GS: } W = O(N^2), \tag{2.8}$$

$$\text{SOR: } W = O(N^{3/2}). \tag{2.9}$$

These results are found to hold in practice not only for the Poisson equation, but for elliptic equations in general, as long as A is an M-matrix. In a general situation, the estimate for SOR may be optimistic, because the optimal ω is not known.

Obviously, the best one may hope to achieve is $W = O(N)$. There exist MG methods with this property, as has been shown rigorously for general elliptic equations. For CG no theoretical results with the same degree of generality as for MG are available. For the Poisson equation Gustaffson (1978) has proved for a certain preconditioned CG method that $W = O(N^{5/4})$. Practical experience indicates that this holds for a wider class of equations.

Before further discussing MG and CG we first introduce a number of iterative methods that have special significance in the context of MG and CG.

Red-black Gauss-Seidel (RBGS) relaxation is Gauss-Seidel relaxation with a certain ordering of the grid-points. These are divided in red and black points in a checkerboard fashion. First the points of one colour are relaxed simultaneously, then the points of the other colour.

Horizontal zebra (HZ) relaxation is Gauss-Seidel relaxation by horizontal lines, taking first the odd and then the even lines, assuming that the boundary lines are odd.

Alternating zebra (AZ) relaxation is Gauss-Seidel relaxation by lines, taking successively the odd, even horizontal, odd, even vertical lines.

Hackbusch (1980) and Foerster et al. (1981) have shown that RBGS and AZ in combination with MG result in efficient iterative methods.

Incomplete LU (ILU) (or incomplete Crout, or incomplete Cholesky) decompositions have been introduced as preconditionings for CG by Meijerink and van der Vorst (1977), and as smoothing processes for MG by Wesseling and Sonneveld (1980). ILU has been found useful in transonic flow computations, cf. van der Wees et al. (1983), Nowak and Wesseling (1983). More recently, incomplete line LU (ILLU) or incomplete block factorizations have been proposed by Underwood (1976), Concus, Golub and Meurant (1982), Axelsson (1983) and Meijerink, see Kettler (1982), Meijerink (1983).

For completeness we will give a description of ILU and ILLU decompositions. Let P be a set of 2-tuples representing a matrix sparsity pattern. Then a class of ILU decompositions of the matrix A can be defined as follows. L and U are lower and upper triangular matrices satisfying

$$\ell_{ij} = 0,\ (i,j) \notin P;\ u_{ij} = 0,\ (i,j) \notin P;$$
$$(LU)_{ij} = a_{ij},\ (i,j) \in P. \tag{2.10}$$

Note that L and U now have a different meaning than in equation (2.3). The ILU decomposition may be made unique by requiring, for example,

$$\ell_{ii} = 1. \tag{2.11}$$

In many cases ILU-decompositions can be computed simply by means of (incomplete) Crout formulae. For example, assume that the given problem (2.1) is a discretization of a partial differential equation on an m*n grid, and let the grid-points be enumerated as in Fig. 2.1.

1+(n-1)m	.	.		nm			
.				.			
.				.	f	g	
.				.	c	d	e
1+2m	.	.		.		a	b
1+m	2+m	.		2m			
1	2	3	.	m			

Fig. 2.1 Enumeration of computational grid-points, and difference molecule

Let A be a 7-point discretization of a second order elliptic partial differential equation with the difference molecule abcdefg of Fig. 2.1 (the atoms b and f are needed if a mixed derivative is present). Then the sparsity pattern of A is:

$$\{(i,i-m),(i,i-m+1),(i,i-1),(i,i),(i,i+1),(i,i+m-1),(i,i+m)\}.$$

For brevity the following notation is introduced:

$$a_i = a_{i,i-m};\ b_i = a_{i,i-m+1};\ c_i = a_{i,i-1};\ d_i = a_{ii};$$
$$e_i = a_{i,i+1};\ f_i = a_{i,i+m-1};\ g_i = a_{i,i+m}. \tag{2.12}$$

Let the sparsity pattern P of L and U be chosen identical to that of A, and let the elements of L and U be called α_i, β_i, γ_i, δ_i, ε_i, ζ_i, η_i. The locations of these elements are identical to those of $a_i, b_i, \ldots, g_i$, respectively. The diagonal of L is specified to be unity; δ_i are the elements of diag(U). Then L and U can be conveniently computed by means of the following Crout formulae:

$$\alpha_i = a_i/\delta_{i-m}, \qquad \beta_i = (b_i - \alpha_i\varepsilon_{i-m})/\delta_{i-m+1},$$

$$\gamma_i = (c_i - \alpha_i\zeta_{i-m})/\delta_{i-1}, \quad \delta_i = d_i - \gamma_i\varepsilon_{i-1} - \beta_i\zeta_{i-m+1} - \alpha_i\eta_{i-m}, \tag{2.13}$$

$$\varepsilon_i = e_i - \beta_i\eta_{i-m+1}, \quad \zeta_i = f_i - \gamma_i\eta_{i-1}, \quad \eta_i = g_i.$$

Quantities that are not defined because their subscript is outside the range [1,nm] are to be replaced by zero. This is but one example of an ILU-decomposition of A. Other possibilities are described, for example, by Meijerink and van der Vorst (1981).

Sometimes it pays to add certain neglected entries (compared to the full LU-decomposition) to the diagonal element or to other non-neglected entries in the same row. Then we no longer have

$$(LU)_{ij} = a_{ij}, \quad (i,j) \in P. \tag{2.14}$$

For details see Axelsson (1982). We will not go into this here.

An ILU decomposition can be used in an iterative method by choosing $B = (LU)^{-1}$ in (2.2), obtaining

$$LUy^{n+1} = b + (LU-A)y^n. \tag{2.15}$$

The cost of one iteration can be reduced by means of the following simple device. With L and U computed by means of (2.13) we have

$$LU = A + C. \tag{2.16}$$

The only non-zero elements of C are given by

$$c_{i,i-m+2} = \beta_i\varepsilon_{i-m+1}, \quad c_{i,i+m-2} = \gamma_i\zeta_{i-1}. \tag{2.17}$$

With (2.16), (2.14) becomes

$$LUy^{n+1} = b + Cy^n \tag{2.18}$$

which is cheaper than (2.15), because C is more sparse than A.

It is easily verified that the construction of L and U according to (2.13) takes 17 flops (floating point operations) per grid-point. L and U are stored in place of A, and if C is generated, no extra storage beyond that for A is needed.

The solution of $LUy = q$ is obtained by back-substitution:

$$y_i := q_i - \gamma_i y_{i-1} - \beta_i y_{i-m+1} - \alpha_i y_{i-m},$$
$$y_i := (y_i - \varepsilon_i y_{i+1} - \zeta_i y_{i+m-1} - \eta_i y_{i+m})/\delta_i. \tag{2.19}$$

Hence, the solution for y^{n+1}, the computation of $b+Cy^n$ (generating C) and the execution of one iteration require 13, 6 and 19 flops per grid-point, respectively.

We will not discuss existence of ILU decompositions. Meijerink and van der Vorst (1977) prove existence for M-matrices, but often ILU is applied successfully to more general matrices.

ILLU decomposition can be described as follows. With the computational grid and the finite difference molecule of Fig. 2.1 the matrix A has the following structure:

$$A = \begin{vmatrix} B_1 & U_1 & & & \\ L_2 & B_2 & U_2 & & \\ & L_3 & B_3 & U_3 & \\ & & \cdot & \cdot & \cdot \\ & & & \cdot\ \cdot & \cdot \\ & & & L_n & B_n \end{vmatrix} \tag{2.20}$$

with L_i, B_i and U_i $m \times m$ matrices; B_i are triangular matrices; L_i and U_i are lower and upper triangular, with sparsity patterns $\{(j,j-1),(j,j)\}$ and $\{(j,j),(j,j+1)\}$ respectively. We try to find a matrix D such that

$$A = (L+D)D^{-1}(D+U), \tag{2.21}$$

where

$$L = \begin{pmatrix} 0 & & & & & \\ L_2 & 0 & & & & \\ & L_3 & 0 & & & \\ & & \cdot & \cdot & & \\ & & & \cdot & \cdot & \\ & & & & L_n & 0 \end{pmatrix}, \quad U = \begin{pmatrix} 0 & U_1 & & & & \\ & 0 & U_2 & & & \\ & & \cdot & \cdot & & \\ & & & \cdot & \cdot & \\ & & & & 0 & U_{n-1} \\ & & & & & 0 \end{pmatrix},$$

$$D = \begin{pmatrix} D_1 & & & & \\ & D_2 & & & \\ & & \cdot & & \\ & & & \cdot & \\ & & & & D_n \end{pmatrix}.$$

We call (2.21) a line LU decomposition of A, because the blocks in L, D and U correspond to (in our case horizontal) lines of the computational grid. Given the decomposition (2.21), solving (2.1) is just as simple as with a classical LU decomposition. Equation (2.21) can be rewritten as

$$A = L + D + U + LD^{-1}U. \tag{2.22}$$

One finds that $LD^{-1}U$ is the following block-diagonal matrix:

$$LD^{-1}U = \begin{pmatrix} 0 & & & & \\ & L_2D_1^{-1}U_1 & & & \\ & & \cdot & & \\ & & & \cdot & \\ & & & & L_nD_{n-1}^{-1}U_{n-1} \end{pmatrix}. \tag{2.23}$$

From (2.22) and (2.23) we deduce the following algorithm for the computation of D:

$$D_1 = B_1, \; D_i = B_i - L_iD_{i-1}^{-1}U_{i-1}, \; i = 2,3,\ldots,n. \tag{2.24}$$

The matrix D_i^{-1} is full, which causes the cost of a line LU decomposition to be $O(nm^3)$, as for standard LU-decomposition. An incomplete line LU decomposition is obtained if we replace $L_i D_{i-1}^{-1} U_{i-1}$ by its tridiagonal part. Thus, algorithm (2.24) is replaced by:

$$\tilde{D}_1 = B_1, \ \tilde{D}_i = B_i - \text{tridiag}(L_i \tilde{D}_{i-1}^{-1} U_{i-1}), \quad i = 2,3,\ldots,n. \tag{2.25}$$

The ILLU decomposition of A is now defined to be

$$A = (L+\tilde{D})\tilde{D}^{-1}(\tilde{D}+U) + E, \tag{2.26}$$

with E the error matrix, and $\tilde{D}$ the block diagonal matrix with blocks $\tilde{D}_i$.

We will now show how $\tilde{D}$ and $\tilde{D}^{-1}$ may be computed. Consider tridiag $(L_i \tilde{D}_{i-1}^{-1} U_{i-1})$, or, temporarily dropping the subscript, tridiag $(L\tilde{D}^{-1}U)$. Let the elements of $\tilde{D}^{-1}$ be s_{ij}; we shall see shortly how to compute them. The elements t_{ij} of tridiag$(L\tilde{D}^{-1}U)$ can be computed as follows:

$$\begin{aligned}
&\sigma_{-1} = \ell_{i,i+1} s_{i+1,i-1} + \ell_{ii} s_{i,i-1}, \quad \sigma_0 = \ell_{i,i+1} s_{i+1,i} + \ell_{ii} s_{ii},\\
&\sigma_1 = \ell_{i,i+1} s_{i+1,i+1} + \ell_{ii} s_{i,i+1}, \quad \sigma_2 = \ell_{i,i+1} s_{i+1,i+2} + \ell_{ii} s_{i,i+2},\\
&t_{i,i-1} = \sigma_{-1} u_{i-1,i-1} + \sigma_0 u_{i,i-1}, \quad t_{ii} = \sigma_0 u_{ii} + \sigma_1 u_{i+1,i},\\
&t_{i,i+1} = \sigma_1 u_{i+1,i+1} + \sigma_2 u_{i+2,i+1}.
\end{aligned} \tag{2.27}$$

The inverse of a tridiagonal matrix can be determined as follows. Let

$$T = \begin{vmatrix} a_1 & c_1 & & & & \\ b_2 & a_2 & c_2 & & & \\ & \cdot & \cdot & \cdot & & \\ & & \cdot & \cdot & \cdot & \\ & & & b_{m-1} & a_{m-1} & c_{m-1} \\ & & & & b_m & a_m \end{vmatrix}.$$

Let the triangular factorization of T be

$$T = (L+I)D^{-1}(I+U), \tag{2.28}$$

where L, D and U are not to be confused with the matrices occurring in (2.20). The only non-zero elements of L, D and U are $\ell_{i,i-1}$, d_{ii} and $u_{i,i+1}$, respectively. Call these elements ℓ_i, d_i, u_i for brevity. They can be computed by means of the following recursion formulae:

$$d_1^{-1} = a_1, \; u_1 = c_1 d_1, \text{ for } i > 1:$$

$$\ell_i = b_i d_{i-1}, \; d_i^{-1} = a_i - \ell_i d_{i-1}^{-1} u_{i-1}, \tag{2.29}$$

$$u_i = c_i d_i.$$

The elements of T^{-1} can be calculated as follows. From (2.28) we have

$$T^{-1} = (I-U)^{-1} D(L+I)^{-1}. \tag{2.30}$$

Let λ_{ij} be the elements of $(L+I)^{-1}$. By requiring $(L+I)^{-1}(L+I) = I$ and proceeding row by row, we find the following recursion formulae:

$$\lambda_{ij} = 0, \; j > i; \; \lambda_{ii} = 1; \; \lambda_{ij} = -\ell_{j+1}\lambda_{i,j+1}, \; j < i. \tag{2.31}$$

Similarly, by requiring $(U+I)(U+I)^{-1} = I$ and proceeding column by column, we find for the elements μ_{ij} of $(I+U)^{-1}$:

$$\mu_{ij} = 0, \; j < 1; \; \mu_{ii} = 1; \; \mu_{ij} = -u_i \mu_{i+1,j}, \; j > i. \tag{2.32}$$

Using (2.30) we find for the elements s_{ij} of T^{-1}:

$$s_{mm} = d_{mm}, \tag{2.33}$$

$$\begin{aligned} s_{kk} &= \sum_{i=k}^{m} \mu_{ki} d_{ii} \lambda_{ik} = d_{kk} + \sum_{i=k+1}^{m} u_k \mu_{k+1,i} d_{ii} \ell_{k+1} \lambda_{i,k+1} \\ &= d_{kk} + u_k \ell_{k+1} s_{k+1,k+1}, \end{aligned} \tag{2.34}$$

$$\begin{aligned} s_{k,k-j} &= \sum_{i=k}^{m} \mu_{ki} d_{ii} \lambda_{i,k-j} = -\sum_{i=k}^{m} \mu_{ki} d_{ii} \ell_{k-j+1} \lambda_{i,k-j+1} \\ &= -\ell_{k-j+1} s_{k,k-j+1}, \end{aligned} \tag{2.35}$$

$$s_{k-j,k} = \sum_{i=k}^{m} \mu_{k-j,i} d_{ii} \lambda_{ik} = - \sum_{i=k}^{m} u_{k-j} \mu_{k-j+1,i} d_{ii} \lambda_{ik}$$
$$= - u_{k-j} s_{k-j+1,k}. \tag{2.36}$$

This completes our description of the computation of tridiag $(L_i \tilde{D}_{i-1} U_{i-1})$.

The complete algorithm for the computation of the ILLU decomposition (2.26) can be summarized as follows. We compute $\tilde{D}$ and its triangular decomposition.

$$D_1 := B_1 ;$$

for $i = 2,3,\ldots,n$ do (i) - (iv):

(i) Compute the triangular decomposition of $\tilde{D}_{i-1}$ according to (2.29);

(ii) Compute the five main diagonals of $\tilde{D}_{i-1}^{-1}$ according to (2.33) - (2.36);

(iii) Compute tridiag $(L_i \tilde{D}_{i-1}^{-1} U_{i-1})$ according to (2.27);

(iv) Compute $\tilde{D}_i$ with (2.25);

Finally, compute the triangular decomposition of $\tilde{D}_n$ according to (2.29).

The number of flops required is given by:

Step (i): 5m; step (ii): 7m; step (iii): 21m; step (iv): 3m.

Hence, the total cost of computing $\tilde{D}$ and its triangular decomposition is 36mn. Storage to the extent of 3mn reals is needed for the triangular decomposition of $\tilde{D}$.

When using ILLU, the iterative method (2.2) becomes:

$$r := b - Ay^n, \tag{2.37}$$

$$(L+\tilde{D})\tilde{D}^{-1}(\tilde{D}+U)y^{n+1} = r, \tag{2.38}$$

$$y^{n+1} := y^{n+1} + y^n. \tag{2.39}$$

Equation (2.38) is solved as follows:

$$(L+\tilde{D})y^{n+1} = r, \tag{2.40}$$

$$r := \tilde{D}y^{n+1}, \tag{2.41}$$

$$(\tilde{D}+L)y^{n+1} = r. \tag{2.42}$$

With the block partitioning used before, and with y_i^{n+1} and r_i denoting m-dimensional vectors corresponding to the i-th block, equation (2.40) is solved as follows:

$$\tilde{D}_1 y_1^{n+1} = r_1, \quad \tilde{D}_i y_i^{n+1} = r_i - L_{i-1} y_i^{n+1}, \quad i > 1. \tag{2.43}$$

Equation (2.42) is solved in similar fashion. The solution of an $m \times m$ tridiagonal system, with triangular decomposition available, takes 5m flops. The cost of the right side of (2.43) is 4m (for the 7-point difference molecule assumed here). The total cost of (2.40), and of (2.42) as well, is therefore 9mn, so that the cost of (2.38) is 23mn. The cost of (2.37) is 14mn. The total cost of one ILLU iteration is therefore 37mn.

For other ILLU variants, see Concus et al. (1982) and Meijerink (1983), who prove existence of ILLU decompositions for M-matrices. For remarks on vectorization, see Meijerink (1983), Meurant (1983), and Hemker, Wesseling and de Zeeuw (1983).

For future reference we note that the cost of RBGS is 12mn flops, assuming a 7-point difference molecule. In HZ, tridiagonal system solving takes 5mn flops, assuming that the necessary triangular decompositions have been computed beforehand, at a cost of 5mn flops, respectively. Residue evaluation takes 8mn flops, so that the total cost of one iteration with HZ is 13mn flops. For AZ, these figures should be doubled.

For RBGS, AZ, ILU and ILLU rate of convergence estimates are not available in the literature, but the number of iterations required certainly increases as the grid is refined. Therefore the computational cost of these methods is $O(N^{\alpha})$ with $\alpha > 1$. In the following sections we will discuss how the convergence of iterative methods such as those just discussed can be accelerated with CG or MG methods.

3. CONJUGATE GRADIENT METHODS

For an introduction to CG (and Chebyshev) acceleration of iterative methods, see Hageman and Young (1981). Within the confines of this paper we can only give a brief discussion.

When A in (2.1) is large and sparse it is attractive, because of efficiency and simplicity, to use A only as a multiplier. This means that we can build polynomials in A. At the start of the iterations the only special vectors available are b and the residue $r^0 = b - Ay^0$, with y^0 the starting iterand. A rather general form of possible algorithms would be

$$y^{n+1} = y^n + \alpha_n p^n, \tag{3.1}$$

$$p^n = \theta_n(A) r^0 + \tilde{\theta}_n(A) b. \tag{3.2}$$

Here θ_n and $\tilde{\theta}_n$ are polynomials, whose degree is increased by one at each iteration. At present, only the case where one chooses $\tilde{\theta}_n \equiv 0$ seems to have been investigated. It may not be worthwhile to allow $\tilde{\theta}_n \not\equiv 0$. For example, it seems reasonable to require that the sequence $\{y^{n+1}-y^n\} = \{\alpha_n p^n\}$ is identical for the following two cases:

$$\text{case 1 : } Ay = b, \quad \text{starting iterand } y^0$$

$$\text{case 2 : } Ay = \bar{b}, \quad \text{starting iterand } \bar{y}^0$$

with $\bar{b} = b + A(\bar{y}^0 - y^0)$. With overbars referring to case 2, we have

$$\bar{\alpha}_n \bar{p}^n = \bar{\alpha}_n \bar{\theta}_n (A) \bar{r}^0 + \bar{\alpha}_n \bar{\tilde{\theta}}_n (A) \bar{b}. \tag{3.3}$$

Since $\bar{r}^0 = r^0$ we can have $\bar{\alpha}_n \bar{p}^n = \alpha_n p^n$ for all b only if $\tilde{\theta}_n \equiv 0$, $\bar{\tilde{\theta}}_n \equiv 0$. Assuming henceforth $\tilde{\theta}_n \equiv 0$, we have

$$\begin{aligned} r^{n+1} &= b - Ay^{n+1} = b - Ay^n - \alpha_n A p^n \\ &= r^n - A\alpha_n \theta_n (A) r^0 = \text{(induction)} \\ &= r^0 - A\{\alpha_n \theta_n (A) + \alpha_{n-1}\theta_{n-1}(A) + \ldots + \alpha_0 \theta_0 (A)\} r^0 \\ &= \phi_{n+1}(A) r^0, \end{aligned} \tag{3.4}$$

where ϕ_{n+1} is a polynomial of degree n+1 with the following property:

$$\phi_n(0) = 1. \tag{3.5}$$

Because of (3.4) we would like to choose ϕ_n such that $||\phi_n(A) r^0||$ is minimized, under the constraint (3.5). For SPD (symmetric positive definite) A this aim is achieved by CG methods. Let us define

$$\Pi_n^1 = \{\psi_n | \psi_n(0) = 1, \ \psi_n \text{ is polynomial of degree} \leq n\}. \tag{3.6}$$

Then we want to construct $\phi_n \in \Pi_n^1$ such that

$$||\phi_n(A) r^0|| \leq ||\psi_n(A) r^0||, \qquad \forall \ \psi_n \in \Pi_n^1. \tag{3.7}$$

If we choose the following norm:

$$||r||^2 \equiv r^T A^{-1} r, \tag{3.8}$$

then the following CG method solves (3.7):

$$\begin{aligned} &p^{-1} = 0,\ r^0 = b - Ay^0, \\ &p^n = r^n + \beta_n p^{n-1},\ \beta_n = \rho_n/\rho_{n-1},\ \rho_n = r^{n^T} r^n, \\ &y^{n+1} = y^n + \alpha_n p^n,\ \alpha_n = \rho_n/\sigma_n,\ \sigma_n = p^{n^T} Ap^n, \\ &r^{n+1} = r^n - \alpha_n Ap^n. \end{aligned} \tag{3.9}$$

For a proof see for example Hageman and Young (1981). The name of the method derives from the fact that the search vectors are conjugate:

$$p^{k^T} Ap^n = 0,\quad k = 0,1,2,\ldots,n-1. \tag{3.10}$$

By making different choices for the norm $||\cdot||$ in (3.7), different CG methods are obtained.

For many practical applications the restriction of CG to SPD systems is a severe drawback. Several ways to generalize CG have been proposed, but at the moment it is not yet clear what are the best CG variants for non symmetric or indefinite systems. We present a promising new method.

First, we rewrite the CG method (3.9) in terms of the polynomials ϕ_n and θ_n introduced before. One easily obtains:

$$\theta_{-1} \equiv 0,\quad \phi_0 \equiv 1, \tag{3.11a}$$

$$\theta_n = \phi_n + \beta_n \theta_{n-1}, \tag{3.11b}$$

$$\phi_{n+1} = \phi_n - \alpha_n \psi \theta_n, \tag{3.11c}$$

with ψ the polynomial $\psi(\tau) = \tau$,

$$\begin{aligned} &\beta_n = \rho_n/\rho_{n-1},\quad \alpha_n = \rho_n/\sigma_n,\quad \rho_n = (\phi_n,\phi_n), \\ &\sigma_n = (\phi_n,\psi\phi_n), \end{aligned} \tag{3.12}$$

where the bilinear form (.,.) is defined by

$$(\phi,\theta) = r^{o^T}\phi(A^T)\theta(A)r^o. \tag{3.13}$$

We will now abandon the assumption that A is SPD, so that the algorithm no longer minimizes the residual in the sense of (3.7); $\|\cdot\|$ no longer has the properties of a norm. We replace (3.13) by

$$(\phi,\theta) = \tilde{r}^{o^T}\phi(A)\theta(A)r^o, \tag{3.14}$$

with $\tilde{r}_o$ a vector to be chosen. In general, this is not an inner product. Then for arbitrary A, (.,.) has the following properties:

$$(\phi,\theta) = (\theta,\phi), \tag{3.15}$$

$$(\phi,\zeta\theta) = (\zeta\phi,\theta), \tag{3.16}$$

for every triple of polynomials ϕ, θ, ζ. The following theorem suggests that the algorithm (3.11) might still be of use for solving Ay = b:

Theorem 3.1 The algorithm defined by (3.11), (3.12) and (3.14) has the following property:

$$(\phi_{n+1},\phi_k) = 0,\ k < n+1;\ (\theta_n,\psi\theta_k) = 0,\quad k < n. \tag{3.17}$$

Hence, if A happens to be such that (.,.) is an inner product, then the residual lies in a subspace the dimension of which is reduced by one at each iteration, just as for the classical CG method.

Proof of theorem 3.1 Obviously

$$\theta_o = \phi_o. \tag{3.18}$$

With (3.11c), (3.12), (3.16), (3.18):

$$(\phi_1,\phi_o) = (\phi_o,\phi_o) - \alpha_o(\psi\theta_o,\phi_o) = 0. \tag{3.19}$$

Using (3.11b,c):

$$\begin{aligned}(\theta_1,\psi\theta_o) &= (\phi_1,\psi\theta_o) + \beta_1(\theta_o,\psi\theta_o) = \\ &(\phi_1,\phi_o-\phi_1)/\alpha_o + \beta_1\sigma_o = -\rho_1/\alpha_o + \beta_1\sigma_o = 0.\end{aligned} \tag{3.20}$$

Similarly,

$$(\phi_2,\phi_o) = (\phi_1,\phi_o) - \alpha_1(\psi\theta_1,\phi_o) = -\alpha_1(\theta_1,\psi\theta_o) = 0,$$

$$\begin{aligned}(\phi_2,\phi_1) &= (\phi_1,\phi_1) = \alpha_1(\psi\theta_1,\phi_1) \\ &= \rho_1 - \alpha_1(\psi\theta_1,\theta_1-\beta_1\theta_o) = \rho_1 - \alpha_1\sigma_1 = 0.\end{aligned} \tag{3.21}$$

This establishes the validity of (3.17) for n = 1. Proceeding by induction, for $k < n$,

$$(\phi_{n+1},\theta_k) = (\phi_n,\theta_k) - \alpha_n(\psi\theta_n,\theta_k) = (\phi_n,\theta_k). \tag{3.22}$$

From (3.11b) it follows that there exist constants c_{kj} such that

$$\theta_k = \sum_{j=1}^{k} c_{kj}\,\phi_j. \tag{3.23}$$

Hence, with (3.22) and the induction hypothesis,

$$(\phi_{n+1},\theta_k) = (\phi_n, \sum_{j=1}^{k} c_{kj}\phi_j) = 0, \quad k < n. \tag{3.24}$$

Furthermore

$$\begin{aligned}(\phi_{n+1},\theta_n) &= (\phi_n,\theta_n) - \alpha_n(\psi\theta_n,\theta_n) \\ &= (\phi_n,\phi_n) + \beta_n(\phi_n,\theta_{n-1}) - \alpha_n\sigma_n \\ &= \rho_n + \beta_n(\phi_n, \sum_{j=1}^{n-1} c_{n-1,j}\phi_j) - \rho_n = 0.\end{aligned} \tag{3.25}$$

It follows that

$$(\phi_{n+1},\phi_k) = (\phi_{n+1},\theta_k) - \beta_k(\phi_{n+1},\theta_{k-1}) = 0, \quad k \leqslant n, \tag{3.26}$$

establishing the first part of the induction hypothesis.

For $k \leqslant n$ we have

$$(\theta_{n+1},\psi\theta_k) = (\phi_{n+1},\psi\theta_k) + \beta_{n+1}(\theta_n,\psi\theta_k)$$

$$= (\phi_{n+1},\phi_k-\phi_{k+1})/\alpha_k + \beta_{n+1}\sigma_n\delta_{nk}$$

$$= (-\rho_{n+1}/\alpha_n + \beta_{n+1}\sigma_n)\delta_{nk} = 0.$$

with δ_{nk} the Kronecker delta. This completes the proof.

The algorithm (3.11) - (3.13) can be put in a form suitable for computation as follows. Define

$$r^n = \phi^n(A)r^0,\ p^n = \theta_n(A)r^0,$$
$$\tilde{r}^n = \phi_n(A^T)\tilde{r}^0,\ \tilde{p}^n = \theta_n(A^T)\tilde{r}^0, \tag{3.27}$$

with r^0 the starting residue and $\tilde{r}^0$ some vector to be specified by the user of the algorithm. Then we have according to (3.12) and (3.13)

$$\rho_n = \tilde{r}^{n^T} r^n,\quad \sigma_n = \tilde{p}^{n^T} Ap^n, \tag{3.28}$$

and we obtain the following algorithm:

$$r^0 = b - Ay^0,\quad \text{choose } \tilde{r}^0,\ p^{-1} = \tilde{p}^{-1} = 0,$$

$$p^n = r^n + \beta_n p^{n-1},$$

$$\tilde{p}^n = \tilde{r}^n + \beta_n \tilde{p}^{n-1},$$

$$r^{n+1} = r^n - \alpha_n Ap^n,$$

$$\tilde{r}^{n+1} = \tilde{r}^n - \alpha_n A^T\tilde{p}^n, \tag{3.29}$$

$$y^{n+1} = y^n + \alpha_n p^n,$$

$$\alpha_n = \rho_n/\sigma_n,\ \beta_0 = 0,\ \beta_n = \rho_n/\rho_{n-1},\ \rho_n = \tilde{r}^{n^T} r^n,$$

$$\sigma_n = \tilde{p}^{n^T} Ap^n.$$

The vectors r^n, $\tilde{r}^n$, p^n, $\tilde{p}^n$ satisfy (3.27). According to theorem 3.1 we have

$$\tilde{r}^{n^T} r^k = 0, \quad k < n, \tag{3.30a}$$

$$\tilde{p}^{n^T} A p^k = 0, \quad k < n. \tag{3.30b}$$

According to (3.30b) the sets $\{\tilde{p}^k\}$ and $\{p^k\}$ are conjugate with respect to A, which is why the algorithm is called the bi-CG method. It has first been proposed by Fletcher (1976).

The bi-CG method can be accelerated appreciably (roughly by a factor 2), by the following stratagem. The idea is to construct an algorithm for which the residue is $\phi_n(A)^2 r^0$ instead of $\phi_n(A) r^0$, which turns out to be possible at hardly any extra cost, and eliminates the need to work with A^T. If bi-CG converges, $\phi_n(A)$ will be a contraction, and $\phi_n(A)^2$ will be smaller than $\phi_n(A)$. A suitable algorithm is obtained by squaring (3.11). We call the resulting method the CGS (conjugate gradients squared) method. From (3.11) we obtain

$$\begin{aligned} \theta_n^2 &= \phi_n^2 + \beta_n^2 \theta_{n-1}^2 + 2\beta_n \phi_n \theta_{n-1}, \\ \phi_{n+1}^2 &= \phi_n^2 + \alpha_n^2 \psi^2 \theta_n^2 - 2\alpha_n \phi_n \psi \phi_n. \end{aligned} \tag{3.31}$$

Using (3.11b),

$$\begin{aligned} \theta_n \phi_n &= \phi_n^2 + \beta_n \phi_n \theta_{n-1}, \\ \theta_n^2 &= \theta_n \phi_n + \beta_n (\phi_n \theta_{n-1} + \beta_n \theta_{n-1}^2), \\ \phi_{n+1} \theta_n &= \phi_n \theta_n - \alpha_n \psi \theta_n^2, \\ \phi_{n+1}^2 &= \phi_n^2 - \alpha_n \psi (\phi_n \theta_n + \phi_{n+1} \theta_n), \end{aligned} \tag{3.32}$$

with α_n, β_n given by (3.12), where ρ_n and σ_n can now be evaluated as follows:

$$\rho_n (\phi_n, \phi_n) = (1, \phi_n^2),$$

$$\sigma_n = (\phi_n, \psi\phi_n) = (1, \psi\phi_n^2).$$

This is transformed into a workable algorithm with the aid of the following vectors:

$$f^n = \phi_n(A)^2 r^o, \quad g^n = \theta_n(A)^2 r^o,$$
$$h^n = \phi_n(A)\theta_{n-1}(A) r^o. \tag{3.33}$$

Equations (3.32) are equivalent to (u corresponds to $\theta_n\phi_n$):

<u>CGS method</u>:

$$\begin{aligned}
f^o &= b - Ay^o, \quad g^{-1} = h^o = 0, \\
u &= f^n + \beta_n h^n, \\
g^n &= u + \beta_n(\beta_n g^{n-1} + h^n), \\
h^{n+1} &= u - \alpha_n A g^n, \\
y^{n+1} &= y^n + \alpha_n(u + h^{n+1}), \\
f^{n+1} &= f^n - \alpha_n A(u + h^{n+1}),
\end{aligned} \tag{3.34}$$

where we have used that $y^{n-1}-y^n$ follows directly from the difference in the residues $f^{n+1}-f^n$. In (3.34) we have

$$\alpha_n = \rho_n/\sigma_n, \quad \beta_o = 0, \quad \beta_n = \rho_n/\rho_{n-1},$$
$$\rho_n = \tilde{r}^{o^T} f^n, \quad \sigma_n = \tilde{r}^{o^T} A g^n. \tag{3.35}$$

We usually choose

$$\tilde{r}^o = b - Ay^o. \tag{3.36}$$

The cost of CGS is about the same as the cost of bi-CG. The correspondence between bi-CG and CGS is that the residues after n iterations are $\phi_n(A)r^o$ and $\phi_n(A)^2 r^o$, respectively.

Another type of method that seems promising for the indefinite case is Chebyshev iteration, for example the version proposed by Manteuffel

(1977, 1978). This method works well if certain parameters related to the spectrum of A can be estimated accurately. An important advantage of CGS is that no parameters need be estimated. A thorough comparison between CGS and Chebyshev iteration has not yet been made.

Bi-CG and CGS are but two examples of extensions of CG to non-SPD systems. We will not review other extensions that have been proposed, but restrict ourselves to mentioning the publications of Concus and Golub (1976), Vinsome (1976), Widlund (1978) and Axelsson (1980).

4. CONJUGATE GRADIENT ACCELERATION OF ITERATIVE METHODS: PRECONDITIONING

Until further notice A is assumed to be SPD. For a stationary iterative method (2.2) it follows from (2.4), that

$$e^n = \psi_n(BA)e^0, \quad \psi_n(x) = (1-x)^n \in \Pi_1^n. \tag{4.1}$$

Assuming that B is SPD we can write

$$B = E^TE. \tag{4.2}$$

For arbitrary powers of E^TEA we have

$$(E^TEA)^k = E^T(EAE^T)^kE^{-T}, \tag{4.3}$$

so that (4.1) can be rewritten as

$$E^{-T}e^n = \psi_n(EAE^T)E^{-T}e^0. \tag{4.4}$$

If we apply CG not to (2.1) but to the following preconditioned version:

$$(EAE^T)(E^{-T}y) = Eb \tag{4.5}$$

then in (4.4) ψ_n is replaced by ϕ_n satisfying the optimality condition (3.7), so that we may say that CG accelerates (2.2). Of course it is equally true that CG is accelerated by preconditioning.

We will now study the rate of convergence that can be obtained with preconditioned CG. In the SPD case the rate of convergence of CG methods can be estimated in an elegant way cf. Axelsson (1977). From (3.4), (3.7) and (3.8) it follows that

$$||r^n||^2 = \min_{\psi\in\Pi_1^n} r^{0^T}\psi(A)^2A^{-1}r^0, \tag{4.6}$$

$\Rightarrow$ choosing $\|z\|^2 = z^TA^{-1}z$.

Let the set of eigenvalues of A be

$$\mathrm{Sp}(A) = \{\lambda_1,\lambda_2,\ldots,\lambda_N\}, \tag{4.7}$$

with corresponding eigenvectors $x_1,x_2,\ldots,x_N$ satisfying $x_i^T x_j = \delta_{ij}$. Let

$$r^0 = \sum_{i=1}^{N} \xi_i x_i, \tag{4.8}$$

then

$$\begin{aligned} \|r^n\|^2 &= \min_{\psi\in\Pi_1^n} \sum_{i=1}^{N} \xi_i^2 \psi(\lambda_i)^2/\lambda_i \\ &\leq \min_{\psi\in\Pi_1^n} \max_{\lambda\in \mathrm{Sp}(A)} \psi(\lambda)^2 \sum_{i=1}^{N} \xi_i^2/\lambda_i \\ &= \|r^0\| \min_{\psi\in\Pi_I^n} \max_{\lambda\in \mathrm{Sp}(A)} \psi(\lambda)^2. \end{aligned} \tag{4.9}$$

Rate of convergence estimates are obtained by making a choice for $\psi(\lambda)$. For example,

$$\begin{aligned} \psi(\lambda) &= T_n(z)/T_n\left(\frac{\bar\lambda+\underline\lambda}{\bar\lambda-\underline\lambda}\right), \\ z &= (\bar\lambda+\underline\lambda-2\lambda)/(\bar\lambda-\underline\lambda), \end{aligned} \tag{4.10}$$

with T_n the Chebyshev polynomial of degree n, and $\bar\lambda$, $\underline\lambda$ the largest and smallest eigenvalue of A. Because

$$\max_{|z|\leq 1} T_n(z)^2 = 1,$$

we obtain

$$\|r^n\|^2/\|r^0\|^2 \leq 1/T_n\left(\frac{\bar\lambda+\underline\lambda}{\bar\lambda-\underline\lambda}\right)^2. \tag{4.11}$$

A well-known property of Chebyshev polynomials is

$$1/T_n\left(\frac{\bar{\lambda}+\underline{\lambda}}{\bar{\lambda}-\underline{\lambda}}\right)^2 \leqslant 4\left\{\frac{1-(\underline{\lambda}/\bar{\lambda})^{\frac{1}{2}}}{1+(\underline{\lambda}/\bar{\lambda})^{\frac{1}{2}}}\right\}^{2n}. \tag{4.12}$$

For $|z| < 1$ the following holds:

$$\left(\frac{1-z}{1+z}\right)^n = \exp\left\{-2n\left(z+\frac{z^3}{3}+\frac{z^5}{5}+\ldots\right)\right\} \leqslant e^{-2nz}. \tag{4.13}$$

Using (4.13) in (4.12) and noting that $\bar{\lambda}/\underline{\lambda} = \mathrm{cond}_2(A)$ we obtain

$$||r^n||/||r^o|| \leqslant 2e^{-2n\,\mathrm{cond}_2(A)^{-\frac{1}{2}}}.$$

Requiring a residue reduction ε the required number of iterations n is

$$n \geqslant \frac{1}{2}\left|\ln\frac{\varepsilon}{2}\right|\mathrm{cond}_2(A)^{\frac{1}{2}}. \tag{4.14}$$

For discretizations of second order elliptic equations we usually have

$$\mathrm{cond}_2(A) = O(1/h^2), \tag{4.15}$$

so that

$$W = O(N^{3/2}), \tag{4.16}$$

as for SOR. In practice CG tends to be somewhat more expensive than SOR, but it is parameter free, and if for SOR the optimal overrelaxation factor is not accurately known, CG is faster.

The efficiency of CG by itself is not very impressive, but the interest of CG derives from the possibility of convergence acceleration by preconditioning. Rewriting the CG algorithm (3.9) for (4.5) one obtains

$$\begin{aligned}
&p^{-1} = 0,\ r^o = Eb - EAy^o,\\
&p^n = r^n + \beta_n p^{n-1},\ \beta_n = \rho_n/\rho_{n-1},\ \rho_n = r^{n^T} r^n,\\
&E^{-T}y^{n+1} = E^{-T}y^n = \alpha_n p^n,\ \alpha_n = \rho_n/\sigma_n,\ \sigma_n = p^{n^T} EAE^T p_n,\\
&r^{n+1} = r^n - \alpha_n EAE^T p^n.
\end{aligned} \tag{4.17}$$

Replacing E^Tp by p and redefining $r = b - Ay$ this can be rewritten as follows:

Preconditioned CG algorithm:

$$p^{-1} = 0,\ r^0 = b - Ay^0$$

$$p^n = E^TEr^n + \beta_n p^{n-1},\ \beta_n = \rho_n/\rho_{n-1},\ \rho_n = r^{n^T}E^TEr^n,$$

$$y^{n+1} = y^n + \alpha_n p^n,\ \alpha_n = \rho_n/\sigma_n,\ \sigma_n = p^{n^T}Ap^n, \tag{4.18}$$

$$r^{n+1} = r^n - \alpha_n Ap^n.$$

It has been found by Meijerink and van der Vorst (1977), that an effective preconditioning is obtained with incomplete Cholesky decomposition, given by

$$LL^T = A + C, \tag{4.19}$$

the symmetric (Cholesky) variant of ILU decomposition discussed in section 2. We choose $E = L^{-1}$ in (4.18). The eigenvalue distributions of $L^{-1}AL^{-T}$ and A are compared for a few examples by Meijerink and van der Vorst (1977) and Kershaw (1978); it is found that $\text{cond}_2(L^{-1}AL^{-T}) \ll \text{cond}_2(A)$. For a full explanation of the acceleration effect of preconditioning not only the condition number but the eigenvalue distribution should be taken into account, but there is no general theory available concerning the influence of preconditioning on the eigenvalue distribution or even the condition number. For a special case, the 5-point discretization of the Poisson equation, Gustafsson (1978) shows that preconditioning with a certain type of ("modified") incomplete LL^T decomposition results in

$$\text{cond}_2(L^{-1}AL^{-T}) = O(1/h), \tag{4.20}$$

so that according to (4.14) the required number of iterations is $O(h^{-\frac{1}{2}})$, resulting in a computational cost of $O(N^{5/4})$. This result seems to hold approximately quite generally for CG with preconditioning by approximate decomposition. One finds that the number of iterations required increases slowly as the grid is refined. The modified incomplete LL^T decomposition seems in general to provide a somewhat better preconditioning than the version described here.

In the preconditioned CG algorithm (4.18) the matrix is needed only for multiplication with r^n. If one does not want to form E or E^{-1}

explicitly, but wants to define E implicitly by means of the iterative method (2.2), one can obtain Br^n from

$$Br^n = y^* - y^n, \tag{4.21}$$

with y^* the result of one iteration (2.2), starting with y^n.

A preconditioned version of CGS can be obtained as follows. Application of CGS to the following preconditioned version of (2.1)

$$BAy = Bb \tag{4.22}$$

results in an algorithm given by (3.34), (3.35) with A and b replaced by BA and Bb. By replacing $B^{-1}f$ by f we obtain:

<u>Preconditioned CGS algorithms</u>:

$$\begin{aligned}
&f^0 = b - Ay^0,\ g^{-1} = h^0 = 0,\\
&u = Bf^n + \beta_n h^n,\\
&g^n = u + \beta_n(\beta_n g^{n-1} + h^n),\\
&h^{n+1} = u - \alpha_n BAg^n,\\
&y^{n+1} = y^n + \alpha_n(u + h^{n+1}),\\
&f^{n+1} = f^n - A\alpha_n(u + h^{n+1}),
\end{aligned} \tag{4.23}$$

with

$$\alpha_n = \rho_n/\sigma_n,\ \beta_0 = 0,\ \beta_n = \rho_n/\rho_{n-1},$$

$$\sigma_n = \tilde{r}^{0^T} Bf^n,\ \sigma_n = \tilde{r}^{0^T} BAg^n.$$

If B is not explicitly available, as for instance when the iterative method to be accelerated is a MG method, then Bf^n and BAg^n can be obtained as follows. Carry out an iteration with the method to be accelerated (2.2), with starting iterand y^n:

$$y^* = y^n + B(b - Ay^n) = y^n + Bf^n. \tag{4.24}$$

Next, carry out an iteration with starting iterand g^n and right-hand-side $b = 0$:

$$g^* = g^n - BAg^n. \tag{4.25}$$

It follows that

$$Bf^n = y^* - y^n, \quad BAg^n = g^n - g^*. \tag{4.26}$$

In this way one may try to use CG or CGS to accelerate the convergence of any iterative method, with the restriction that for CG the matrix B must be symmetric. Kettler (1982) has used CG to accelerate MG. To make B symmetric, he used incomplete LL^T decomposition for smoothing, and the V-cycle multigrid schedule (see the next section). One might say that CG accelerates MG which accelerates incomplete LL^T. Behie and Forsyth (1983) have used Orthomin, a non-symmetric CG variant (Vinsome 1976) to accelerate MG using the sawtooth cycle.

For CGS applied to a general system there is no guarantee that convergence will be rapid, but a rule of thumb is that a good rate of convergence may be expected with ILU and ILLU preconditioning if A satisfies

$$a_{ii} \geqslant - \sum_{j=1} a_{ij}, \quad a_{ij} \leqslant 0, \quad j = i. \tag{4.27}$$

(This makes A an M-matrix).

In order to obtain a rough idea of the computational cost of CGS we count flops (per grid-point) in (4.23). Preconditioning takes place with ILU or ILLU. Assume that multiplication with B or BA takes place using (4.24)-(4.26). Using in this case the explicitly available matrix B one can obtain slightly lower operation counts than those obtained below, but we will neglect this possibility here. Note that in (4.24) the residue $b - Ay^n = f^n$ is already available, so that $(b+Cy^n)$ or (2.37) need not be carried out for ILU or ILLU, respectively. Using the ILU and ILLU operation counts of section 2, we find that the cost of y^* is 13 flops (ILU) or 23 flops (ILLU). Similarly, the cost of g^* is found to be 18 flops (ILU) or 36 flops (ILLU). We assume that A has 7 non-zero elements per row. Hence, multiplication with A takes 18 flops. In addition to matrix multiplications, CGS needs 18 flops, as is easily seen from (4.23), including of course the cost of α_n, β_n. Therefore the total cost of one preconditioned CGS iteration is 60 flops (ILU) or 88 flops (ILLU).

5. MULTIGRID ACCELERATION OF ITERATIVE METHODS

The basic ideas of MG methods are quite general and have a wide range of application. They can be used not only to accelerate iterative methods, but also, for example, to formulate novel ways to solve non-linear problems, or to devise algorithms that construct adaptive discretizations. The volume edited by Hackbusch and Trottenberg (1982) represents a useful survey.

We restrict ourselves here to the one aspect of MG mentioned in the title of this chapter. This makes it possible to simplify MG, and to distinguish situations where its effectiveness is guaranteed. The significance of MG as accelerating technique derives from the fact that, in principle, a computational complexity of O(N) can be achieved, with N the number of unknowns. This has been proved rigorously by Fedorenko (1964) for a finite difference approximation of the Poisson equation and by Bakhvalov (1966), Hackbusch (1980), Wesseling (1980) for finite difference approximations to general second order elliptic partial differential equations. For a survey of MG rate of convergence theory, including finite element discretizations, see Hackbusch (1982). These general theories result in O(N) but pessimistic, and fortunately unrealistic, computational complexity estimates. The papers by Brandt (1977) and Hackbusch (1978) showed the great potential of MG for practical applications. The theoretical work just mentioned assumes a W-cycle. More recently, work on rate of convergence theory for the V-cycle has appeared, such as Musy (1982), Maitre and Musy (1983), McCormick (1983), Braess and Hackbusch (1983). The terms V- and W-cycle are explained for example by Stüben and Trottenberg (1982). These terms refer to the MG schedule, i.e. the switching strategy between the grids. For special equations, notably Poisson's equation, work on realistic rate of convergence predictions is underway, see for example Braess (1981, 1982), Stüben and Trottenberg (1982). We will not discuss these theoretical aspects here.

Equation (2.1) is assumed to represent a discretization of a partial differential equation. If the basic iterative method (2.2) converges, it usually (but not always) has the property, exploited by MG, that the non-smooth part of error and residue is annihilated rapidly, whereas it takes many iterations to get rid of the smooth part. A precise definition of smoothness will be given shortly. The fundamental MG idea is to approximate the problem with smooth error and residue on coarser grids. In the MG context (2.2) is called a <u>smoothing process.</u>

One way of discriminating between smooth and non-smooth parts of grid functions is by means of Fourier analysis, as proposed by Brandt (1977). Let the computational grid G associated with the discretization (2.1) be defined by (we restrict ourselves for simplicity to two-dimensional problems):

$$G := \{(x_1,x_2) \mid x_i = 0, h_i, 2h_i, \ldots, m_i h_i\}. \tag{5.1}$$

Any grid-function $e : G \to \mathbb{R}$ can be represented by a Fourier series as follows:

$$e_{mn} = \sum_{s=-m_1/2}^{m_1/2} \sum_{t=-m_2/2}^{m_2/2} c_{st} \exp(im\theta_s + in\phi_t). \tag{5.2}$$

with

$$\theta_s = (2s-1)\pi/(m_1+1), \ \phi_t = (2t-1)\pi/(m_2+1), \tag{5.3}$$

where we have assumed that m_i is even. By e_{mn} we mean the value of the grid-function e in the grid-point with coordinates (mh_1,nh_2). Let $\bar{G}$ be a coarse grid with step-size doubled, i.e.

$$\bar{G} := \{(x_1,x_2) \mid x_i = 0, 2h_i, 4h_i, \ldots, \tfrac{1}{2}m_i 2h_i\}. \tag{5.4}$$

Then we call those Fourier components that cannot be represented without aliasing on $\bar{G}$ non-smooth. That is, the set of non-smooth Fourier components is given by

$$\begin{aligned} &\{\exp(im\theta+in\phi) \mid (\theta,\phi) \in F\}, \\ &F := \{(\theta,\phi) \mid -\pi \leqslant \theta,\phi \leqslant \pi, \ |\theta| \geqslant \pi/2 \text{ and/or } |\phi| \geqslant \pi/2\}, \end{aligned} \tag{5.5}$$

where for convenience we do not restrict (θ,ϕ) to the discrete set occurring in (5.3).

For periodic boundary conditions and constant coefficients in the differential equation, many iterative processes (but not for example RBGS and AZ) have the property that, if the error e before iteration is given by (5.2), then the error $\bar{e}$ after iteration is given by

$$\bar{e}_{mn} = \sum_{s=-m_1/2}^{m_1/2} \sum_{t=-m_2/2}^{m_2/2} \bar{c}_{st} \exp(im\theta_s + in\phi_t), \tag{5.6}$$

with

$$\bar{c}_{st} = \rho(\theta_s,\phi_t)c_{st}. \tag{5.7}$$

The annihilation of the non-smooth part of the error can be measured by the quantity defined below (Brandt (1977)):

Definition 5.1 The Fourier smoothing factor is

$$\rho_F := \sup_{(\theta,\phi)\in F} |\rho(\theta,\phi)|.$$

Note that ρ_F does not depend on the mesh-size h_i, in contrast to the rate of convergence of (2.2). A large catalogue of Fourier smoothing factors for various equations and smoothing processes has been compiled by Kettler (1982). It turns out that simple point-wise smoothing processes, such as damped Jacobi or Gauss-Seidel relaxation, have a good smoothing factor (i.e. ρ_F well below 1) for Poisson's equation, but not for equations with strong coupling in a certain direction, such as the convection-diffusion equation at high Péclet number, or anisotropic diffusion problems. In such cases more robust smoothing processes are called for, such as block Gauss-Seidel relaxation, AZ, ILU or ILLU (AZ does not work for convection-diffusion problems with upwind differences).

As noted before, Fourier smoothing analysis as just described assumes periodic boundary conditions and constant coefficients in the differential equation. The Fourier smoothing factor may be expected to be a good indicator of the quality of a smoothing process in more general circumstances, provided the coefficients vary smoothly, and provided the influence of perturbations of the boundary conditions attenuates as one moves into the interior of the region. However, MG is applied successfully to problems with discontinuous coefficients and problems where perturbations of the boundary conditions are felt in the interior, such as convection-diffusion and anisotropic diffusion problems. Apart from these limitations, Fourier smoothing analysis has the disadvantage, that the performance of the coarse grid corrections in the no-man's land between the smooth and non-smooth parts of the error is not taken into account. A different type of smoothing analysis that does not suffer these disadvantages is as follows.

Let the sets of grid-functions $G \to \mathbb{R}$ and $\bar{G} \to \mathbb{R}$ be defined by Y and $\bar{Y}$ respectively. Let the coarse grid approximation of (2.1) be given by

$$\bar{A}\bar{y} = \bar{b}. \tag{5.8}$$

Furthermore, let there be given a prolongation operator P and a restriction operator R:

$$P : \bar{Y} \to Y, \quad R : Y \to \bar{Y}. \tag{5.9}$$

A two-grid method for the acceleration of the iterative method (2.2) can be formulated as follows. Let y^j be the current iterand, and let $y^{j+\frac{1}{2}}$ be the result of applying a coarse grid correction to y^j:

$$y^{j+\frac{1}{2}} = y^j - P\bar{A}^{-1}R(Ay^j-b), \tag{5.10}$$

where we assume for the time being that the coarse grid problem is solved exactly. For the residue $r^j := Ay^j-b$ we find:

$$r^{j+\frac{1}{2}} = (I-AP\bar{A}^{-1}R)r^j. \tag{5.11}$$

We now make the following choice for $\bar{A}$, called

Galerkin approximation:

$$\bar{A} = RAP. \tag{5.12}$$

Then it follows from (5.11) that

$$r^{j+\frac{1}{2}} \in \mathrm{Ker}(R), \tag{5.13}$$

as noted by Hemker (1982) and McCormick (1982). In other words, $r^{j+\frac{1}{2}} \perp \mathrm{Ker}^{\perp}(R)$, which justifies the appellation "Galerkin approximation" for (5.12). Following Hemker (1982A) we will relate the concept of smoothness to the kernel and range of R and P.

Definition 5.2 The set of R-smooth grid-functions is $\mathrm{Ker}^{\perp}(R)$.

Whether the grid-functions just defined are also what one would call physically smooth or smooth in the sense of the Fourier analysis presented above, depends on the choice made for R.

It remains to annihilate the non-smooth part of r, and this is done in the second part of the two-grid iteration, called smoothing. This is done with an iterative method of type (2.2):

$$y^{j+1} = y^{j+\frac{1}{2}} + B(b-Ay^{j+\frac{1}{2}}), \tag{5.14}$$

and we find:

$$r^{j+1} = (I-AB)r^{j+\frac{1}{2}}. \tag{5.15}$$

The projection operator on Ker(R) is given by $I-R^T(RR^T)^{-1}R$, and we may conclude from (5.13) and (5.15) that

$$||r^{j+1}|| \leq ||(I-AB)(I-R^T(RR^T)^{-1}R)|| \; ||r^{j+\frac{1}{2}}||. \tag{5.16}$$

This leads us to the following definition:

Definition 5.3 The R-smoothing factor of the smoothing process (2.2) is

$$\rho_R := ||(I-AB)(I-R^T(RR^T)^{-1}R)||.$$

Whether ρ_R will be approximately equal to ρ_F depends on whether $\mathrm{Ker}^{\perp}(R)$ approximately equals the space spanned by the Fourier components (5.5), and on the applicability of Fourier smoothing analysis. An advantage of Fourier smoothing analysis is that ρ_F is usually easier to compute than ρ_R.

We now take the dual viewpoint of considering the error instead of the residue, and define:

Definition 5.4 The set of P-smooth grid-functions is Range(P).

Let the error e^j be defined by $e^j := y^j - y$. Then it follows from (5.10) that

$$e^{j+\frac{1}{2}} = (I - P\bar{A}^{-1}RA)e^j. \qquad (5.17)$$

A streamlined reasoning is obtained if we now assume that smoothing precedes coarse grid correction, so that (5.14) is replaced by

$$y^j = y^{j-\frac{1}{2}} + B(b - Ay^{j-\frac{1}{2}}). \qquad (5.18)$$

Let $e^j = e_1^j + e_2^j$ with $e_1^j \in \text{Range}(P)$, $e_2^j \in \text{Range}^{\perp}(P)$. Again choosing $\bar{A}$ according to (5.12) we see that

$$(I - P\bar{A}^{-1}RA)e_1^j = 0, \qquad (5.19)$$

(write $e_1^j = P\hat{e}$ for some $\hat{e}$), so that

$$e^{j+\frac{1}{2}} = (I - P\bar{A}^{-1}RA)e_2^j. \qquad (5.20)$$

In general $e^{j+\frac{1}{2}}$ will be small only if e_2^j is small, which motivates the following definition:

Definition 5.5 The P-smoothing factor of the smoothing process (2.2) is

$$\rho_P := ||(I-P)(P^TP)^{-1}P^T)(I-BA)||$$

(Note that the projection operator on $\text{Range}^{\perp}(P)$ is given by $I-P(P^TP)^{-1}P^T$.) In the special case that $R = P^T$ we have that the sets of R-smooth and of P-smooth grid-functions are identical, since $\text{Ker}^{\perp}(R) = \text{Range}(R^T)$, but ρ_R and ρ_P will in general not be identical. The quantity ρ_P is defined and studied by McCormick (1982)

The two-grid method defined by (5.10) and (5.14) or (5.18) can be regarded as an acceleration technique for the iterative method (2.2). The striking efficiency of MG methods is due to the fact that there exist simple iteration methods of type (2.2) for which ρ_F, ρ_R and ρ_P

are well below 1 for a large class of problems, independent of the mesh-size.

Prolongation and restriction operators can be chosen in various ways. Examples of prolongations are (denoting grid-functions in $\bar{Y}$ by an overbar): 9-point prolongation:

$$(P\bar{y})_{2p,2q} = \bar{y}_{pq},\quad (P\bar{y})_{2p+1,2q} = \tfrac{1}{2}(\bar{y}_{pq}+\bar{y}_{p+1,q})$$

$$(P\bar{y})_{2p,2q+1} = \tfrac{1}{2}(\bar{y}_{pq}+\bar{y}_{p,q+1}), \tag{5.21}$$

$$(P\bar{y})_{2p+1,2q+1} = \tfrac{1}{2}\{(P\bar{y})_{2p+1,2q} + (P\bar{y})_{2p,2q+1}\}.$$

7-point prolongation: as 9-point prolongation, except

$$(P\bar{y})_{2p+1,2q+1} = \tfrac{1}{2}(\bar{y}_{p+1,q}+\bar{y}_{p,q+1}). \tag{5.22}$$

Examples of restrictions are:

Injection:

$$(Ry)_{pq} = y_{2p,2q}, \tag{5.23}$$

9-point restriction or full weighting:

$$(Ry)_{pq} = y_{2p,2q} + \tfrac{1}{2}(y_{2p+1,2q}+y_{2p,2q+1}+y_{2p-1,2q}+y_{2p,2q-1}) + \\ + \frac{1}{4}(y_{2p+1,2q+1}+y_{2p-1,2q+1}+y_{2p+1,2q-1}+y_{2p-1,2q-1}), \tag{5.24}$$

7-point restriction:

$$(Ry)_{pq} = y_{2p,2q}+\tfrac{1}{2}(y_{2p+1,2q}+y_{2p,2q+1}+y_{2p-1,2q} \\ +y_{2p,2q-1}+y_{2p+1,2q-1}+y_{2p-1,2q+1}). \tag{5.25}$$

We call (5.24), (5.25) 9-point or 7-point restriction because a weighted average is taken of 9 or 7 grid-function values, and we call (5.21), (5.22) 9-point or 7-point prolongation, because they are closely related to (5.24) and (5.25) respectively: with P, R according to (5.21), (5.24) or according to (5.22), (5.25) we have

$$R = P^T, \tag{5.26}$$

i.e. as matrices, P and R are adjoint.

In Fig. 5.1 we define a 5-point, 7-point and 9-point difference molecule. With a 5-point or a 7-point molecule, 7-point prolongation and restriction can be used. With a 9-point molecule, 9-point prolongation and restriction is more accurate. With a 7-point molecule one can construct finite difference approximations to any second order partial differential equation in two dimensions, including mixed derivatives. A 7-point molecule is also obtained with finite elements, using Courant triangulation (cf. Fig. 5.1). If one desires exactly symmetric numerical solutions to symmetric problems a 9-point molecule should be used, an example being symmetric flow around a symmetric airfoil.

Fig. 5.1 Difference molecules: 5-point, 7-point, 9-point. Courant triangle.

In a loose sense, P is accurate if, given that $\bar{y}$ is a good discrete approximation to the exact solution of the differential equation, Py is also a good approximation. For prolongations based on linear interpolation, such as (5.21) and (5.22), this is certainly the case when the exact solution is smooth. An important case when the exact solution is not smooth occurs when the coefficients of the differential equation are discontinuous. In that case <u>matrix-dependent prolongation</u> should be used. In order to define this type of prolongation we use the grid-point enumeration of Fig. 2.1. A grid-point with coordinates (ph_1, qh_2) has the number $1+p+qm$. Indicating elements of the matrix A by a_{ij} in the usual way corresponding with this enumeration, we define:

<u>Matrix-dependent prolongation</u>:

$$(P\bar{y})_{2p,2q} = \bar{y}_{pq}, \tag{5.27a}$$

$$(P\bar{y})_{2p+1,2q} = (A_{i,i-1}\bar{y}_{pq}+A_{i,i+1}\bar{y}_{p+1,q})/(A_{i,i-1}+A_{i,i+1}), \tag{5.27b}$$

$$(P\bar{y})_{2p,2q+1} = (A_{j,j-m}\bar{y}_{pq}+A_{j,j+m}\bar{y}_{p,q+1})/(A_{j,j-m}+A_{j,j+m}), \tag{5.27c}$$

$$(P\bar{y})_{2p+1,2q+1} = -\sum_{j\neq k} A_{kj}y_j/A_{kk}, \tag{5.27d}$$

where $i = 2+2p+2qm$, $j = 1+2p+(2q+1)m$, $k = 2+2p+(2q+1)m$. In (5.27d) y-values obtained with (5.27a-c) are used. It is possible to use the right-hand-side in (5.27d); sometimes this enhances the rate of convergence. A matrix dependent restriction is obtained with (5.26). Matrix-dependent prolongations of this and related type have been proposed by Alcouffe, Brandt, Dendy and Painter (1981), Kettler (1980, 1982), Kettler and Meijerink (1981).

The coarse grid problem (5.10) is not solved exactly of course, but approximately. In the MGD-family of MG codes we do this with one two-grid iteration employing an additional coarser grid with doubled mesh-size, and so on recursively, until the coarsest grid (usually a 3 × 3 grid) is reached, where a few iterations (usually one) are performed according to (2.2). Smoothing is the costliest part of the algorithm. Therefore we choose to let coarse grid correction precede smoothing (i.e. we have (5.10), (5.14)), so that the first time that smoothing takes place on the finest grid we already have a first approximation available. The resulting MG method is said to be of sawtooth type, because its schedule is represented in a natural way by the schematic of Fig. 5.2, which is a sawtooth curve.

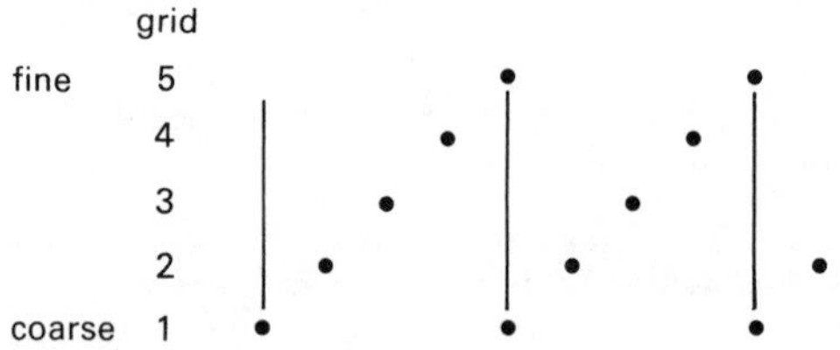

Fig. 5.2 Sawtooth multigrid schedule. A dot represents a smoothing step.

Various more general MG schedules have been described, see for example Brandt (1977), Stüben and Trottenberg (1982). Some comparative experiments are described in Wesseling (1982A). The sawtooth schedule is the simplest possible MG schedule. One may wonder whether such a simple fixed schedule can handle a sufficiently large variety of cases. Experience indicates that the answer is affirmative, see e.g. the experiments carried out by Wesseling and Sonneveld (1980), Kettler (1982), Wesseling (1982A,B), Hemker, Kettler, Wesseling and de Zeeuw (1983), McCarthy (1983). In transonic potential flow computation an MGD-type method has proved reliable, see Nowak and Wesseling (1983). We think that with an effective smoother and accurate coarse grid approximation, a simple MG schedule suffices for linear problems.

The sawtooth schedule can be programmed in a simple way without using recursion. Let the computational grids employed be denoted by $G^1, G^2, \ldots, G^\ell$, with G^1 the coarsest and G^ℓ the finest grid. Let a superscript k indicate grid-functions and operators on G^k. Let one application of the smoothing process (5.14) be executed by a subroutine SMOOTHING (y,b,k). Then a quasi-FORTRAN outline of MG algorithms using the sawtooth schedule is given by:

```
C     MULTIGRID PROGRAM, SAWTOOTH SCHEDULE
C     INITIAL GUESS IS y^ℓ = 0
      r^ℓ = b^ℓ
      DO 10 k = ℓ-1(-1)1
      CALL RESTRICTION (r,k)          r^k = R^{k+1} r^{k+1}
   10 CONTINUE
C     START OF maxit MULTIGRID ITERATIONS
      DO 50 n = 1(1) maxit
      IF (n.EQ.1) GO TO 30
      CALL RESIDUE (r,b,y,ℓ)          r^ℓ = b^ℓ-A^ℓ y^ℓ
      DO 20 k = ℓ-1(-1)1
      CALL RESTRICTION (r,k)          r^k = R^{k+1} r^{k+1}
   20 CONTINUE
   30 y^1 = 0
      CALL SMOOTHING (u,r,1)          y^1 = y^1+B^1(r^1-Ay^1)
      DO 40 k = 2(1)ℓ-1
      CALL PROLONGATION (y,y,k)       y^k = P^k y^{k-1}
      CALL SMOOTHING (y,r,k)          y^k = y^k+B^k(r^k-A^k y^k)
   40 CONTINUE
      CALL PROLONGATION (v,y,ℓ)       v^ℓ = P^ℓ y^{ℓ-1}
      y^ℓ = y^ℓ+v^ℓ
      CALL SMOOTHING (y,b,ℓ)          y^ℓ = y^ℓ+B^ℓ(b-A^ℓ y^ℓ)
   50 CONTINUE
```

Based on this algorithm, the MGD family of codes is being developed. Two portable FORTRAN codes have been implemented, called MGD1 and MGD5. They can be obtained by sending a magnetic tape to the second author. In these codes, prolongation and restriction are of 7-point type. For smoothing MGD1 and MGD5 use ILU and ILLU, respectively. Versions MGD1V and MGD5V have been designed for auto-vectorization on vector computers, such as the CYBER-205 and the CRAY-1, without sacrificing much on sequential machines. They are easily changed to versions MGD1S and MGD5S, which are slightly faster on sequential machines. More details, and CPU-time measurements on CYBER-170, CYBER-205 and CRAY-1 can be found in Hemker et al. (1983, 1983), Hemker and de Zeeuw (1984). Extensive tests of MGD1 have been carried out by McCarthy (1983). Other MG-software that is generally available is the collection of multigrid solution modules MGOO, see Foerster and Witsch (1982).

In order to facilitate comparison with other methods, especially CGS, for which we only have a research code in another programming language, we will estimate the cost in flops per finest grid-point. The cost of one MGD1 iteration is 30 flops per finest grid-point, see Wesseling (1982B). In MGD1 the cost of a smoothing step on one grid is 19 flops, for MGD5 it is 37 flops per grid-point, as shown in section 2.

The number of grid-points on all grids taken together is about 4/3 times the number of grid-points of the finest grid. Hence the total smoothing work for one MG iteration for MGD1 or MGD5 is about 25 or 49 flops per finest grid-point, respectively. Since the only difference between the two codes is the smoothing process, we estimate that the cost of one MGD5 iteration is 30-25+49=54 flops per finest grid-point. The measured CPU-time ratio on a CYBER-170 is 1.6.

Some design considerations concerning the MGD codes can be found in Wesseling (1982B). These codes have been constructed such that they are perceived by the user just like any other code for solving linear systems of algebraic equations. The user has only to give the matrix and the right-hand-side in a prescribed data structure. The matrix should have a sparsity pattern corresponding to a 7-point finite difference discretization. The user remains unaware of the underlying multigrid algorithm, and cannot make any choices or decisions, since the code is completely autonomous (black box MG, cf. Dendy (1982)).

The use of coarse grid Galerkin approximation (CGGA) (5.12) greatly facilitates the realization of the design goals just mentioned, since by using (5.12) the algorithm can set up the coarse grid operators independently from the user, using as input only the fine grid matrix. This would be less easy to achieve with the popular alternative of coarse grid finite difference approximation, (CGFDA), in which the coarse grid matrices are finite difference approximations of the given differential equation, usually of the same type as the fine grid matrix. CGFDA has another disadvantage, namely, that the approximations obtained on the coarsest grids make little sense if the coefficients of the differential equation are sampled pointwise. This may lead to divergence; Wesseling (1982B) gives an example. Of course the user can avoid this by using suitably averaged values of the coefficients of the differential equation on the coarsest grids. Using CGGA leads automatically to an accurate type of averaging. A disadvantage of CGGA can be the cost, which in practice equals the cost of about two MG iterations with our codes. If the coefficients of the differential equation are not expensive it is cheaper to set up finite difference approximations. For a few experiments comparing CGFDA and CGGA and a few remarks on efficient programming of (5.12), see Wesseling (1982A,B). The total work for computing A^k, $k = \ell-1(-1)$ is found to be about 64 flops per finest grid point, for a 7-point finite difference approximation on the finest grid.

Additional preliminary work is required for setting up the incomplete decompositions before iteration starts. The total cost of preliminary work is equivalent to about 3 iterations for MGD1 and 2 iterations for MGD5, cf. Hemker and de Zeeuw (1984). This preliminary work is considerable in view of the fact that convergence is usually so rapid that only a few iterations are needed. This price buys robustness. For self-adjoint problems with smoothly varying coefficients of the same order of magnitude, one obtains good rates of convergence with pointwise relaxation processes for smoothing and CGFDA, which require little preliminary work. But with these MG-ingredients convergence will deteriorate if the problem is strongly anisotropic, or strongly non-self-adjoint (convection-diffusion at high Péclet number). Under these circumstances the MGD codes continue to converge fast, with some exceptions for MGD1.

Under what conditions may MG methods be expected to converge rapidly? A priori theoretical results are not available except for the Poisson equation, but a rule of thumb is that good smoothing processes can be found and rapid convergence may be expected if A satisfies (4.27). If (4.27) is strongly violated, deterioration of the rate of convergence may occur.

However, for convection-diffusion problems at high Péclet numbers MG methods have not performed well. This situation has improved only recently. In order to satisfy (4.27) upwind differencing must be used, or a sufficient amount of artifical viscosity must be added. Nevertheless, when (5.12) is used with 7-point or 9-point prolongation and restriction, the coarse grid matrices do not satisfy (4.27). This is illustrated by the transformation that an upwind difference undergoes by repeated application of (5.12) with 7-point prolongation and restriction:

$$\begin{array}{ccc} 0 & 0 & \\ -1 & 1 & 0 \\ & 0 & 0 \end{array} \qquad \begin{array}{ccc} -1 & 1 & \\ -5 & 4 & 1 \\ & -1 & 1 \end{array} \qquad \begin{array}{ccc} -5 & 5 & \\ -15 & 8 & 7 \\ & -5 & 5 \end{array} \qquad \begin{array}{ccc} -21 & 21 & \\ -51 & 16 & 35 \\ & -21 & 21 \end{array}$$

$$\begin{array}{ccc} -85 & 85 & \\ -187 & 32 & 155 \\ & -85 & 85 \end{array} \qquad \begin{array}{ccc} -341 & 341 & \\ -751 & 64 & 651 \\ & -341 & 341 \end{array} \; .$$

Scaling factors have been omitted. As the number of grids increases, the diagonal becomes weaker. Because the coarse grid matrices do not satisfy (4.27) the smoothing process does not perform well, and furthermore, the coarse grid solution may show wiggles. The situation becomes worse as the number of grids increases. Hence, convergence is not rapid or divergence occurs, unless the smoothing process is almost an exact solver on the finest grid. In that case, the bad coarse grid approximations are corrected on the finest grid and convergence is rapid; in fact the coarse grids are superfluous. ILLU has this property: ILLU-decomposition is almost exact for the convection-diffusion equation at high Péclet number with upwind differences. Therefore MGD5 works for convection-diffusion equations. For MGD1 cases of divergence have been found. If one does not want to use ILLU smoothing, an easy way out would seem to be not to use CGGA but CGFDA with upwind differences or artificial viscosity on all grids. But then convergence is found to become disappointingly slow. A very good way to handle the convection-diffusion equation turns out to be the use of matrix-dependent prolongation and restriction, with CGGA or CGFDA. With this prolongation and restriction CGGA leaves upwind differences invariant, so that the coarse grid matrices satisfy (4.27), and do not differ much from the coarse grid matrices obtained with upwind differencing and CGFDA. Good rates of convergence are obtained with smoothing processes less formidable than ILLU. For a more extensive treatment of the ideas just discussed and numerical experiments, see van Asselt (1982), de Zeeuw and van Asselt (1985), Hemker, Kettler, Wesseling and de Zeeuw (1983).

A rather different MG approach to the convection-diffusion equation, proposed by Brandt, is not to use upwind differencing, which has inherent anisotropic numerical viscosity, but to use isotropic artificial viscosity. This makes point-wise relaxation processes applicable for smoothing. The accuracy and probably also the rate of convergence is improved by what is called double discretization, which amounts to applying defect correction on every grid. For this approach see Brandt (1982) section 10.2. A disadvantage is that the method is especially designed for convection-diffusion problems, so that for other problems one would perhaps prefer other MG ingredients. No definitive results with this approach have been published as yet for the convection-diffusion equation.

The accuracy of upwind or artificial viscosity discretizations can also be improved by applying defect-correction on the finest grid only. See Hemker (1982) for an application of this idea to the convection-diffusion equation.

For ease of programming of MG methods it is very convenient if coarser grids can be obtained by mesh doubling. Therefore, the number of grid points of the finest grid in the x_i-direction should be given by $1+2^{\ell}(m_i-1)$, with m_i a small integer. Sometimes it is awkward to achieve this, for example when a system of partial differential equations is solved on a staggered grid. One can then change the number of grid points by either eliminating Dirichlet boundaries or not, or by increasing the number of discretized equations by adding artificial equations, for example the identity. This is called padding. Padding can also be used to make the shape of the computational region rectangular. Of course, the computational complexity is influenced unfavourably by padding.

6. NUMERICAL EXPERIMENTS

Realistic estimates of the performance in practice of CG and MG by purely theoretical means are possible only for very simple problems. Therefore, numerical experiments are necessary to obtain insight and confidence in the efficiency and robustness of a particular method. Numerical experiments can be used only to rule out methods that fail, not to guarantee good performance of a method for problems that have not yet been attempted. Nevertheless, one strives to build up confidence by carefully choosing test problems, trying to make them representative for large classes of problems, taking into account the nature of the mathematical models that occur in the field of application that one has in mind. For the development of CG and MG, in particular the subject areas of computational fluid dynamics, petroleum reservoir engineering and neutron diffusion are pace-setting. We will list here the most significant test problems, and discuss a few numerical results.

Only the case of a single second order elliptic equation in two dimensions is discussed, although the applicability of CG and MG is not restricted to this case. The general form of our problem then is, in Cartesian tensor notation,

$$-(a_{ij}\phi_{,j})_{,i} + (b_i u)_{,i} + cu = f. \qquad (6.1)$$

Important constant coefficient test problems are the following special cases of (6.1):

$$-(\varepsilon c^2+s^2)\phi_{,11} - 2(\varepsilon-1)sc\phi_{,12} - (\varepsilon s^2+c^2)\phi_{,22} = f \tag{6.2}$$

and

$$-\varepsilon\phi_{,ii} + c\phi_{,1} + s\phi_{,2} = f, \tag{6.3}$$

with $c = \cos\alpha$, $s = \sin\alpha$. Equation (6.2) is obtained by a coordinate rotation over an angle α for the anisotropic diffusion equation:

$$-\varepsilon\phi_{,11} - \phi_{,22} = f. \tag{6.4}$$

Equation (6.3) is the convection diffusion equation. Equation (6.2) is self-adjoint, and can be handled such that the matrix arising from discretization is SPD.

Problems with constant coefficients are thought to be representative of problems with smoothly varying coefficients. Of course, in the code to be tested the fact that the coefficients are constant should not be exploited. As pointed out by Curtiss (1981), one should keep in mind that for constant coefficient problems the spectrum of the matrix resulting from discretization can have very special properties, that are not present when the coefficients are variable. Therefore one should also carry out tests with variable coefficients, especially with CG, for which the properties of the spectrum are very important. For MG, constant coefficient test problems are often more demanding than variable coefficient problems, because it may happen that the smoothing process is not effective for certain combinations of ε and α. This fact goes easily unnoticed with variable coefficients, where the unfavourable values of ε and α perhaps occur only in a small part of the domain.

In petroleum reservoir engineering and neutron diffusion problems quite often equations with strongly discontinuous coefficients appear. For these equations (6.2) and (6.3) are not representative. Suitable test problems with strongly discontinuous coefficients have been proposed by Stone (1968) and Kershaw (1978); a definition of these test problems may also be found in Kettler (1982). In Kershaw's problem the domain is non-rectangular, but is a rectangular polygon. The matrix for both problems is SPD. For the parameter p in Stone's problem we choose p=5 (cf. Kettler (1982)).

The four test problems just mentioned, i.e. (6.2), (6.3), and the problems of Stone and Kershaw, are gaining acceptance among CG and MG practitioners as standard test problems. Given these test problems, the dilemma of robustness versus efficiency presents itself. Should one try to devise a single code to handle all problems (robustness), or develop codes that handle only a subset, but do so more efficiently than a robust code? This dilemma is not novel, and just as in other parts of numerical mathematics, we expect that both approaches will be fruitful, and no single "best" code will emerge.

For CG methods a natural subdivision of the problems presents itself, namely in self-adjoint and non-self-adjoint problems. The former lead to SPD matrices, to which the applicability of classical CG is restricted. In non-self-adjoint cases, a non-symmetric CG variant should be used, for example CGS. Of course, CGS can be used also for SPD matrices, at little extra cost compared with classical CG.

The robustness and efficiency of CG and MG are determined to a large extent by the preconditioning and the smoothing process respectively. Pointwise relaxation methods, such as RBGS, are easy to implement and require no preliminary work before iterations start, and are efficient for (6.2), (6.3) for $\varepsilon \approx 1$. But these methods fail for ε differing widely from 1, and for the problems of Stone and Kershaw. In these cases suitable block relaxation methods are called for. We have not yet found a case where ILLU fails. ILU is found to fail in certain cases where property (4.27) is violated. AZ may fail also when (4.27) holds, for convection-diffusion problems. These findings will be amplified in the sequel.

Property (4.27) is violated in the case (6.2) for certain combinations of ε and α, for which the coefficients of the mixed derivative are relatively large. However, in practical applications the mixed derivative coefficient is often small. When the mixed derivative is introduced by a non-orthogonal coordinate transformation its coefficient is usually small, because for accuracy reasons one prefers coordinate transformations that do not deviate much from orthogonality. In anisotropic diffusion problems there is usually a preferred direction, along which one aligns one of the coordinate axes, so that sc = 0, and no mixed derivative is present.

Property (4.27) is also violated for (6.3) when $\varepsilon < h/2$ and central differences are used. With upwind differences it can still be violated on the coarse grids, as discussed in the preceding section.

Apart from a robust smoothing process, an MG method for the problems of Stone and Kershaw needs matrix-dependent prolongation and restriction, because of the occurrence of discontinuous coefficients.

Numerical experiments with MG concerning special cases of (6.2) (notably Poisson's equation) have been reported by Brandt (1977), Hackbusch (1978), Nicolaides (1979), Foerster et al. (1981), Foerster and Witsch (1982), Kettler (1982), Wesseling (1982A,B), Hemker et al. (1983, 1983). We will not list here experiments with CG, of which there are many more. Hackbusch (1978) and Foerster and Witsch (1982) include examples of Poisson's equation in non-rectangular regions. In the last mentioned publication also an MG method specially designed for Poisson's equation is presented. Computing times are reported similar to those obtained with fast Poisson solvers using the fast Fourier transform and cyclic reduction, and about 15 times as fast as a certain CG method (ICCG, Meijerink and van der Vorst (1977)) on a 257×257 mesh. It is to be noted that ICCG is much more generally applicable than the MG method concerned, which uses RBGS smoothing. This method would fail for example for the problems of Stone and Kershaw, for which ICCG performs well.

For test problem (6.3) MG results have been reported by Wesseling and Sonneveld (1980), Hemker (1982), Wesseling (1982A,B), Hemker,

Kettler, Wesseling and de Zeeuw (1983), de Zeeuw and van Asselt (1985). Because the discretization matrix is not SPD, classical CG cannot be applied. Chebyshev iteration has been used for this type of problem by Manteuffel (1977, 1978) and van der Vorst (1981). This method is not parameter free, unlike CGS.

We will present results for the general case of (6.2) and (6.3), letting α vary with intervals of 15°, and choosing $\varepsilon << 1$, ε intermediate, and $\varepsilon = 1$. The following methods will be tested:

- MGD1 and MGD5, described in section 5;
- CGS1 and CGS5, the CGS method described in section 4 with ILU and ILLU preconditioning, respectively;
- MGHZ and MGAZ, which are MGD1 with ILU smoothing replaced by HZ and AZ smoothing, respectively.

For easy reference, in the following table we give the operation counts for the various methods, as determined before. For MGHZ and MGAZ the operation count is determined by noting that the work for MGD1 excluding smoothing is 5 flops per finest grid-point. The smoothing work with HZ and AZ is 4/3 times the work of a single grid iteration. Here we neglect certain savings that are possible because the residue is zero in half the number of grid-points after application of HZ and AZ. From the results reported by Hemker, Wesseling and de Zeeuw (1983) we deduce that both for PW and IW the measured CP-time ratio on a CYBER-170 is MGD1 : MGHZ = 1.24. From Hemker and de Zeeuw (1984) we deduce that on the same machine MGD5 : MGD1 = 1.10 for PW and MGD5 : MGD1 = 1.62 for IW. These figures are roughly consistent with table 6.1.

Table 6.1

Flops per (finest) grid-point for one iteration (IW) and preliminary work (PW)

	MGD1	MGD5	MGHZ	MGAZ	CGS1	CGS5
PW	87	114	69	74	17	29
IW	30	54	22	40	60	88

We have run test problems (6.2) and (6.3) on a uniform 65 × 65 computational grid in the unit square. The initial guess is given by $-\sin\pi x_1 \sin\pi x_2 + \sin 48\pi x_1 \sin 48\pi x_2$, and the boundary conditions, which are eliminated, are given by

$$\phi\big|_{\partial\Omega} = x_i x_i . \qquad (6.5)$$

For the MG methods, the termination criterion was that the ℓ_2-norm of the residue should be less than 10^{-10}, with a maximum of 10 iterations. The CG iterations were terminated after a residue reduction factor of 10^{-8} had been reached, with a maximum of 15 iterations.

For test problem (6.2), the discretization of the mixed derivative is as given in Fig. 6.1.

$$\begin{array}{ccc} \frac{1}{2} & -\frac{1}{2} & \\ -\frac{1}{2} & 1 & -\frac{1}{2} \\ & -\frac{1}{2} & \frac{1}{2} \end{array}$$

Fig. 6.1 Difference molecule for $-h^2\phi_{,12}$.

The following table specifies the problems that were treated. For problems 5, 6, upwind difference were used, for the other problems, central differences.

Table 6.2

Specification of test problems

Problem	1	2	3	4	5	6
Equation	(6.2)	(6.2)	(6.3)	(6.3)	(6.3)	(6.3)
ε	10^{-2}	10^{-8}	10^{-1}	$h/2$	10^{-3}	10^{-8}

In Figs. 6.2-6.5 we give a graphical representation of the number of iterations needed to reduce the ℓ_2-norm of the residue by a factor 10, or, roughly speaking, to gain a decimal figure in accuracy. This number is given by

$$N = n/\log_{10} \{||\text{initial residue}||/||\text{final residue}||\} \tag{6.6}$$

where n is the number of iterations that were performed.

Computations were performed with α a multiple of 15°. If for a value of α no symbol appears for one of the methods, this means that more than 10 iterations are necessary to gain one decimal, or that the method diverges.

The results clearly show that the rate of convergence can strongly depend on α, and that performing experiments for just a few values of α can be misleading.

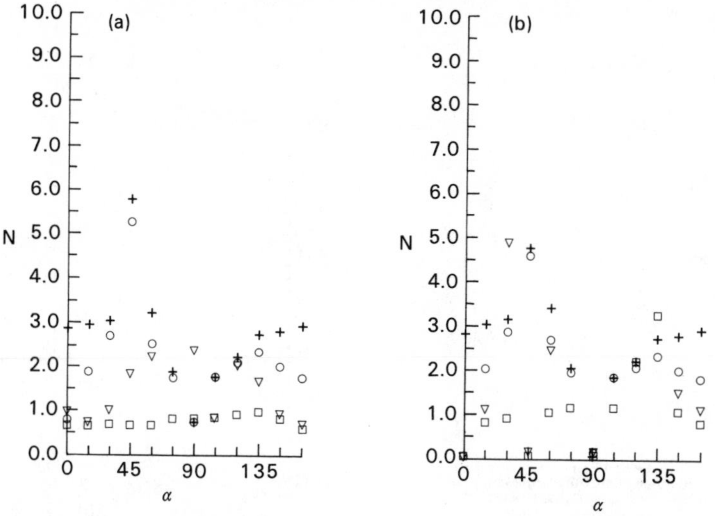

Fig. 6.2 Multigrid results for equation (6.2). (a): Problem 1. (b): Problem 2. ∇: MGD1; □: MGD5; +: MGHZ; O: MGAZ.

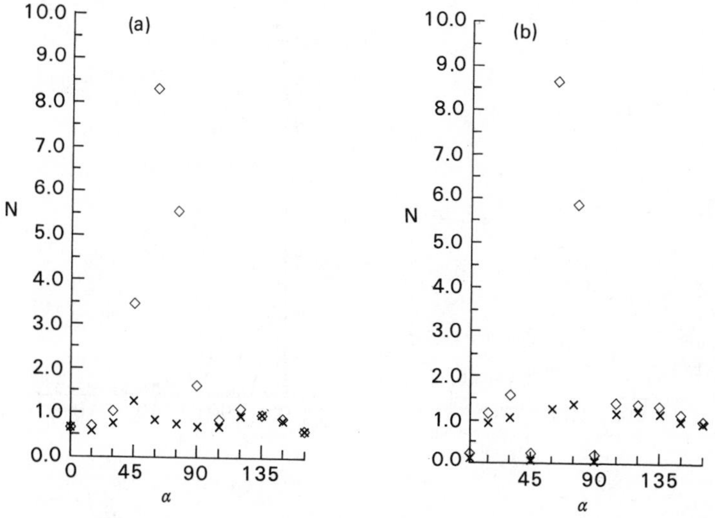

Fig. 6.3 CGS results for equation (6.2). (a): Problem 1; (b): Problem 2; ◊: CGS1; ×: CGS5.

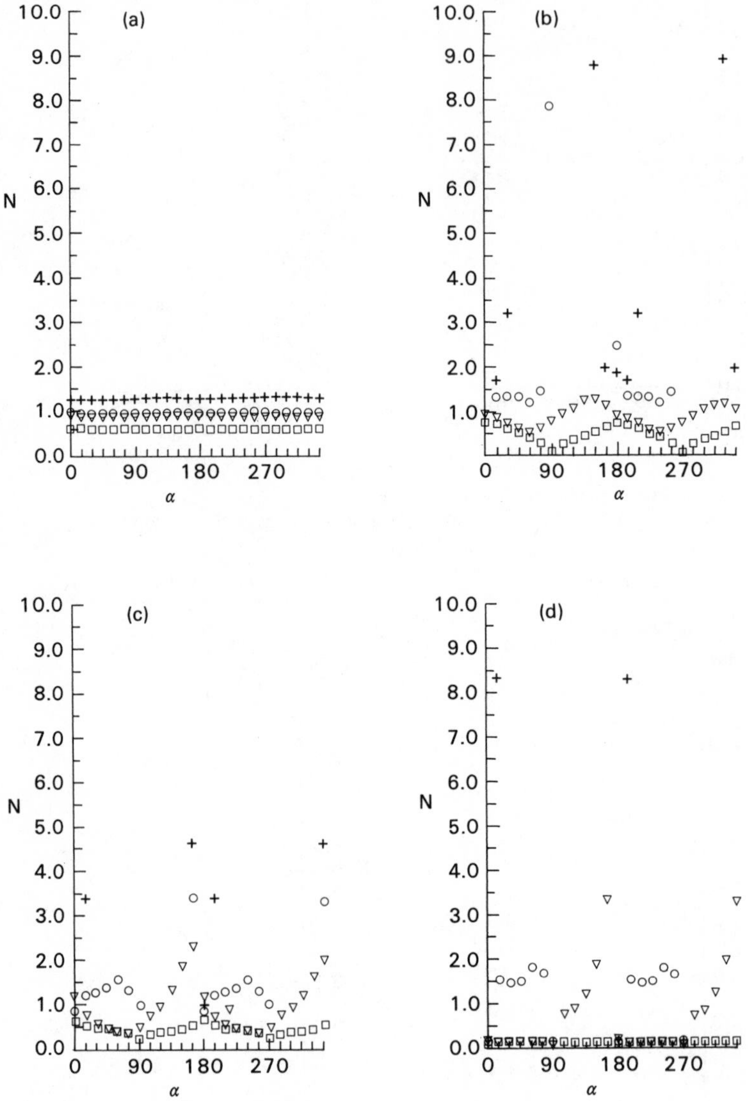

Fig. 6.4 Multigrid results for equation (6.3). (a)-(d): Problems 3-6, respectively. Symbols as in Fig. 6.2

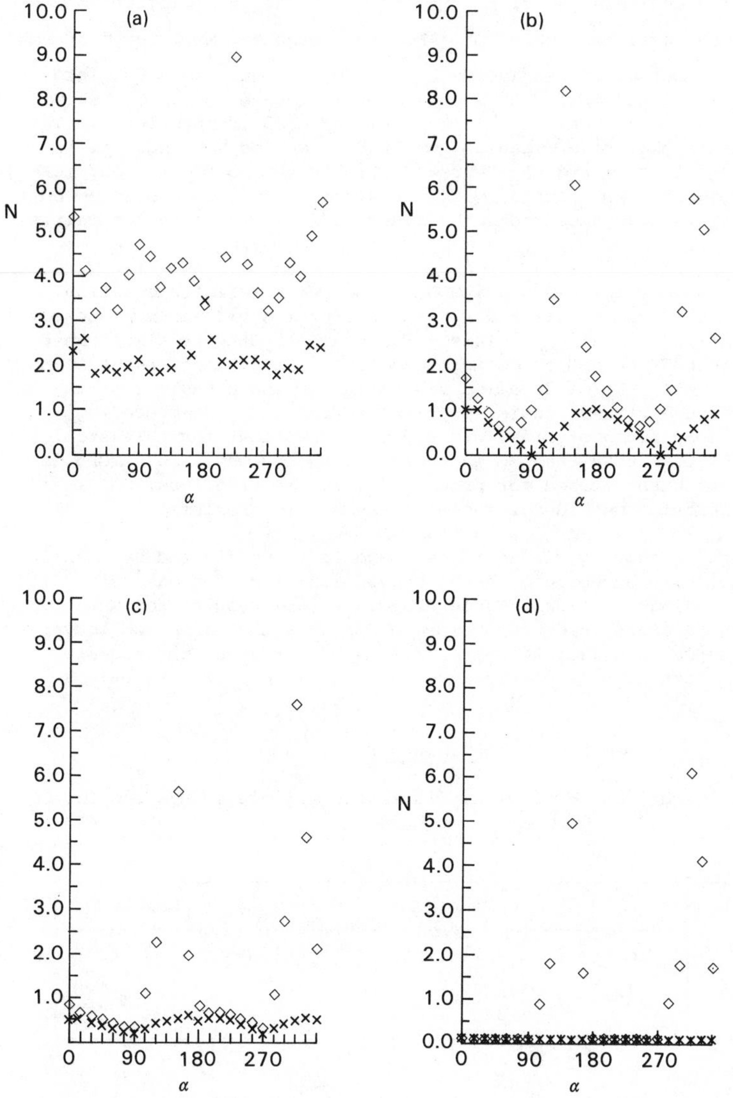

Fig. 6.5 CGS results for equation (6.3). (a)-(d): Problems 3-6, respectively. Symbols as in Fig. 6.3

Fig. 6.2 shows that for the anisotropic diffusion problem (6.2) MGD1 does not work well for α slightly different from 90°. This is predicted by smoothing analysis (Kettler (1982)), which does not explain, however, why it works for α precisely 90°. For this problem, MGD1 out-performs the other MG methods for α around 0° or 180°, taking table 6.1 into account. For general α, MGD5 is the best MG method. Fig. 6.3 shows that CGS1 and CGS5 behave much like MGD1 and MGD5, respectively. This means that in that case, when ILU or ILLU is a good smoother, it is a good preconditioner, and vice-versa. CGS5 is the most efficient method for this problem. Of course, with classical CG one would even be better off, and it is guaranteed to work (hence, CGS also), since the matrix is SPD.

Fig. 6.4 shows that MGHZ and MGAZ do not work well for equation (6.3). This is because HZ and AZ are ineffective smoothing processes for convection-diffusion problems. For small ε, ILLU is almost an exact solution method, and MG or CG acceleration is in fact not needed. MGD1 also works well. For a detailed discussion of the behaviour of the MGD codes for convection-diffusion problems we refer to the preceding section. Comparison of Figs. 6.4 and 6.5 shows that for this problem CGS is less effective than MG in accelerating ILU and ILLU. MGD5 is the most efficient method for problem (6.3). Nevertheless, CGS is a good acceleration method for these non-symmetric problems.

The rate of convergence of MG is found to be unaffected by mesh-refinement, with exceptions in the convection-diffusion case discussed in the preceding section. Table 6.3 gives some results for N as defined in equation (6.6) for CGS1 and CGS5, as the mesh-size is varied. The dependence on h is not clear-cut. As h decreases, the required number of iterations generally increases, but there are exceptions.

Table 6.3

Number of iterations per decimal figure for various mesh-sizes for CG methods

1/h		CGS1	CGS5		CGS1	CGS5
33	Problem 2	1.151	.857	Problem 5	.899	.369
65	$\alpha=120^{\circ}$	1.322	1.186	$\alpha=165^{\circ}$	1.967	.598
129		3.340	1.842		div	.233

In the case of problem 5, $\alpha=120^{\circ}$, like CGS1, MGD1 is found to diverge see Hemker, Kettler, Wesseling and de Zeeuw (1983). The explanation and the remedy has been given in the preceding section. In the case of CGS1 all we can say at present is, that apparently ILU is not a good preconditioning for this problem.

Next, we turn to the test-problems of Stone and Kershaw. Because the matrix is SPD, CGS will behave more or less like classical CG, which has been applied to these problems by several authors. Therefore CGS will not be used. The MGD1 and MGD5 codes are not applicable, because matrix-dependent prolongation and restriction is necessary. We show some of the results obtained by Kettler (1982). The MG methods used are similar to MGD1 and MGD5, but matrix-dependent prolongation and restriction is used, as defined by (5.27) and (5.26), and the sawtooth-cycle is replaced by the V-cycle. Fig. 6.6 gives $\log_{10}||\text{residue}||_2$ as a function of the estimated number of flops per finest grid-point. The ILU and ILLU iteration methods are accelerated by MG and classical CG. As explained in section 4, CG can be used to accelerate any iterative method (which corresponds to a symmetric preconditioned matrix), and Kettler (1982) has used CG to accelerate MG (MGCG). Of course, one could also try to use MG for acceleration of CG. Then one would expect fast convergence if CG is a good smoother; therefore one would have to tailor CG such that it works primarily on the non-smooth components of the error. We have not pursued this avenue.

The figures show no systematic trend. All methods converge very rapidly compared with older methods, see Kershaw (1978). CG is not inferior to MG in these tests, but when the grid is refined, CG would probably start lagging behind. The dimension of the computational grid was 31×31 for Stone and 51×51 for Kershaw. By padding (see the preceding section) this was increased to 33×33 and 57×57, respectively, in order to make construction of 4 or 3 coarse grids possible by mesh doubling. Padding was also used to fill in the L-shaped region in Kershaw's problem to a square, to facilitate MG programming. For more results, including the use of several other smoothing processes, see Kettler (1982).

Fig. 6.6 shows that CG acceleration of MG is effective when ILU smoothing is used, but with ILLU it does not help, although it does no harm either. CG and MG have also been applied to the test problems of Stone and Kershaw (and two other similar problems) by Behie and Forsyth (1983). They advocate acceleration of MG in its non-symmetric sawtooth variant by means of Orthomin (Vinsome (1976)), a CG-variant for non-symmetric problems.

7. CONCLUSIONS

Multigrid and conjugate gradient type techniques for the acceleration of iterative methods have been discussed. A detailed discussion has been given of incomplete factorizations (ILU and ILLU), which lend themselves especially well for MG or CG acceleration.

A brief review has been given of the theoretical background of classical CG and preconditioning. Classical CG methods are restricted to SPD matrices, but generalization is possible. One such generalized algorithm, called the CGS method, has been presented. Preconditioning of CG type methods has been discussed.

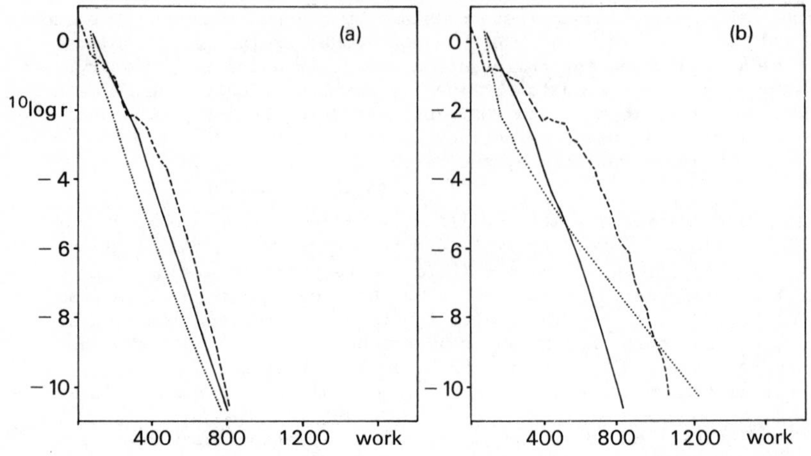

Fig. 6.6 Results for test problems of Stone (above) and Kershaw (below)
(a) : ILLU; (b) : ILU.
—— : MGCG; --- : CG; ... : MG.

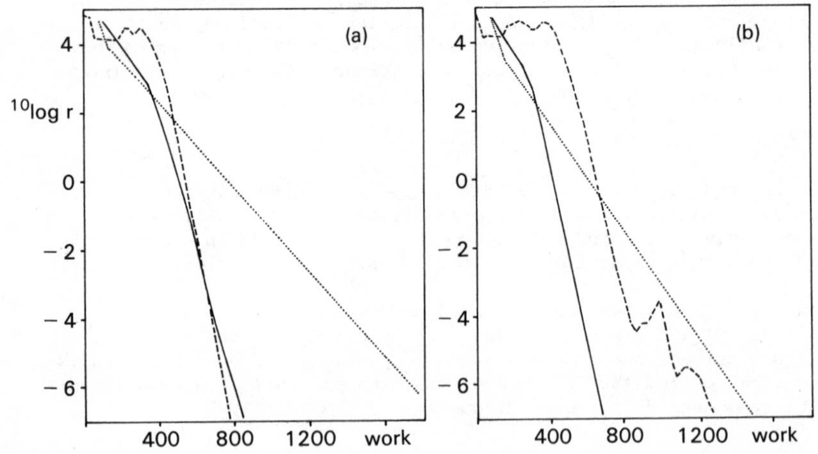

Multigrid methods have been discussed within the framework of acceleration techniques. Various ways of looking at the smoothing factor have been discussed. Prolongation, restriction and coarse grid approximation methods have been reviewed. Two portable, autonomous multigrid codes, MGD1 and MGD5, have been introduced. MG treatment of convection-diffusion problems entails special difficulties, and ways to overcome these have been outlined.

For the general class of problems that we have treated no rate of convergence theory of practical utility is available. Therefore, numerical experiments are necessary for comparison and validation purposes. The choice of a suitable set of test problems has been discussed, and experiments described with several CG and MG methods, including a combination of MG and CG, both self-adjoint and non-self-adjoint problems, and problems with strongly discontinuous coefficients. These problems are of medium size, and roughly speaking, CG is about as efficient as MG, but as the mesh is refined, we would expect CG to lag behind. It should be remembered that CG is easier to program than MG.

Generally speaking, the use of incomplete factorizations leads to more robust and efficient methods than the use of line-relaxations with a zebra pattern, for the test problems considered. With incomplete line factorization (ILLU) one can handle all problems considered with a single code (MGD5 or CGS5), without requiring user-provided adaptations.

ACKNOWLEDGEMENT

The authors are indebted to Mr. W. Lioen for performing the calculations with the MGHZ and MGAZ methods.

REFERENCES

Alcouffe, R.E., Brandt, A., Dendy, Jr., J.E. and Painter, J.W. (1981) The multigrid method for the diffusion equation with strongly discontinuous coefficents. *SIAM J. Sci. Stat. Comp.*, **2**, pp. 430-454.

Asselt, E.J. van (1982) The multigrid method and artificial viscosity. In: Hackbusch and Trottenberg, pp. 313-326.

Axelsson, O. (1977) Solution of linear systems of equations: iterative methods. In: "Sparse Matrix Techniques", V.A. Barker (ed.), Lecture Notes in Math. 572, Springer-Verlag, Berlin, pp. 1-51.

Axelsson, O. (1980) Conjugate gradient type methods for unsymmetric and inconsistent systems of linear equations. *Lin. Algebra and its Applications*, **29**, pp.1-16.

Axelsson, O. (1982) Numerical Integration of Differential Equations and Large Linear Systems. In: J. Hinze (ed.), Proceedings, Bielefeld 1980. Lecture Notes in Mathematics 968, Springer-Verlag, Berlin, pp. 310-322.

Axelsson, O. (1983) A General Incomplete Block-Matrix Factorization Method. Report 8337, Catholic University, Nijmegen, The Netherlands.

Bakhvalov, N.S. (1966) On the convergence of a relaxation method with natural constraints on the elliptic operator. *USSR Comp. Math. Math. Phys.*, **6**, No. 5, pp. 101-135.

Behie, A. and Forsyth, Jr., P. (1983) Comparison of Fast Iterative Methods for Symmetric Systems. *IMA J. of Numer. Anal.*, **3**, pp. 41-63.

Braess, D. (1981) The contraction number of a multigrid method for solving the Poisson equation. *Numer. Math.*, **37**, pp. 387-404.

Braess, D. (1982) The convergence rate of a multigrid method with Gauss-Seidel relaxation for the Poisson equation. In: Hackbusch and Trottenberg, pp. 368-386.

Braess, D. and Hackbusch, W. (1983) A new convergence proof for the multigrid method including the V-cycle. *SIAM J. Num. Anal.*, **20**, pp. 967-975.

Brandt, A. (1977) A Multi-level adaptive solutions to boundary-value problems. *Math. Comp.*, **31**, 333-390.

Brandt, A. (1982) Guide to Multigrid Development. In: Hackbusch and Trottenberg, pp. 220-312.

Concus, P. and Golub, G.H. (1976) A generalized conjugate gradient method for nonsymmetric systems of linear equations. In: R. Glowinski and J.L. Lions (eds.), Proc. of the Second Int. Symposium on Computer Methods in Applied Sciences and Engineering, Paris, 1975. Lecture Notes in Economics and Mathematical Systems, 134, Springer-Verlag, Berlin.

Concus, P., Golub, G.H. and Meurant, G. (1982) Block Preconditioning for the Conjugate Gradient Method. Report LBL-14856, Lawrence Berkeley Laboratory, Un. of California.

Curtiss, A.R. (1981) On a property of some test equations for finite difference or finite element methods. *IMA J. Numer. Anal.*, **1**, pp. 369-375.

Dendy, Jr. J.E. (1982) Black Box Multigrid. *J. Comp. Phys.*, **48**, pp. 366-386.

Fedorenko, R.P. (1964) The speed of convergence of one iterative process. *USSR Comp. Math. Math. Phys.*, **4**, no. 3, pp. 227-235.

Fletcher, R. (1976) Conjugate gradient methods for indefinite systems. In: G.A. Watson (ed.): Numerical analysis. Proceedings, Dundee 1975, Lect. Notes in Math., 506, Springer-Verlag, Berlin, pp. 73-89.

Foerster, H., Stüben, K. and Trottenberg, U. (1981) Non-standard multigrid techniques using checkered relaxation and intermediate grids. In: M. Schultz (ed.): Elliptic Problem Solvers, Academic Press, New York, pp. 285-300.

Foerster, H. and Witsch, K. (1982) Multigrid software for the solution of elliptic problems on rectangular domains: MG00 (Release 1). In: Hackbusch and Trottenberg, pp. 427-461.

Gustafsson, I. (1978) A class of first order factorization methods. *BIT,* **18**, pp. 142-156.

Hackbusch, W. (1978) On the multigrid method applied to difference equations. *Computing,* **20**, pp. 291-306.

Hackbusch, W. (1980) Convergence of multigrid iterations applied to difference equations. *Math. Comp.,* **34**, pp. 425-440.

Hackbusch, W. and Trottenberg, U., eds. (1982) Multigrid Methods. Proceedings, Köln-Porz, 1981. Lecture Notes in Mathematics 960. Springer-Verlag, Berlin.

Hageman, L.A. and Young, D.M. (1981) Applied Iterative Methods. Academic Press, New York.

Hemker, P.W. (1982A) A note on defect correction processes with an approximate inverse of deficient rank. *J. Comp. Appl. Math.,* **8**, pp. 137-139.

Hemker, P.W. (1982) Mixed Defect Correction Iteration for the Accurate Solution of the Convection Diffusion Equation. In: Hackbusch and Trottenberg, pp. 485-501.

Hemker, P.W., Kettler, R., Wesseling, P. and de Zeeuw, P.M. (1983) Multigrid methods: development of fast solvers. *Appl. Math. and Comp.,* **13**, pp. 311-326.

Hemker, P.W., Wesseling, P. and de Zeeuw, P.M. (1983) A portable vector-code for autonomous multigrid modules. Report NW 154/83, Mathematical Centre, Kruislaan 413, 1098 SJ Amsterdam, The Netherlands.

Hemker, P.W. and de Zeeuw, P.M. (1984) Some implementations of multigrid linear systems solvers. In this volume.

Kershaw, D.S. (1978) The incomplete Choleski-conjugate gradient method for the iterative solution of systems of linear equations. *J. Comp. Phys.,* **26**, pp. 43-65.

Kettler, R. (1980) A study of the applicability of the multiple grid method in reservoir simulation. Part II. Master's thesis, Delft University of Technology, Nov. 1980.

Kettler, R. (1982) Analysis and Comparison of Relaxation Schemes in Robust Multigrid and Preconditioned Conjugate Gradient Methods. In: Hackbusch and Trottenberg, pp. 502-534.

Kettler, R. and Meijerink, J.A. (1981) A multigrid method and a combined multigrid-conjugate gradient method for elliptic problems with strongly discontinuous coefficients in general domains. Publication 604, Shell Research B.V., Kon. Shell Expl. and Prod. Lab., Rijswijk, The Netherlands.

Maitre, J.-F. and Musy, F. (1983) Méthodes multigrilles: opérateur associé et estimations du facteur de convergence; le cas du V-Cycle. C.R. Acad. Sc. Paris, 296, Série I, pp. 521-524.

Manteuffel, T.A. (1977) The Tchebychev Iteration for Nonsymmetric Linear Systems. *Numer. Math.*, **28**, pp. 307-327.

Manteuffel, T.A. (1978) Adaptive Procedure for Estimating Parameters for the Nonsymmetric Tchebychev. *Numer. Math.*, **31**, pp. 183-208.

McCarthy, G.J. (1983) Investigations into the Multigrid Code MGD1. Report AERE R 10889, Harwell, U.K.

McCormick, S.F. (1982) An algebraic interpretation of multigrid methods. *SIAM J. Numer. Anal.*, **19**, pp. 548-560.

McCormick, S.F. (1983) Multigrid Methods for Variational Problems: the V-cycle. *Math. and Comp. in Sim.*, **25**, pp. 63-65.

Meurant, G. (1983) Vector Preconditionings for the Conjugate Gradient Method. To be submitted to BIT. Private Communication.

Meijerink, J.A. and van der Vorst, H.A. (1977) An iterative solution method for linear systems of which the coefficient matrix is a symmetric M-matrix. *Math. Comp.*, **31**, pp. 148-162.

Meijerink, J.A. (1983) Iterative Methods for the Solution of Linear Equations based on Incomplete Factorization of the Matrix. Publication 643, Shell Research B.V., Kon. Shell Expl. and Prod. Lab., Rijswijk, The Netherlands, July 1983.

Meijerink, J.A. and van der Vorst, H.A. (1981) Guidelines for the usage of incomplete decompositions in solving sets of linear equations as they occur in practical problems. *J. Comp. Phys.*, **44**, pp. 134-155.

Musy, F. (1982) Sur les méthodes multigrilles: formalisation algébrique et démonstration de convergence. C.R. Acad. Sc. Paris 295, Serie I, pp. 471-474.

Nicolaides, R.A. (1979) On some theoretical and practical aspects of multigrid methods. *Math. Comp.*, **33**, pp. 933-952.

Nowak, Z. and Wesseling, P. (1983) Multigrid acceleration of an iterative method with application to transonic potential flow. In: INRIA, Proceedings Sixth International Conference on Computing Methods in Applied Sciences and Engineering, Versailles, France, Dec. 1983.

Stone, H.L. (1968) Iterative solution of implicit approximations of multidimensional partial differential equations, *SIAM J. Numer. Anal.*, **5**, pp. 530-558.

Stüben, K. and Trottenberg, U. (1982) Multigrid Methods: Fundamental Algorithms, Model Problem Analysis and Applications. In: Hackbusch and Trottenberg, pp. 1-176.

Underwood, R.R. (1976) An approximate factorization procedure based on the block Cholesky decomposition and its use with the conjugate gradient method. Report NEDO-11386, General Electric Co., Nuclear Energy Div., San Jose, CA.

Vinsome, P.K.W. (1976) ORTHOMIN, an iterative method for solving sparse sets of simultaneous linear equations. Society of Petroleum Engineers, paper SPE 5729.

Van der Vorst, H.A. (1981) Iterative Solution Methods for Certain Sparse Linear Systems with a Non-Symmetric Matrix Arising from PDE-Problems. *J. Comp. Phys.*, **44**, pp. 1-19.

Van der Wees, A.J., van de Vooren, J. and Meelker, J.H. (1983) Robust calculation of 3D transonic potential flow based on the nonlinear FAS multigrid method and incomplete LU-decomposition. AIAA paper 83-1950.

Wesseling, P. (1980) The rate of convergence of a multiple grid method. In: G.A. Watson (Ed.), Numerical Analysis. Proceedings, Dundee 1979. Lect. Notes in Math. 773, Springer-Verlag, Berlin, pp. 164-184.

Wesseling, P. (1982A) Theoretical and practical aspects of a multigrid method. *SIAM J. Sci. Stat. Comput.*, **4**, pp. 387-407.

Wesseling, P. (1982B) A robust and efficient multigrid method. In: Hackbusch and Trottenberg, pp. 614-630.

Wesseling, P. and Sonneveld, P. (1980) Numerical experiments with a multiple grid and a preconditioned Lanczos type method. In: R. Rautmann (ed.), Approximation methods for Navier-Stokes problems. Proceedings, Paderborn 1979. Lecture Notes in Math. 771, Springer-Verlag, pp. 543-562.

Widlund, O. (1978) A Lanczos method for a class of nonsymmetric systems of linear equations. *SIAM J. Numer. Anal.*, **15**, pp. 801-812.

Young, D.M. (1971) Iterative solution of large linear systems. Academic Press, New York.

Zeeuw, P.M. de and van Asselt, E.J. (1985) The convergence rate of multi-level algorithms applied to the convection-diffusion equation. To appear, *SIAM J. Sci. Stat. Comp.*

EFFICIENT SOLUTION OF FINITE DIFFERENCE AND FINITE ELEMENT EQUATIONS

J. Ruge and K. Stüben

(Gesellschaft für Mathematik und Datenverarbeitung, W. Germany)

INTRODUCTION

Algebraic multigrid (AMG) methods are designed for the automatic solution of certain sparse linear systems

$$A\,u = f \text{ with } A = (a_{ij})_{N,N},\ u = (u_1,\ldots,u_N)^T,\ f = (f_1,\ldots,f_N)^T \qquad (1.1)$$

using multigrid principles. One goal behind the development of AMG is to obtain black box solvers for elliptic partial differential equations, discretized by either finite differences or finite elements. The characteristics of AMG solvers are, however, quite different from those of "geometric" multigrid solvers. Geometric multigrid solvers (even if they are called black box solvers) assume some underlying geometrical structure: Operators are defined on (mostly regular uniform) grids, a fixed predefined type of grid coarsening is assumed, etc. On the contrary, AMG in its most general form makes no use of geometrical or physical properties which may be inherent to the problem at hand: only the matrix A is known to AMG. In the context of AMG, "coarser grids" may be regarded as subsets of unknowns, and "coarse-grid-correction equations" are merely considerably smaller systems of linear equations which play the same role as the coarse-grid-correction equations in geometric multigrid. These equations (and corresponding inter-grid transfer operators) are constructed fully automatically, based solely on algebraic information contained in the matrix A, mainly in terms of the strength of connection between unknowns.

More generally, AMG characterizes the attempt to apply multigrid ideas to linear systems (1.1) independent of their origins. As a consequence, corresponding solvers can - in contrast to geometric multigrid solvers - easily be incorporated into existing non-multigrid software which requires efficient solution of linear systems. Up to now, however, there is no concept of how to solve arbitrary systems (1.1) efficiently by AMG. Extensive experience, however, has been gained under the assumptions

$$A \text{ symmetric and } a_{ij} \leqslant 0 \ (j \neq i),\quad \sum_{j=1}^{N} a_{ij} \geqslant 0. \qquad (1.2)$$

Assuming this, the effect of smoothing (by Gauss-Seidel point-relaxation) is understood rather well and the strength of connection between unknowns is reflected by the size of the corresponding matrix coefficients. If not stated otherwise in this paper, we will assume (1.2) to be satisfied. We point out, however, that the AMG code described in this paper makes no use of this assumption, and (1.2) does not have to be satisfied strictly for the code to work efficiently (see, e.g., Sections 5.3 and 5.4). Some remarks on the limitations of the code will be given later.

An AMG code (AMG01) has been developed in Stüben (1983), the main goal of which was to demonstrate the robustness of AMG compared to geometric multigrid. Details on the program design, in particular on the general and flexible data structure, are contained in Stüben, Trottenberg and Witsch (1983). According to its test character, AMG01 was far from being optimized. Compared to fast geometric multigrid solvers, it was not really competitive in terms of efficiency. The main reasons for this were complexity and speed of convergence per cycle, which was - although remarkably good - not satisfying. Also, the preparation phase needed to compute all necessary AMG components was quite expensive, not so much in terms of floating point operations but in terms of CP-time.

In this paper we want to report on the further development and improvements of AMG01 which were directed towards obtaining higher efficiency. Major changes in the previous algorithm were performed in the construction of the coarser grids and the definition of interpolation. As a result, the new algorithm has become considerably more efficient than the previous one. The final work on the algorithm, as well as the numerical investigations reported in this paper, were carried out under project MATAN at the Gesellschaft für Mathematik und Datenverarbeitung (GMD) in St. Augustin, FRG, through its Institut für Methodische Grundlagen (F1).

The concept of AMG was first introduced in Brandt, McCormick and Ruge (1982) and more recently in Brandt, McCormick and Ruge (1984). Several theoretical results on smoothing and two-grid convergence are contained in Brandt (1981). Although the theorems given there are not yet tailored to the AMG algorithms used in practice, they greatly influenced the particular choices of the multigrid components used in AMG01. In Chapters 2 and 3 we summarize the essential ideas and experiences which led to the development of the recent AMG01 code. In particular, Chapter 3 recalls the most important algorithmic components and the essential features of the old algorithm described in Stüben (1983). Certain improvements are discussed and a corresponding algorithm is described in Chapter 4. In addition, a general discussion on the efficiency of AMG01 is included.

The performance of AMG01 and its efficiency are demonstrated in Chapter 5. This chapter also includes examples which do not satisfy (1.2), for instance, some non-symmetric problems. Although the range of applications of the code is more general, in this paper, we focus on finite difference (Sections 5.1 through 5.3) and finite element (Section 5.4) discretizations of typical (scalar) elliptic PDE's. In particular, we make a comparison with some generally available geometric multigrid solvers for finite difference equations in Section 5.2, namely with BOXMG (Dendy (1982)), MGD1 (Hemker, Kettler, Wesseling and de Zeeuw (1983)) and MG00 (Foerster and Witsch (1982)). This comparison will show that the efficiency of AMG01, measured in terms of convergence speed and CP-time per cycle, is quite comparable to that of robust geometric solvers. For the different situations which may occur within the corresponding range of problems (isotropic problems, anisotropic problems, problems with strongly oscillating or even discontinuous coefficients, etc.), all AMG components, even the grids, are automatically constructed by the program in an adaptive and suitable way which gives fast convergence. Clearly, from a practical point of view, an advantage of AMG01 over the above solvers is that it is a "pitch black" box, which can be applied directly to more general situations as, for example, to arbitrary domains, three-dimensional problems, non-uniform grids, and even to situations for which geometric multigrid methods are hardly applicable at all, for example, to randomly structured grids.

Geometrically more complex structures are treated in Section 5.4 by means of some finite element discretizations. In particular, we show that the performance of AMG01 does not deteriorate in case of "stretched" or locally refined triangulations. Also, we give examples of discretizations for which the application of geometric multigrid would be quite troublesome.

The general applicability and the flexibility of AMG01 is, of course, not free of cost. One has to pay for it by setup time which is - although cheaper than that of the old algorithm by around 80% - still quite high: Typically, the setup costs roughly between 5 and 9 corresponding cycles. Taking into account, however, the ease of use of the code and the savings of human effort in designing multigrid methods for complex situations, this extra cost seems to be reasonable. Furthermore, there are still many possibilities to improve and optimize AMG01 even further, in particular with respect to the setup time. Some remarks on the limitations and the further development of AMG01 are contained in Chapter 6.

Finally, we want to mention that - from a general point of view - the AMG concept has some similarities to the so-called "method of aggregation and disaggregation" (see e.g., Chatelin and Miranker (1982)). Two crucial differences between the two concepts are, first, AMG is to yield fully automatic algorithms and, second, AMG is to give typical multigrid efficiency.

2. THE GENERAL IDEA OF AMG01; NOTATION

In this chapter we first give the most important notation, in particular Section 2.1 introduces a (formal) grid terminology. Section 2.2 contains the main ideas and briefly points out the essential differences between AMG01 and geometric multigrid methods. Some further notation is given in Section 2.3.

2.1 Grid terminology

Given a problem (1.1), AMG01 will automatically construct a sequence of increasingly smaller systems of (residual) equations which play the same role as coarse-grid-correction equations in usual multigrid methods:

$$\sum_{j=1}^{N_m} a_{ij}^{(m)} u_j^{(m)} = f_i^{(m)} \qquad (i = 1,\dots,N_m) \tag{2.1}$$

for m = 1,...,M with m = 1 being just the given system (1.1). The M-th system of equations will be very small (usually one or two equations only) and can easily be solved exactly. Before we describe how this construction is actually done, we want to introduce a formal grid terminology in the following.

The way the above systems will be constructed, each of the "coarser-level" unknowns $u_j^{(m+1)}$ $(m < M)$ will represent a certain variable $u_{i(j)}^{(m)}$ on the next finer level (the corresponding fine-level variable), i.e., in the multigrid correction process $u_j^{(m+1)}$ will be used to correct $u_{i(j)}^{(m)}$. According to this relationship, the set of unknowns on each level m $(m < M)$ can be split into subsets: a set of unknowns which are also represented on the next coarser level m+1 (C-variables of level m), and a set of unknowns which are not represented on level m+1 (F-variables of level m).

It is now very convenient to renumber the unknowns (and all related quantities) such that coarser-level variables have the same index as their corresponding finer-level variables. According to this, we re-write (2.1) in the form

$$\sum_{j\in\Omega^m} a_{ij}^{(m)} u_j^{(m)} = f_i^{(m)} \qquad (i \in \Omega^m) \tag{2.2}$$

where $\Omega^1 := \{1,...,N\}$ and $\Omega^{m+1} \subset \Omega^m$ is recursively defined by

$$\Omega^{m+1} := \{i \in \Omega^m : u_i^{(m)} \text{is C-variable on level m}\} \qquad (m = 1,...,M-1).$$

In the following, we will usually interpret $u^{(m)}$ and $f^{(m)}$ as grid functions on the (fictitious) grid Ω^m with grid points $i \in \Omega^m$. The equation (2.2) for a fixed i is regarded to correspond to the unknown $u_i^{(m)}$ or, alternatively, to the point i. The full system (2.2) will be interpreted as an operator equation on the space $\mathbb{G}(\Omega^m)$ of grid functions on Ω^m:

$$A^{(m)}u^{(m)} = f^{(m)} \text{ with } A^{(m)} : \mathbb{G}(\Omega^m) \to \mathbb{G}(\Omega^m).$$

More generally, we may think in terms of directed graphs instead of grids. Connections between grid points of Ω^m are given by the coefficients $a_{ij}^{(m)}$: We define a point $i \in \Omega^m$ to be connected to a point $j \in \Omega^m$ if $a_{ij}^{(m)} \neq 0$.

Thinking in terms of grids, grid functions, etc. is convenient, in particular, as the formal relation of AMG01 to geometric multigrid methods then becomes quite obvious. Occasionally, however, we will also use matrix terminology if this is more convenient. No confusion should arise.

In analogy to the definition of C- and F-variables, we introduce the notation of C- and F-points:

$$C^m := \Omega^m \cap \Omega^{m+1}: \text{ set of C-points (of grid } \Omega^m),$$

$$F^m := \Omega^m - \Omega^{m+1}: \text{ set of F-points (of grid } \Omega^m).$$

Although C^m and Ω^{m+1} formally contain the same points, it is reasonable to distinguish between these sets: Connections of points in C^m are always understood with respect to $a_{ij}^{(m)}$ while connections of points in Ω^{m+1} are understood with respect to $a_{ij}^{(m+1)}$.

Using the terminology introduced above, the following multigrid components (cf. Stüben and Trottenberg (1981), Section 2.3) have to be constructed by the program recursively for $m = 1,\ldots,M-1$ in order to set up a multigrid procedure:

(C1) a coarser grid $\Omega^{m+1} \subset \Omega^m$;

(C2) a smoothing process for the equations on Ω^m and the numbers ν_1 and ν_2 of smoothing steps to be performed before and after the coarse-grid corrections, respectively;

(C3) a coarse-to-fine transfer operator (interpolation) $I_{m+1}^m : \mathbb{G}(\Omega^{m+1}) \to \mathbb{G}(\Omega^m)$;

(C4) a fine-to-coarse transfer operator (restriction) $I_m^{m+1} : \mathbb{G}(\Omega^m) \to \mathbb{G}(\Omega^{m+1})$;

(C5) a coarse-grid operator $A^{(m+1)} : \mathbb{G}(\Omega^{m+1}) \to \mathbb{G}(\Omega^{m+1})$.

Once these components are constructed by the program in a separate preparation phase, a standard multigrid cycle can be defined as, e.g., described in Stüben and Trottenberg (1981), Chapter 4. We will usually use the simple V-cycle.

2.2 General idea

The assumptions (1.2) are satisfied for many (scalar) second order elliptic difference equations. In order to motivate the main ideas of AMG01, let us for the moment think in terms of such elliptic problems, defined on some regular rectangular grid.

There are two major differences between AMG01 and geometric multigrid solvers. A first difference lies in the different kind of coarser grids and the particular smoothing processes used by these methods: In geometric multigrid, the way in which coarser grids are derived from finer ones is (usually) pre-defined independently of the smoothing process (and of the problem). In most of the available codes, coarsening is simply done by standard coarsening (i.e., by doubling the mesh sizes). The efficiency of a smoothing operator for a given problem, however, depends on the particular type of coarsening chosen: Smoothing means that fast reduction occurs for all those error frequencies which are not "represented" on the next coarser grid. Thus, if coarser grids are pre-defined, a proper choice of the smoothing operator becomes important to obtain efficient smoothing. For instance, assuming standard coarsening, isotropic problems can very efficiently be smoothed by simple pointwise relaxation; anisotropic problems, however, require more complicated relaxation processes as, e.g., line-relaxation or even alternating line-relaxation (Stüben and Trottenberg (1981)).

In algebraic multigrid, the philosophy is the reverse of that of geometric multigrid: AMG01 fixes the smoothing process to simple Gauss-Seidel point-relaxation. According to this choice, the coarser grids are constructed such that smoothing by point-relaxation becomes efficient. For this we have to note that, roughly speaking, point-relaxation smoothes in the direction of strong dependencies. In a matrix of type (1.2), however, the strength of dependencies (connections) between unknowns is reflected by the size of the corresponding matrix coefficients. Thus, by merely looking at the size of these coefficients, it is possible to construct coarser grids automatically in a proper way. This makes AMG01 much more flexible than geometric multigrid solvers, in particular, in treating very general grid structures such as those discussed in Section 5.4.

The second important difference is the way interpolation is defined in AMG01. Clearly, there is no way to define interpolation based on the geometric locations of grid points in a natural way. Instead, interpolation to a fine-grid point is - in general - defined by a weighted average of certain connected coarse-grid values. A proper choice of these coarse-grid points and the corresponding weights has to be closely related to relaxation: Roughly speaking, interpolation has to be "good" for smooth grid functions, i.e., for those grid functions which cannot be reduced efficiently by relaxation. For details on the interpolation, see Sections 3.2 and 4.1.

The definition of the remaining multigrid components, namely the restriction and the coarse-grid operators, is simple in AMG01: For restriction we use just the transpose of interpolation and the coarse-grid operators are defined to be of "Galerkin-type" (see Section 3.1). These choices are quite often used also in geometric multigrid (cf., e.g., Dendy (1982) and Hemker, Kettler, Wesseling and de Zeeuw (1983)).

The advantage of these choices is that the coarse-grid-correction processes satisfy a certain variational condition (see Section 3.2).

2.3 Further notation

We generally use the letters u, v and e to denote approximations, corrections and errors, respectively. Here, the error e is defined by e:= U - u, where U denotes the exact solution corresponding to u. In AMG01, errors will be smoothed using pointwise relaxation in a certain ordering of the grid points: we will use C/F-relaxation, one step of which is defined by the application of one partial Gauss-Seidel step first to those equations which correspond to C-points (C-relaxation) followed by a corresponding partial step over the F-points (F-relaxation).

For any set S, we will denote the number of elements in that set by $|S|$. Two particular sets needed in this paper are given in the following. For a point $i \in \Omega^m$, we define its neighbourhood to be the set of points $j \in \Omega^m$ to which i is connected:

$$N_i^{(m)} := \{j \in \Omega^{(m)} : j \neq i,\ a_{ij}^{(m)} \neq 0\} \qquad (i \in \Omega^m).$$

In AMG01, strong connections will play an essential role. The set of points j to which a point i is strongly connected will generally be denoted by $\mathcal{S}_i$:

$$\mathcal{S}_i^{(m)} := \{j \in N_i^{(m)} : \text{i is strongly connected to j}\} \quad (i \in \Omega^m).$$

Here "strongly connected" means, roughly, that $|a_{ij}^{(m)}|$ is comparable in size to the largest of the $|a_{ik}^{(m)}|$ $(k \in N_i^m)$. For most of our discussion, a general understanding of this term will suffice. A precise definition is given in Section 4.2, where the AMG01 algorithm is described in detail.

The efficiency of an algebraic multigrid method will be measured by its speed of convergence and by its complexity. More precisely, we define its Ω-complexity and its A-complexity by

$$c_\Omega := \left(\sum_{m=1}^{M} N_\Omega^{(m)}\right)/N_\Omega^{(1)}, \qquad c_A := \left(\sum_{m=1}^{M} N_A^{(m)}\right)/N_A^{(1)}$$

respectively, with $N_\Omega^{(m)}$ and $N_A^{(m)}$ being the number of points in Ω^m and the number of (non-zero) coefficients in $A^{(m)}$, respectively. Furthermore, we will use the notation $N_F^{(m)}$ and $N_C^{(m)}$ to denote the number of F- and C-points on Ω^m. These quantities directly affect storage and numerical work (see Section 4.3).

Finally, on $\mathbb{G}(\Omega^m)$ we use the ℓ_2-inner product (,) and the energy inner product

$$(u^{(m)}, v^{(m)})_E := (A^{(m)}u^{(m)}, v^{(m)})$$

along with their corresponding norms.

3. SOME REMARKS ON THE DEVELOPMENT OF AMG01

In the previous section, we have outlined the idea of AMG01 without specifying the choice of all the multigrid components (C1) - (C5). The following Chapters 3 and 4 deal with the question of how to select these components. In particular, proper definitions of the coarse grid point choice and the interpolation are quite crucial for obtaining good multigrid efficiency.

In Section 3.1 we will specify the components such that the resulting method yields a direct solver. This solver is described for reasons of demonstration only; it is useless in practice as it is much too costly both with respect to complexity (see Section 2.3) and numerical work. More "realistic" (iterative) AMG methods such as those discussed below can, however, be regarded as being approximations to this solver. In Section 3.2, we give a short review on older ideas in the development of AMG01 (cf. Brandt, McCormick and Ruge (1982) and Stüben (1983)). The corresponding older codes were - although quite robust - not yet satisfactory with respect to their efficiency (if compared to geometric multigrid methods). The remarks given in Section 3.2 will be used for motivating the most recent version of AMG01 which is described in detail in Section 4.

3.1 An unrealistic direct AMG solver

An extreme choice of the multigrid components (C1) - (C5) leads to a direct solver (i.e., corresponding V-cycles converge in only one iteration step) if A is non-singular and all $a_{ii} \neq 0$. For this "ideal" AMG algorithm, the components (C1) - (C5) for m = 1,2,... are defined recursively as follows:

(C1) Select any disjoint subsets $F^m \subset \Omega^m$, $C^m \subset \Omega^m$ with $\Omega^m = F^m \cup C^m$ such that there are no F-to-F connections, i.e.

$$\text{for all } i \in F^m: \quad \mathbb{N}_i^{(m)} \subset C^m. \tag{3.1}$$

Then the coarser grid is defined by $\Omega^{m+1} := C^m$. Note that F^m can always be chosen to contain at least one point. Clearly, the goal is to choose F^m as large as possible. If (3.1) holds with

$F^m = \Omega^m$ and $C^m = \emptyset$, the algorithm is terminated and M is set to m.

(C2) One smoothing step on Ω^m is defined by the application of one F-point relaxation step (see Section 2.3). The number of smoothing steps to be performed on each level is given by $\nu_1 = 1$, $\nu_2 = 0$.

(C3) Given a coarser-grid function $v^{(m+1)} \in \mathbb{G}(\Omega^{m+1})$, the interpolated function $v^{(m)} := I^m_{m+1} v^{(m+1)}$ is defined by

$$v_i^{(m)} := \begin{cases} v_i^{(m+1)} & (i \in C^m) \\ -\left(\sum_{j \in C^m} a_{ij}^{(m)} v_j^{(m)}\right)/a_{ii}^{(m)} & (i \in F^m). \end{cases} \tag{3.2}$$

(C4) The restriction operator I^{m+1}_m is defined to be the transpose of the interpolation operator

$$I^{m+1}_m := (I^m_{m+1})^T. \tag{3.3}$$

(C5) The coarse-grid operator on $\mathbb{G}(\Omega^{m+1})$ is defined to be of "Galerkin-type", i.e.

$$A^{(m+1)} := I^{m+1}_m A^{(m)} I^m_{m+1}. \tag{3.4}$$

Note, that the above recursive process of constructing the multigrid components is well defined as long as each $A^{(m)}$ is non-singular and $a_{ii}^{(m)} \neq 0$ $(i \in \Omega^m)$. Then, the application of just one V-cycle (starting with any first approximation) yields the exact solution of the given system of equations (cf. Brandt, McCormick and Ruge (1982) and Stüben (1983)). We want to recall the two properties from which this statement follows immediately by a straightforward recursive argument:

(1) On each of the grids Ω^m $(m < M)$, the range of the corresponding smoothing operator in (C2) is equal to the range of the interpolation operator I^m_{m+1}.

(2) On each of the grids Ω^m $(m < M)$, the corresponding (m,m+1)-two-grid correction operator (cf. Stüben and Trottenberg (1981)), Chapter 2)

$$K^{m+1}_m := I_m - I^m_{m+1}[A^{(m+1)}]^{-1} I^{m+1}_m A^{(m)} \quad (I_m \text{ denotes the identity on } \Omega^m) \tag{3.5}$$

is a projection onto the null space of $I_m^{m+1} A^{(m)}$; in particular, K_m^{m+1} eliminates grid functions which are in the range of interpolation.

The latter of these two properties is satisfied due to the choice of the coarse-grid operators (3.4). Concerning the first property, we remark that after the performance of one F-relaxation step on Ω^m, residuals are zero at points $i \in F^m$ (because of (3.1)). Thus the corresponding error satisfies

$$e_i^{(m)} = -\left(\sum_{j \in C^m} a_{ij}^{(m)} e_j^{(m)} \right) / a_{ii}^{(m)} \qquad (i \in F^m). \tag{3.6}$$

As interpolation (3.2) at points $i \in F^m$ is defined by just the same formula, the first property is obviously satisfied, too.

3.2 More realistic iterative AMG solvers

As already mentioned before, the "ideal" multigrid method defined above is ideal only with respect to its "speed of convergence". For use in practice the method is of no significance. Its essential deficiency is a tremendous complexity. Due to the very restrictive definition of the coarser grids in (3.1), coarsening proceeds extremely slowly. As a consequence, the "radius" of interpolation (3.2) and by this the size of the coarse-grid operators grows quickly and the corresponding matrices will be filled very soon.

For the development of more realistic (iterative) algebraic solvers we now assume A to be symmetric and positive definite. The restriction andcoarse-grid operators will always be defined as in (C4) and (C5), respectively, of the previous section. This is because then - independent of the actual definition of the coarser grids and the interpolation (as long as it has full rank) - all coarse-grid operators $A^{(m)}$ are also symmetric and positive definite, and the corresponding two-grid correction operators K_m^{m+1} are orthogonal projectors with respect to the energy inner product (see Section 2.3). As a consequence, two-grid corrections are characterized by a minimization principle: the energy norm of the error after such a correction step is minimum with respect to corrections by grid functions in the range of interpolation I_{m+1}^m.

One goal in the following will be to reduce complexity considerably. For this we will change the coarse grid choice and the interpolation (components (C1) and (C3), respectively). The resulting methods will be iterative methods, the convergence factors of which should be considerably smaller than one and - at least nearly - independent of the size of the problem. To obtain this, we first replace the partial Gauss-Seidel steps which were used in the previous section by complete Gauss-Seidel steps. More precisely, we will use C/F-relaxation (see Section 2.3) for smoothing, usually with $\nu_1 = \nu_2 = 1$. In order to have good smoothing (cf. Brandt (1983) and Brandt, McCormick and Ruge (1984)), we will assume - in addition to symmetry and positive definiteness of A - also the inequalities in condition (1.2) to hold.

With the algorithmic components (C2), (C4) and (C5) fixed as stated above, the remaining design choices are interpolation and coarse grid choice. In order to obtain good coarse-grid corrections, these choices are actually coupled and, furthermore, must be closely related to "smoothness". Generally, our goal will be to define interpolation in such a way that its range is "close" to the range of the smoothing operator (cf. Remark 3.2 below).

Several multigrid procedures have been investigated so far. The first one (Brandt, McCormick and Ruge (1982)) was based on a (two-grid) theory developed in Brandt (1983), also see Brandt, McCormick and Ruge (1984) (cf. Remark 3.1 below). According to this theory, condition (3.1) on the coarse grid choice was weakened by requiring that, for each point $i \in F^m$, only a fixed percentage of its total connections are contained in C^m. More precisely, F^m and C^m are chosen such that with some fixed $0.5 \leq \eta \leq 1$:

$$\text{for all } i \in F^m: \quad \sum_{j \in C^m} |a_{ij}^{(m)}| \geq \eta \sum_{j \in N_i^{(m)}} |a_{ij}^{(m)}|. \tag{3.7}$$

Along with this change of the coarse grid choice, the interpolation formula (3.2) was replaced by

$$v_i^{(m)} := \begin{cases} v_i^{(m+1)} & (i \in C^m) \\ -\left(\sum_{j \in C^m} a_{ij}^{(m)} v_j^{(m)}\right) / \left(\sum_{j \in F^m} a_{ij}^{(m)}\right) & (i \in F^m). \end{cases} \tag{3.8}$$

<u>Remark 3.1</u>: In Brandt (1983) it is shown that, for any $0 < \eta \leq 1$, corresponding two-grid methods (i.e., M = 2) converge with a convergence factor (with respect to the energy norm) which is smaller than and bounded away from 1 independently of the size of the problem (but, of course, depending on η). The choice $\eta \geq 0.5$ guarantees that assumption (1.2) is satisfied for all coarse-grid operators $A^{(m)}$ and that the above statement holds also for all intermediate (m,m+1)-two-grid methods.

Note that $\eta = 1$ yields just the previous definitions (3.1) and (3.2). We do not want to go into the details of how to construct sets C^m and F^m which satisfy (3.7). We want to point out, however, that - in order to obtain fast grid reduction (small Ω-complexity) - a corresponding algorithm has to be arranged in such a way that, whenever a point i is decided to be an F-point, primarily points $j \in S_i^{(m)}$ (cf. Section 2.3) become C-points. This is to make (3.7) satisfied with an amount of C-points which is as small as possible. A heuristic motivation for the definition of interpolation (3.8) is given in the following remark.

Remark 3.2: As already mentioned above, our goal is to define interpolation I^m_{m+1} in such a way that smooth error functions $e^{(m)}$ can be approximated well, i.e., if $e_i^{(m+1)} := e_i^{(m)}$ $(i \in \Omega^{m+1})$ then, roughly, $I^m_{m+1}\, e^{(m+1)} \approx e^{(m)}$. (Here, by smooth functions we mean functions which are "dominant" in the range of relaxation or, in other words, functions which cannot efficiently be reduced by relaxation.) Such functions $e^{(m)}$ are characterized by the inequality

$$([D^{(m)}]^{-1}\, r^{(m)},\, e^{(m)})_E << (e^{(m)},\, e^{(m)})_E \tag{3.9}$$

where $D^{(m)} := \text{diag}(A^{(m)})$ and $r^{(m)} := A^{(m)}\, e^{(m)}$ denotes the residual. This holds for Gauss-Seidel relaxation with any ordering of the grid points and for any matrix of type (1.2) after several relaxation sweeps (see Brandt (1983) and Brandt, McCormick and Ruge (1984)). From (3.9) one can deduce that (at least on the average) for all $i \in \Omega^m$

$$\text{(a)}\ |r_i^{(m)}| << |e_i^{(m)} a_{ii}^{(m)}| \quad \text{(b)}\ |e_i^{(m)} - e_j^{(m)}| << |e_i^{(m)}| \text{ if } |a_{ij}^{(m)}| \text{ is large.} \tag{3.10}$$

The latter inequality is also derived in Brandt (1983) and Brandt, McCormick and Ruge (1984). It states in a more precise way what we mean by "point-relaxation smoothes in the direction of strong connections". Concerning the first inequality, note that in case of C/F-relaxation residuals are particularly small at points $i \in F^m$. Due to the above properties of a smooth function, we have approximately

$$(r_i^{(m)} =)\ a_{ii}^{(m)}\, e_i^{(m)} + \sum_{\substack{j \in F^m \\ j \neq i}} a_{ij}^{(m)} e_j^{(m)} + \sum_{j \in C^m} a_{ij}^{(m)} e_j^{(m)} \approx 0 \quad (i \in F^m). \tag{3.11}$$

In order to get an approximation of how $e_i^{(m)}$ and depends on $e_j^{(m)}$ $(j \in C^m)$, we may now replace $e_j^{(m)}$ in the second part of the sum by $e_i^{(m)}$: either $a_{ij}^{(m)}$ is large in which case we can - due to the second inequality in (3.10) - expect that $e_j^{(m)} \approx e_i^{(m)}$, or $a_{ij}^{(m)}$ is a small in which case the above replacement does not change the left hand side of (3.12) essentially. This gives an approximate description of smooth error functions:

$$e_i^{(m)} \approx - \Big(\sum_{j \in C^m} a_{ij}^{(m)} e_j^{(m)} \Big) / \Big(\sum_{j \in F^m} a_{ij}^{(m)} \Big) \qquad (i \in F^m). \qquad (3.12)$$

From the definition (3.8) of interpolation we see that the goal stated at the beginning of this remark is reached.

The AMG code corresponding to the above components (with η around 0.5) turned out to be quite satisfactory as far as convergence was concerned; complexity, however, was still a problem. The main reason for this is that the "radius" of interpolation (3.8) is usually still quite large on coarser grids which leads to an unacceptable filling of the coarser-grid operators.

In order to obtain a better complexity, another algorithm was investigated (Stüden (1983)) in which condition (3.7) has been weakened further: Instead of a fixed percentage, only a fixed number p of (strong) connections of every point $i \in F^m$ was required to be contained in C^m. Accordingly, only these points were used in interpolation (p-point interpolation). These definitions of the coarse grid choice and the interpolation are also discussed in Brandt (1983). p-point interpolation was motivated by geometrical considerations; It was reasoned that, if the origin of the matrix A was a d-dimensional problem, then a good interpolation could be obtained using at most d + 1 points. In fact, for most two-dimensional problems, taking p = 3 gave good convergence factors over a wide variety of problems and a much better complexity than the previous algorithm (cf. Stüben (1983)).

Although the results with the last algorithm (we will refer to it in the following as p-point algorithm) were quite impressive, its efficiency was still not satisfactory: Compared to geometric multigrid methods, complexity was still too high and speed of convergence (of the V-cycle) depended sometimes on the number M of grids used. A detailed examination of these two deficiencies shows that they have essentially the same origin. For a brief discussion let us assume a geometrical environment and the operator given to be derived from some isotropic problem (e.g., Poisson's equation on a uniform grid; for problems which exhibit a strong directional dependence, the p-point algorithm behaves considerably better):

- The unsatisfactory complexity of the p-point algorithm has a different reason than that of the other algorithm: The algorithm has no knowledge of the geometric locations of grid points. Instead, decisions on which p points are actually to be used for the coarser grids (and by this also for interpolation) are made by some optimization criterion the goal of which is to minimize the number of coarse-grid points. It is essentially by this optimization that interpolation tends to become "one-sided". Roughly speaking, the algorithm forces a coarse grid point choice in which many F-points have actually p strong connections to C-points, and tends to pick F-points in groups surrounded by C-points (for an example, see Fig. 3.1a in Stüben (1983)). The effect of this is to give a coarse-grid operator in which there are connections between each of the C-points around that group (and each C-point is usually next to several of these F-point groupings). As a consequence, the resulting coarse-grid operator may be one with a large stencil although the "radius" of interpolation is - by definition - small. This effect becomes worse the larger the number of strong connections present in A.

- Another consequence of the one-sidedness of interpolation is the fact that speed of convergence may depend on the number of grids used. This difficulty occurs also (in a weaker form) with the other algorithm and it has been discussed by means of a very simple example in Brandt (1983). There it is shown that - if interpolation is very one-sided - coarse-grid corrections become worse. This is an important effect, as it may arise for very smooth grid functions (eigenfunctions to small eigenvalues). As a consequence, although two-grid convergence is still independent of the size of the problem, V-cycles tend to accumulate the errors from different levels. (This shows, in particular, that - assuming nothing more than (3.7) and (3.8) - the theorems given in Brandt (1983) do not carry over to theorems on the convergence of complete V-cycles.) Clearly, an easy way to overcome the problem of convergence would be to use more precise multigrid cycles, for example W-cycles. Such cycles, however, are not efficient here: coarsening by the methods discussed above usually reduces the number of grid points from one grid to the next coarser one by, roughly, 50%.

It is apparent from the above discussion that for more efficient solution, the interpolation must be improved. This includes both the interpolation formula and the way in which the coarser grids are chosen. An improved algorithm which essentially overcomes the above difficulties, is given in the next section.

4. THE PRESENT AMG01 ALGORITHM

In this chapter, we improve the coarse grid point choice and the interpolation formula. Together these produce much better complexities and convergence factors for a variety of problems. Section 4.1 states the main objectives of the coarse grid point choice, and gives the new interpolation formula. Section 4.2 gives the coarse grid point choice algorithm in some detail. Finally, Section 4.3 discusses the aspects of work and storage involved in the setup and solution processes.

4.1 Interpolation and the coarse grid point choice

The interpolation formulas used with the previous algorithms, in particular (3.8), were derived from the residual equation, and required (3.10b) concerning the error at points not used in interpolation. A similar method is used below, but a more refined assumption is made about the error at the unused points.

In the following, let us assume the level, m, to be fixed, and all matrices, vectors and sets written without the superscript m belong to that level. Now, denote the set of points j to be used in interpolation to an F-point i by S_i^I. This will always be a subset of $C \cap S_i$. Furthermore, a point $j \notin S_i^I$ is said to depend strongly on S_i^I if the sum of the connections from j to points in this set is large in some sense. (We define this more precisely in the next section. For now, a general idea of its meaning will suffice).

Now assume that sets C - and by this, F - and S_i^I ($i \in F$) have been chosen in some way. (The requirements for choosing such sets will be examined below.) With the same arguments as in Section 3.2, we can assume that after relaxation:

$$a_{ii}e_i + \sum_{\substack{j \notin S_i^I \\ j \neq i}} a_{ij}e_j + \sum_{j \in S_i^I} a_{ij}e_j \approx 0 \qquad (i \in F). \qquad (4.1)$$

In order to derive from this an interpolation formula to points $i \in F$, we would like to express the error at i only in terms of that at points in S_i^I. This is possible if the error at the other neighbours of i is known as a function of the error at points in $S_i^I \cup \{i\}$. For instance, in Section 3.2, the "smoothing property" (3.10b) was used, and for each neighbour j of point i not used in interpolation, e_j was simply replaced by e_i. It is, however, not generally true that the error is constant along strong connections, so the accuracy of the resulting interpolation is limited. In order to make an improvement in interpolation possible, we will now suppose that C has been constructed so that it satisfies the following criterion, which also implicitly defines S_i^I:

> C should be chosen such that for each point i in F, there is a set $S_i^I \subseteq C \cap S_i$ such that every point $j \in S_i$ is either in S_i^I, or strongly depends on it.

(CG1)

Once a set C has been found which satisfies the above criterion, the sets S_i^I are not necessarily uniquely defined. In the interest of complexity, we would like to have C, and thus S_i^I ($i \in F$), small. In order to obtain this, we state another objective (which, in practice, will be only approximately satisfied since the computational cost of strict enforcement would be prohibitive):

> The overall dependence between C-points should be as small as possible.

(CG2)

Instead of replacing e_j for $j \in N_i - S_i^I$ in (4.1) by e_i, a better approximation can be made when C satisfies (CG1). Since the error is smoothed along strong connections, the error at these points j can be approximately expressed as a weighted average of the errors at i and points in S_i^I as follows:

$$e_j \approx (a_{ji}e_i + \sum_{k \in S_i^I} a_{jk}\, e_k)/(a_{ji} + \sum_{k \in S_i^I} a_{jk}) \qquad (j \in N_i - S_i^I). \tag{4.2}$$

This can be used in (4.1) to obtain an approximation for the error at point i in terms of that at the points in S_i^I, which then yields the following interpolation formula:

$$v_i := \begin{cases} v_i^{(m+1)} & (i \in C) \\ -[\sum_{j \in S_i^I} (a_{ij} + c_{ij})v_j]/(a_{ii} + c_{ii}) & (i \in F) \end{cases} \tag{4.3}$$

where

$$c_{ij} := \sum_{\substack{k \notin S_i^I \\ k \neq i}} a_{ik}\, a_{kj}/[a_{ki} + \sum_{\ell \in S_i^I} a_{k\ell}] \qquad (j \in S_i^I \cup \{i\}).$$

Remark 4.1: For geometrically based problems, (CG1) and (CG2) ensure, in particular, that a more "balanced" set of interpolation points is chosen. However, it is the combination of the better C-point choice and interpolation, and not the coarse grid choice alone, which is responsible for the improvement in performance for such problems.

The main focus here is on matrices which satisfy (1.2). However, when positive off-diagonals occur (which is possible for coarser grid operators, even if (1.2) holds for the fine-grid operator), a slight modification of (4.2) is used: e_j is written as a weighted average of only those e_k, $k \in S_i^I \cup \{i\}$, for which a_{jk} is negative. Small positive connections, such as those which may occur on coarser levels, then do not affect convergence. In problems for which these are considerably larger, this change helps, but convergence will generally deteriorate as the size of the positive connections grows.

4.2 The coarse grid point choice algorithm

This section will be of interest mainly to the readers who want to know the more technical details of the algorithm used in this paper. In order to look at this new algorithm, some more definitions and notation are required. All rely on the concept of strength of connection, which has been used extensively before, but has never been specified precisely. Strength of connection of a point i to a point j is measured against the largest connection of i, which is denoted by s_i in the following (cf. Remark 4.2 below):

$$s_i := \max \{-a_{ik} : k \neq i\}.$$

Now, the dependence of a point i on a point $j \neq i$ and on a set of points P with $i \notin P$ are denoted by

$$d(i,j) := -a_{ij}/s_i \text{ and } d(i,P) := \sum_{j \in P} d(i,j),$$

respectively. We say that a point i strongly depends on (or is strongly connected to) a point j with respect to $0 \leqslant \alpha \leqslant 1$ if $d(i,j) \geqslant \alpha$, and we define the set of strong connections of i by:

$$\mathcal{S}_i(\alpha) := \{j \in N_i : d(i,j) \geqslant \alpha\}.$$

We also refer to the set $\mathcal{S}_i^T(\alpha)$ of points which strongly depend on the point i:

$$\mathcal{S}_i^T(\alpha) := \{j : i \in \mathcal{S}_j(\alpha)\}.$$

(Note that even for a symmetric operator, strong dependence defined in the above way need not be a symmetric relation.) We omit the parameter α in the following and simply write $\mathcal{S}_i$ and $\mathcal{S}_i^T$.

Remark 4.2: When the row corresponding to point i is strongly diagonally dominant (i.e. the absolute value of the sum of the off-diagonals is small compared to the size of a_{ii}, say one tenth the size), then the point i is given special consideration, and, in particular, is forced to be an F-point. Such exceptional cases are excluded from the following discussions for reasons of simplicity.

There are many possible algorithms for picking coarse grids which satisfy (CG1) while approximately satisfying (CG2). We primarily considered a method which would give a fast setup process without sacrificing complexity. Our coarse grid point choice is a two part process. Aside from α, another parameter $0 \leqslant \beta \leqslant 1$, is used which determines strong connections to sets as defined later. (Concerning reasonable values for α and β, see Remark 4.4.) In the first part, a tentative coarse grid is quickly decided. Here, C-points are distributed over the grid as "densely" as possible while minimizing the number of strong C-C connections. (The connection between $i, j \in C$ is called a strong C-C connection if $i \in \mathcal{S}_j$ or $j \in \mathcal{S}_i$.) Then, in the second part, a final sweep is made over the F-points to ensure that for each point the property (CG1) holds. This is done by forcing additional C-points when necessary. After the first part of the algorithm is performed, few extra C-points will need to be forced, so (CG2) is approximately satisfied.

First part of the process

The algorithm for the initial choice of C- and F-points is shown in Fig. 4.1. Initially, C and F are empty, and at any stage in the algorithm, all points not yet in C or F are "undecided" points. Now, some undecided point i is picked to become a C-point. This point i is chosen so that $|\$_i^T| + |\$_i^T \cap F|$ is maximal over all undecided points. The first term guarantees an initial bias towards the points that could be used to interpolate to many others. The second increases the probability that undecided points on which the largest number of F-points strongly depend (i.e., points which are most necessary for interpolation to existing F-points) will become C-points. If this sum is zero, all remaining undecided points, including i, are put into F, and the process ends. Otherwise, i becomes a C-point, and all undecided points $j \in \$_i^T$ are put into F. The process is repeated, picking C-points and forcing F-points, until all points are in C or F.

Remark 4.3: Note that, at any stage in the process, no undecided point strongly depends on any point in C, so each new C-point chosen does not depend strongly on the previous ones. However, in problems with many non-symmetric strong connections (i.e. those for which $j \in \$_i$, but $i \notin \$_j$), the previously chosen C-point can depend strongly on undecided points. So it is possible that many strong C-C connections can result, strongly violating (CG2) (and in extreme cases, give very large coarse grids). In such cases, in the above description, the sum $|\$_i^T| + |\$_i^T \cap F|$ is replaced by $|\$_i^T| + |\$_i^T \cap F| - |\$_i^T \cap C|$. (This modification is used in the code.) The extra term ensures that points with strong connections from existing C-points will be less likely to become C-points themselves.

Second part of the process

Fig. 4.2 shows the algorithm used for the final choice of C-points. When testing dependence of a point $k \in \$_i - \$_i^I$ on $\$_i^I$ $(i \in F)$ according to (CG1), we define strong dependence by

$$d(k,\$_i^I) \geq \beta\, d(i,k). \tag{4.4}$$

For a good interpolation to a point i, it is most important to get an accurate representation for the error at the points on which i most strongly depends. Thus, the above dependence is measured relative to d(i,k), so the strongest connections of i will be more likely to become C-points, or if they are not, a high dependence on $\$_i^I$ is ensured.

Now, we will enforce (CG1). For this, each F-point resulting from the previous part is tested. Letting the current F-point be i, $\$_i^I$ is

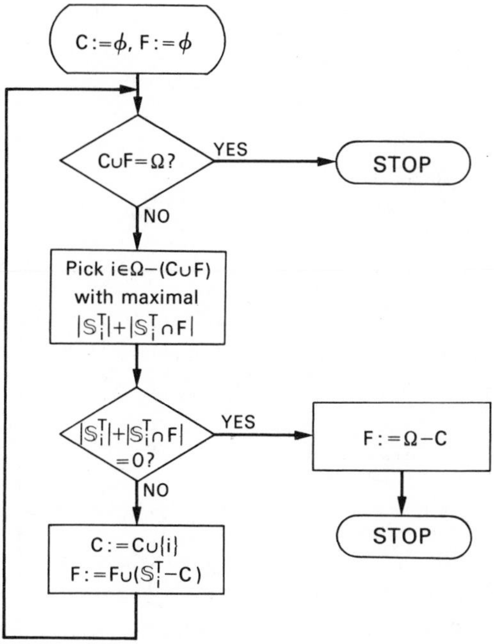

Fig. 4.1 The first part of the coarse grid point choice.

set initially to $\mathbb{S}_i \cap C$. Then the points in $\mathbb{S}_i - \mathbb{S}_i^I$ must be tested for strong dependence on $\mathbb{S}_i^I$ using (4.4). These are examined one at a time. If a point is found which does not strongly depend on $\mathbb{S}_i^I$, it becomes a conditional C-point. That is, it is put into $\mathbb{S}_i^I$, and the process continues, but its final status depends on whether any additional points are found which do not strongly depend on the new $\mathbb{S}_i^I$. We expect that the initial C-point choice was such that only a few isolated F-points must be changed to C-points. Thus, if more than one additional C-point is required in order to be able to interpolate to the point i, then i itself becomes a C-point, and the conditional C-point remains in F. If no additional C-points are needed, the conditional C-point is actually put into C. Now, if all points of $\mathbb{S}_i - \mathbb{S}_i^I$ have been tested, and at most one has been made a C-point, the interpolation weights to point i are computed according to (4.3). The process continues with the next F-point. This continues until all the F-points have been tested.

<u>Remark 4.4:</u> The optimal choice of the values α and β may vary somewhat depending on the problem being solved, but the exact values chosen are

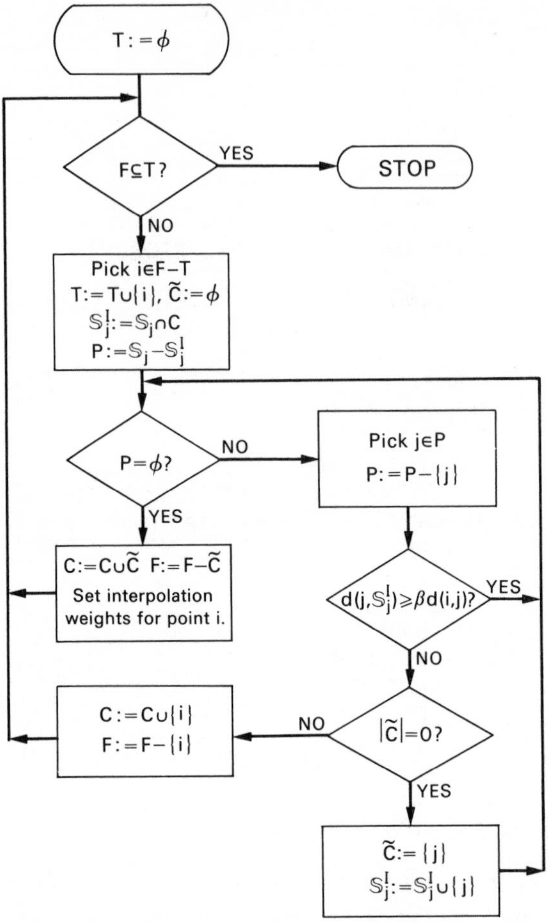

Fig. 4.2 The second part of the coarse grid point choice. T is the set of points which have been tested for qualification as F-points according to (CG1). $\tilde{C}$ denotes the set of conditional C-points for the point i.

not critical, so they can be held fixed, and good performance can be obtained for a wide variety of problems. For obtaining the results reported on in Chapter 5, we chose $\alpha = .25$ and $\beta = .35$. The value of .25 for α turned out to be reasonable, and in order to ensure sufficient accuracy of the approximation (4.2), a slightly higher value for β was picked.

4.3 Work and storage: efficiency considerations

Here, the amount of work, in terms of floating-point operations, of the setup and solution phases of AMG01 are examined in some detail, and the storage requirements are given. We use the quantities C_A, C_Ω, $N_A^{(m)}$, $N_\Omega^{(m)}$, $N_C^{(m)}$ and $N_F^{(m)}$ as introduced in Section 2.3. In addition, let the average number of entries per row of the given matrix be a (that is, $a := N_A^{(1)}/N_\Omega^{(1)}$), and let the average number of interpolation points per F-point over all grids be p. The operation counts computed below depend directly on the quantities C_A, C_Ω, p, and a. Because of the fully automatic performance of the setup phase of AMG01, the quantities C_A, C_Ω and p are extremely difficult to compute a priori. In order to obtain an O(N) method, the program must construct the coarser grids and coarse-grid operators in such a way that these numbers are reasonably bounded.

Setup phase

This phase is the most expensive part of the AMG01 algorithm. Using the quantities given above, floating-point operation counts can be derived for the coarse-grid operator construction and the computation of the interpolation weights. Precise calculation of these is not possible, but approximations can be found by assuming that certain average values hold throughout. A coarse-grid matrix is constructed using the Galerkin formulation (3.4), so connections from one coarse-grid point to others are generated by multiplying the restriction operator, the matrix itself, and the interpolation operator. Taking advantage of the 1's in the grid-transfer operators, the average number of operations required to compute a row of the coarse-grid matrix can be computed, and used to approximate W_A, the total number of floating point operations per fine-grid point invested in the coarse-grid operator computations. In order to approximate W_I, the total number of floating point operations per fine-grid point required for computing the interpolation weights, it is sufficient to compute the average amount of work at one point, since the total number of F-points on all grids is approximately equal to (but less than) the number of points on the finest grid. Average operation counts are

$$W_A \approx p(2p(\tilde{a}-p) + 3p + \tilde{a}), \quad W_1 \approx 3(\tilde{a} - (p+1))(p+1) + p$$

where $\tilde{a} := a\, C_A/C_\Omega$ is just the average number of non-zero entries per row of the matrices on all the grids. These approximations are quite realistic.

Example: For a 5-point finite difference discretization of a symmetric second-order elliptic partial differential equation, some typical quantities are $C_A \approx 2.50$, $C_\Omega \approx 1.75$ and $p \approx 3.25$. These values yield $W_A \approx 137$ and $W_I \approx 40$.

Clearly, the larger of these two counts is in the coarse-grid operator definition. W_A and W_I give, however, only part of the setup work. The remainder of the work is in the determination of the structure of the matrix and picking the coarser grids, and the only good measure of that is in terms of actual computation time, since in these processes, most of the work is in the performance of checks and loops. Generally, total setup time is about 2 to 3 times that required to construct the coarse-grid operators, and, typically is about the time needed for the performance of 5 to 9 V-cycles. More exact comparisons are made in Chapter 5, where times are given for a variety of test problems.

Solution phase

For the cycling process, operation counts in terms of C_A, C_Ω, p and a are the same as stated in Stüben (1983), and are given again here. The actual computation times have decreased, though, since lower complexities are obtained with the present algorithm. The total number of operations required for one relaxation sweep on Ω^m, interpolation and correction from Ω^{m+1} to Ω^m, and computation and transfer of residuals from Ω^m to Ω^{m+1} are as follows:

	Floating point operations
Relaxation	$2\ N_A^{(m)}$
Residual transfer	$2\ N_A^{(m)} + 2\ p\ N_F^{(m)}$
Interpolation	$N_C^{(m)} + 2\ p\ N_F^{(m)}$

If ν relaxation sweeps per grid per cycle are performed, then the total number of floating-point operations involved in one AMG01 V-cycle is given by:

$$2(1+\nu)C_A\ N_A^{(1)} + [C_\Omega + (4p-1)]N_\Omega^{(1)}.$$

This can be expressed in terms of total number of operations per fine-grid point, since $N_A^{(1)} = a\ N_\Omega^{(1)}$:

$$\boxed{2(1+\nu)a\ C_A + 4p + C_\Omega - 1}$$

Example: Using the same values given in the last example, and letting $\nu=2$ (one relaxation sweep before and after coarse-grid correction), this gives about 85 floating-point operations per fine-grid point per cycle. For comparison, a robust geometric multigrid algorithm, using standard coarsening, full weighting, and two alternating ZEBRA relaxation steps for smoothing (assuming the decompositions have been stored and do not have to be recomputed), requires 59 operations per fine-grid point per V-cycle (see Stüben and Trottenberg (1981), Table 10.7).

Storage

A complete description of the storage scheme used in AMG01 is given in Stüben, Trottenberg and Witsch (1983). The storage requirements can be divided into three groups: the matrices themselves, the interpolation weights, and the solution approximations and right-hand sides. Since the program is designed for matrix problems with no fixed structure, the matrices for each level are stored in the form used in the Yale Sparse Matrix Package (cf. Eisenstat, Gursky, Schultz and Sherman (1977)). This means that all non-zero entries of each matrix are stored, and an integer pointer vector of the same length is needed. In addition, three other integer vectors, each with a length of the total number of points on all grids, are also needed. These are for accessing the matrix information and for keeping track of the correspondence between the points on different grids. Similarly, all interpolation weights are stored, along with an integer pointer vector of the same length. In addition, another integer vector, with a length of the total number of F-points, is needed. Finally, of course, storage is required for the solution approximations and right-hand sides on all grids. Separated into real and integer variables, the storage requirements per fine-grid point are summarized in the following table:

	Real storage	Integer storage
Matrices	$a\ C_A$	$a\ C_A + 3\ C_\Omega$
Interpolation	p	$p+1$
Solution, R.H.S.	$2\ C_\Omega$	0
Total	$a\ C_A + 2\ C_\Omega + p$	$a\ C_A + 3\ C_\Omega + p + 1$

Example: Using the average values for 5-point operators given in the examples above, this gives 19.25 real variables and 22 integer variables per fine-grid point. In terms of words (where a real variable occupies two words and an integer one word), the total storage is about 3 times that required for the problem itself.

Remark 4.5: When larger stencils are given, complexity is considerably less than those average values shown for the 5-point discretizations in the examples. In practice, for 9-point stencils, the average storage requirements (as well as the operation counts per V-cycle) are about the same as for 5-point stencils.

Remark 4.6: There are techniques for lessening storage and operation counts in special cases. For example, although the main area of application has been symmetric problems, the program has been written to handle more general cases, and the storage scheme picked for the matrices does not assume symmetry. The program could easily be adapted to use a storage scheme in which only the diagonal and non-zero upper off-diagonals are stored. For the solution process, storage would be reduced by something less than 50%. During the setup process, it would be an easy matter to expand the matrix on the current working grid to full form in order to use the present algorithm. The upper off-diagonals of the matrix for the next coarser grid could then be generated, and the finer-grid matrix could then be collapsed. The gain in storage would depend somewhat on the complexity achieved, but the savings over full storage would be considerable. In addition, the work required to compute the coarse-grid matrices would be reduced, since only half the off-diagonals need to be computed.

Remark 4.7: Some storage is needed in order to make the setup process efficient. For example, instead of searching for points with maximal values as required by the initial C-point choice, lists and stacks are maintained which allow immediate access to such points. In addition, once the interpolation matrix is computed, its transpose is stored and is helpful in speeding up the computation of the coarse-grid matrix. However, at any point in the setup process, the storage that will be needed for the grids not yet determined is available. This will suffice for most of the work space needed unless complexity is especially low.

5. NUMERICAL RESULTS

In this section we want to demonstrate the robustness and the efficiency of AMG01. By changing the parameters α and β (see Section 4.2) in a proper way, it is often possible to improve the efficiency of AMG01 even further. According to its black-box nature, however, we are not going to adapt the above parameters to particular situations, but rather fix them to the following values (cf. Remark 4.4)

$$\alpha = 0.25, \quad \beta = 0.35.$$

AMG01 was designed to be a black-box solver primarily for general (sparse) systems (1.1) with (1.2). In order to make some comparisons with geometric multigrid, in this paper we focus on the application of AMG01 to discretized partial differential equations, namely to several (scalar) elliptic boundary value problems of the second order discretized by either finite differences (Sections 5.1 through 5.3) or finite elements (Section 5.4). We want to mention, however, that AMG has also been successfully applied to several problems which do not have a continuous background (see, e.g., Brandt, McCormick and Ruge (1983) and Stüben (1983)). Also, we have already stated that the AMG01 code does not make use of the assumption (1.2). In fact, for AMG01 to work efficiently, these assumptions do not have to be satisfied strictly. We will give examples for such situations later in this section (also cf. the remarks in Section 6).

In Section 5.1 we demonstrate that the new AMG01 code matches our objectives, namely, low complexity and fast convergence. Section 5.2 contains comparisons of AMG01 with three generally available geometric multigrid solvers. It turns out that the performance of AMG01 is favourably comparable to that of robust geometric multigrid solvers. The following Section 5.3 contains some remarks and results for non-symmetric problems. Finally, Section 5.4 gives several examples on finite element discretizations. Most of these examples serve as a demonstration of the performance of AMG01 in case of (geometrically) more complex-structured situations. In particular, we also give examples there for which the direct application of geometric multigrid would be quite difficult.

The quantities shown in the tables of this section are complexity values, convergence factors and CP-times. All computations and time measurements are carried out on the IBM 3083 computer of the GMD. Most of the convergence factors refer to the simple V-cycle (using one C/F-relaxation step before and one after the coarse-grid correction steps). It is only for reasons of comparison that we sometimes include also convergence factors for the V*-cycle (Stüben (1983)). This cycle is distinguished from the V-cycle in that corrections $v^{(m)}$ computed on coarser grids Ω^m $(m > 1)$ are scaled such that the error of the scaled correction minimizes the energy norm with respect to the one-dimensional space spanned by $v^{(m)}$. More precisely, $v^{(m)}$ is multiplied by

$$(f^{(m)}, v^{(m)})/(v^{(m)}, v^{(m)})_E$$

before it is interpolated to the next finer grid. Here, $f^{(m)}$ denotes the actual right hand side on level m. It was already found in Stüben (1983) that the V*-cycle often turns out to improve convergence essentially without increasing numerical work by much.

Finally, we remark that convergence factors of V-cycles are numerically computed by a v. Mises vector iteration power method. "Convergence factors" of V*-cycles are defined as the average residual reduction factor measured over 20 cycles. (In general, of course, the convergence speed of the V*-cycle depends somewhat on the right hand side of (1.1). The values shown below are typically observed ones.)

5.1 Complexity and convergence

The objectives of the coarse-grid point choice and the interpolation as defined in Chapter 4 were to improve both complexity and V-cycle convergence obtained by the old p-point algorithm (Stüben (1983)). As already mentioned at the end of Section 3.2, p-point interpolation may lead to unsatisfactory results, especially for larger stencils with many connections of about the same strength. Therefore, we here focus primarily on simple problems exhibiting this property. We point out, however, that the results given below are typical also for more complicated problems (cf. the following sections).

In Table 5.1, we compare results for several discretizations of the Laplace operator (up to a scaling factor): the usual 5-point stencil, the corresponding skewed stencil, the Mehrstellen operator and an approximation which occurs in connection with finite element discretizations using bilinear finite elements. We assume all stencils to be defined on a uniform square grid of mesh size $h = 1/64$ matching the boundary of the unit square. For simplicity, we assume Dirichlet boundary conditions. Besides complexity values, the table shows convergence factors and CP-times (in seconds and per cycle) for both the V- and the V*-cycle. The upper value in each of the boxes refers to the new AMG01 code, the corresponding lower value refers to the older AMG01 code (Stüben (1983)).

Table 5.1

Typical convergence factors and complexity values of AMG01.

problem	convergence		CP-times		complexity		
	V	V*	V	V*	A	Ω	
$\begin{bmatrix} & -1 & \\ -1 & +4 & -1 \\ & -1 & \end{bmatrix}$	0.054	0.049	0.31	0.35	2.26	1.69	new code
	0.318	0.092	0.44	0.54	3.90	2.01	old code
$\begin{bmatrix} -1 & & -1 \\ & +4 & \\ -1 & & -1 \end{bmatrix}$	0.091	0.048	0.29	0.32	2.13	1.65	new code
	0.385	0.140	0.40	0.49	3.53	1.97	old code
$\begin{bmatrix} -1 & -4 & -1 \\ -4 & 20 & -4 \\ -1 & -4 & -1 \end{bmatrix}$	0.103	0.084	0.30	0.32	1.41	1.38	new code
	0.313	0.120	0.55	0.66	3.00	2.02	old code
$\begin{bmatrix} -1 & -1 & -1 \\ -1 & +8 & -1 \\ -1 & -1 & -1 \end{bmatrix}$	0.119	0.066	0.30	0.32	1.38	1.33	new code
	0.456	0.284	0.48	0.56	2.58	1.74	old code

The results in Table 5.1 show that the new algorithm not only exhibits a much faster convergence than the older one but is also considerably cheaper. It has been confirmed also for many more complex problems that V-cycle convergence is typically below 0.1 for small stencils and around 0.1 for larger stencils. These convergence factors are (nearly) independent of the meshsize. Complexity is quite satisfactory, also: First, Ω-complexity is now much smaller than 2, which means that the (average) reduction factor for the number of grid points is considerably below 1/2; for the 9-point stencils, the reduction factor of grid points is even as low as about 1/4. Second and more important, A-complexity is smaller than that of the old algorithm by up to more than 50%. In particular, for larger stencils it is about equal to the Ω-complexity. This is, in a sense, optimal, as it means that the (average) stencil size is about the same on all grids. Of course, this cannot be expected to be generally true also for 5-point stencils: at least on the first coarser grid the stencil size will usually grow (to typically about 9 points). As a consequence, the A-complexity will become somewhat larger than the Ω-complexity. (This is, of course, similar in geometric multigrid whenever the Galerkin approach is used on coarser grids.)

Remark 5.1: The skewed Laplacian is an interesting example showing the flexibility of AMG01 in automatically picking coarse grids which match the smoothing properties of relaxation. Usual multigrid methods (using

Table 5.2

Convergence factors and complexity values of AMG01 for a large stencil. (Σ denotes the sum of all surrounding stencil elements.)

problem	ε	rates		CP-times		complexity		
		V	V*	V	V*	A	Ω	
$\begin{bmatrix} -\varepsilon & -\varepsilon & -\varepsilon & -\varepsilon & -\varepsilon \\ -\varepsilon & -1 & -1 & -1 & -\varepsilon \\ -\varepsilon & -1 & -\Sigma & -1 & -\varepsilon \\ -\varepsilon & -1 & -1 & -1 & -\varepsilon \\ -\varepsilon & -\varepsilon & -\varepsilon & -\varepsilon & -\varepsilon \end{bmatrix}$	0.2	0.091	0.044	0.58	0.63	1.32	1.35	new code
		0.477	0.299	0.74	0.83	1.68	1.63	old code
	1	0.111	0.062	0.53	0.55	1.17	1.18	new code
		0.707	0.507	0.58	0.62	1.29	1.30	old code
	5	0.129	0.076	0.70	0.78	1.59	1.40	new code
		0.687	0.411	0.73	0.82	1.68	1.49	old code

standard coarsening) have difficulties with this operator because there are actually two sets of points which are not coupled by the equations. The coarse grid points are only contained in one of these sets, and no meaningful reduction of the error at the other set of points can be obtained by means of coarse-grid corrections.

For still larger stencils, A-complexity of the p-point algorithm becomes better (although the increase of stencil sizes on coarser grids is worse than for the previous problems) which is a consequence of the fact that p-point interpolation then enforces (by definition) rapid grid coarsening. The good complexity, however, is obtained at the expense of very unsatisfactory convergence (due to extremely one-sided interpolation). For such cases, the improvement obtained by the new AMG01 code is even more striking than for the above examples. To demonstrate this, we give results for a (typical but admittedly somewhat artificial) 25-point stencil in Table 5.2: In particular, if all connections are of the same strength ($\varepsilon = 1$) or if the connections to points further away are stronger than to direct neighbours ($\varepsilon = 5$), convergence is very unsatisfactory if p-point interpolation is used. On the other hand, convergence of the new AMG01 code is hardly influenced by the size of the stencil given.

5.2 *Comparison with geometric multigrid solvers*

In order to demonstrate the efficiency of AMG01, we make some comparisons with three generally available geometric multigrid solvers, namely with BOXMG (Dendy (1982)), MGD1 (Hemker, Kettler, Wesseling and de Zeeuw (1983)) and MG00 (Foerster and Witsch (1982)). All these codes can be used to solve certain scalar elliptic differential equations of the second order on rectangular domains discretized on a uniform rectangular grid. While MG00 is written for high efficiency and low storage, the other two packages mainly aim at robustness with respect to a somewhat larger range of applications. We want to recall briefly the main characteristics of the above codes. For more details, we refer to the references given above.

- MGOO has been written for the high-speed low-storage solution V boundary value problems

$$- a(x,y)u_{xx} - b(x,y)u_{yy} + c(x,y)u = f$$

or

$$- (a(x,y)u_x)_x - (b(x,y)u_y)_y + c(x,y)u = f$$

with general (including periodic) boundary conditions. Coarser grids are obtained by standard coarsening and coarse-grid operators are defined in the "natural" way, i.e., by using the same difference formula as on the finest grid. Smoothing is done by red-black relaxation in isotropic cases, otherwise by ZEBRA relaxation (alternatingly if necessary). Interpolation is always linear, and restriction is done by half injection (in isotropic cases) or full weighting (otherwise). Due to the goal of high efficiency and low storage, no "Galerkin operators" or more sophisticated interpolation operators are used in MGOO. (This, of course, restricts the range of an efficient application to problems with relatively smooth coefficients a, b and c.) Furthermore, in case of ZEBRA relaxations, decompositions are not stored. This increases CP-time for relaxation by roughly 40% but saves a considerable amount of storage. As a consequence, preparation time for MGOO is negligible.

- MGD1 has been written for arbitrary linear elliptic equations of the second order (including mixed derivatives and first order terms, not, however, equations in divergence form). This code always uses 7-point ILU for smoothing and 7-point interpolation and restriction. The coarse-grid operators are the corresponding Galerkin operators. The setup phase of MGD1 includes computing and storing the ILU-decompositions and the coarse-grid operators.

- BOXMG has been written for symmetric problems. In contrast to MGD1, BOXMG allows the user to adapt the algorithm to certain specific situations in order to increase efficiency. In its most robust form it uses alternating linewise relaxation (in lexicographic order) for smoothing. Interpolation takes the coefficients of the operator into account (cf. Alcouffe, Brandt, Dendy and Painter (1981)) and the coarse-grid operators are of Galerkin-type. This makes the code especially suitable for solving self-adjoint problems with strongly discontinuous coefficients. Setup in BOXMG includes the computations of LU-decompositions needed for the smoothing process and the computation of the weights of interpolation and the coarse-grid operators.

In Table 5.3, we compare CP-times and convergence factors of AMG01 with those of the above solvers. In all examples we use 5-point discretizations on the unit square, Dirichlet boundary conditions and the mesh size $h = 1/64$. The diffusion coefficient of the last example is discontinuous across some grid lines (its definition is given in Fig. 5.1). The corresponding discretization is the same as used in BOXMG (cf. Alcouffe, Brandt, Dendy and Painter (1981)). The following quantities are shown in the table:

ρ - convergence factor per V-cycle;

t - CP-time per V-cycle;

t_{rel} - CP-time per reduction of the error by 0.1;

setup - setup time.

d = 10	d = 100
d = 1	d = 1000

Fig. 5.1 Coefficient d of the last example in Table 5.3

We want to point out that the comparison in Table 5.3 is not meant to be a competition between the geometric solvers. In particular, the results given for MG00 and BOXMG are not comparable: MG00 was meant to be representative of a very efficient solver and accordingly was run in modes suitable for the different examples (note, however, the above remark on the setup phase of MG00). BOXMG, on the other hand, was run in its most robust version for all the examples. The primary purpose of the table is to show that the cycling time and the convergence of AMG01 compares favourably with those of a robust geometric solver. This is seen, for instance, by a comparison of t_{rel} for AMG01 and BOXMG. Setup time is, of course, higher than that of geometric solvers: setup time in AMG01 typically costs between 5 and 9 corresponding cycles. However, this cost seems to be acceptable, considering the very general structure of AMG01 and its direct applicability to much more complicated problems (e.g., arbitrary boundary conditions and variable coefficients, irregular grids on arbitrary domains, larger stencils, three dimensional problems, problems where the application of geometric multigrid is difficult or does not even make sense). Some more complex situations are considered in Section 5.4.

Table 5.3 again shows the flexibility of AMG01. There is little dependence of convergence factors on anisotropy or varying, even discontinuous, diffusion coefficients. MGD1, on the other hand, exhibits a marked dependence on the direction of the anisotropy. (This is because ILU is used for smoothing, see above. Another version of MGD1 (Hemker, Kettler, Wesseling and de Zeeuw (1983)), not yet available, using ILLU will probably change this situation at the expense of additional numerical work.)

Table 5.3

Comparison of AMG01 performance with that of geometrical multigrid solvers for some symmetric problems.

problem		AMG01	BOXMG	MGD1	MG00	
$-u_{xx}-u_{yy}$		0.056	0.015	0.122	0.038	ρ
		0.30	0.33	0.12	0.06	t
		0.24	0.18	0.13	0.04	t_{rel}
		2.26	0.23	0.37	—	setup
$-\varepsilon u_{xx}-u_{yy}$	$\varepsilon=0.001$	0.084	0.043	0.004	0.005	ρ
		0.33	0.33	0.12	0.15	t
		0.31	0.24	0.05	0.07	t_{rel}
		1.83	0.24	0.37	—	setup
	$\varepsilon=0.01$	0.093	0.107	0.139	0.044	
		0.34	0.33	0.12	0.15	t
		0.33	0.34	0.14	0.11	t_{rel}
		2.10	0.24	0.38	—	setup
	$\varepsilon=0.1$	0.058	0.088	0.158	0.040	ρ
		0.38	0.33	0.12	0.15	t
		0.31	0.31	0.15	0.11	t_{rel}
		3.33	0.24	0.38	—	setup
	$\varepsilon=0.5$	0.069	0.036	0.140	0.021	ρ
		0.31	0.32	0.12	0.15	t
		0.27	0.22	0.14	0.09	t_{rel}
		2.31	0.24	0.37	—	setup
	$\varepsilon=2$	0.079	0.037	0.120	0.021	ρ
		0.31	0.32	0.12	0.12	t
		0.28	0.22	0.13	0.07	t_{rel}
		2.25	0.23	0.37	—	setup
	$\varepsilon=10$	0.087	0.084	0.265	0.040	ρ
		0.38	0.32	0.12	0.12	t
		0.36	0.30	0.21	0.09	t_{rel}
		3.26	0.24	0.37	—	setup
	$\varepsilon=100$	0.093	0.104	0.540	0.044	ρ
		0.33	0.32	0.12	0.12	t
		0.32	0.33	0.45	0.09	t_{rel}
		2.08	0.24	0.38	—	setup
	$\varepsilon=1000$	0.083	0.043	0.390	0.005	ρ
		0.32	0.32	0.12	0.12	t
		0.30	0.23	0.29	0.05	t_{rel}
		1.86	0.24	0.37	—	setup
$-\Delta u-(x+y)u$		0.060	0.015	0.122	0.037	ρ
		0.31	0.33	0.12	0.07	t
		0.25	0.18	0.13	0.05	t_{rel}
		2.35	0.23	0.37	—	setup
$-(100^{x+y-1}u_x)_x-u_{yy}$		0.080	0.044	—	0.053	ρ
		0.33	0.32	—	0.32	t
		0.30	0.24	—	0.25	t_{rel}
		2.46	0.24	—	—	setup
$-\nabla(d\nabla u)$ with d defined in Fig. 5.1		0.069	0.044	—	—	ρ
		0.32	0.32	—	—	t
		0.28	0.24	—	—	t_{rel}
		2.36	0.24	—	—	setup

Finally, we want to recall that - usually - by far the most efficient way to apply geometric multigrid methods in elliptic PDE's is to use the full multigrid (FMG) principle, the goal of which is to solve a given problem below truncation error as quickly as possible (cf., e.g., Brandt (1981) or Stüben and Trottenberg (1981)). This principle is incorporated in the solvers MG00 and BOXMG above (not, however, in MGD1). Because of the algebraic nature of AMG, there is no natural way to incorporate the FMG principle (at least not without using more specific information on the problem at hand). Clearly, it is possible to use a formal FMG-type process within AMG01 (by starting the solution process on the coarsest level M using reasonable restrictions of the fine-grid right-hand side as right-hand sides on coarser grids) with the goal to obtain a good first approximation on the finest grid. This process, however, will - if applied to PDE's - not guarantee a solution below truncation error.

5.3 Application to non-symmetric problems

Although AMG01 is designed (and was motivated) primarily for symmetric problems, it works just as well in mildly non-symmetric cases. But even for highly non-symmetric problems, such as convection-diffusion equations, convergence is usually (at least) as good as expected from symmetric problems, although complexity may become worse. To see this, note that, for convection-diffusion equations, strong dependence is directed upstream. Thus, for any point i, the points in S_i will generally not depend strongly on each other: they, in turn, depend on points further upstream. According to the coarse grid point choice objective (CG1) in Section 4.2, for each F-point i, we obtain $S_i^I = S_i$. As a consequence, all points in S_i $(i \in F)$ have to be C-points which leads to an unnecessarily high complexity. On finer grids, this is not a problem, but on coarser grids the number of strong connections to points upstream increases, and the problem compounds causing complexity to grow. (This is somewhat of an over-simplification, since - through the Galerkin operator - connections will also be generated to points which do not lie upstream, but these tend to be weaker.)

Remark 5.2: It is possible to construct a coarsening strategy more suitable to handle such cases. It depends on the fact that points which depend strongly on the same set of points in roughly the same way will tend to have approximately the same value. (This condition is actually equivalent to stating that the points are strongly connected in A^TA.) On coarser grids, points in S_i will tend to have the same domain of dependency (i.e., if j and k are in S_i, then $S_j \cap S_k$ is large). Then, in a different sense, we can say that there is a strong connection between them and not all points in S_i need to be C-points for i to be an F-point. In the simplest case, the values at some of these points can be replaced by the value of others on which they "strongly depend" in order to derive an interpolation formula. In this paper we are not going to modify AMG01 in order to treat non-symmetric problems in the most efficient way. Instead, we want to give some results for such problems by using AMG01 as it stands.

Table 5.4

Comparison of AMG01 performance for a mildly non-symmetric discrete problem

problem	AMG01	MGD1	MG00	
$-100^{x+y-1}u_{xx} - u_{yy} + (x+y)u$	0.094 0.35 0.34 2.63	0.356 0.12 0.27 0.37	0.054 0.27 0.21 -	ρ t t_{rel} setup

In three tables we give some results on the performance of AMG01 for several non-symmetric problems. For comparison, we give the corresponding results obtained by MGD1. (The non-symmetric version of BOXMG (Dendy (1983)) was not available to us. MG00 cannot be applied to problems involving first order derivatives.) The results shown in the tables correspond to those shown in Table 5.3. All differential equations are - for simplicity - again discretized on the unit square with Dirichlet boundary conditions and a uniform mesh size of $h = 1/64$.

The example in Table 5.4 is discretized by the usual 5-point stencil, thus yielding a mildly non-symmetric system of equations. The corresponding results are typical for such systems: there is no degradation compared to symmetric systems. The results in Table 5.5 and 5.6 correspond to highly non-symmetric problems derived by discretizing the convection-diffusion equation

$$- \varepsilon\Delta u + a(x,y)u_x + b(x,y)u_y = f(x,y) \qquad (\varepsilon \text{ small}) \tag{5.1}$$

using the difference formula

$$\frac{1}{h^2}\begin{bmatrix} & -\varepsilon + bh\,\mu_y & \\ -\varepsilon + ah(\mu_x - 1) & -\Sigma & -\varepsilon + ah\,\mu_x \\ & -\varepsilon + bh(\mu_y - 1) & \end{bmatrix}_h$$

with

$$\mu_x = \begin{cases} \varepsilon/2ah & (\text{if } ah > \varepsilon) \\ 1 + \varepsilon/2ah & (\text{if } ah < -\varepsilon) \\ 1/2 & (\text{if } |ah| \leqslant \varepsilon). \end{cases}$$

μ_y is defined analogously with a replaced by b. Σ denotes just the sum of the surrounding stencil elements. (We note that AMG01 works just as well if we used, for example, pure upstream differencing.)

Table 5.5

Comparison of AMGO1 performance with that of MGD1 for the convection-diffusion equation (5.1) with variable coefficients.

problem		AMG01	MGD1	
example (a)	$\varepsilon=.1$	0.068 0.30 0.26 2.27	0.126 0.12 0.13 0.38	ρ t t_{rel} setup
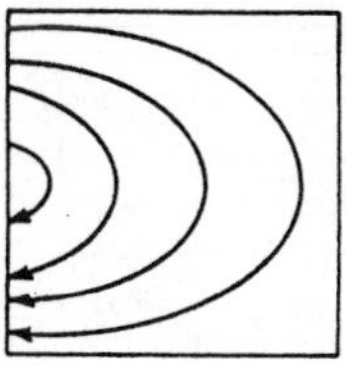	$\varepsilon=.001$	0.050 0.42 0.32 2.96	0.219 0.12 0.18 0.38	ρ t t_{rel} setup
	$\varepsilon=.00001$	0.025 0.40 0.25 2.72	0.317 0.12 0.24 0.38	ρ t t_{rel} setup
example (b)	$\varepsilon=.1$	0.062 0.31 0.26 2.25	0.126 0.12 0.13 0.38	ρ t t_{rel} setup
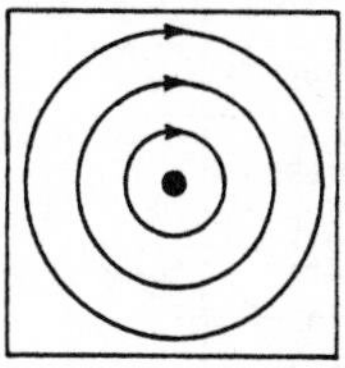	$\varepsilon=.001$	0.076 0.43 0.38 3.05	0.324 0.12 0.25 0.37	ρ t t_{rel} setup
	$\varepsilon=.00001$	0.102 0.43 0.43 3.00	0.785 0.12 1.14 0.37	ρ t t_{rel} setup
example (c)	$\varepsilon=.1$	0.049 0.31 0.24 2.29	0.126 0.12 0.13 0.37	ρ t t_{rel} setup
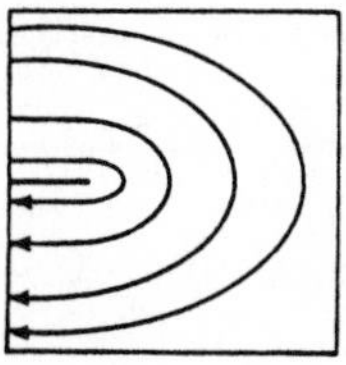	$\varepsilon=.001$	0.064 0.42 0.35 2.91	0.206 0.12 0.17 0.37	ρ t t_{rel} setup
	$\varepsilon=.00001$	0.022 0.37 0.22 2.43	0.284 0.12 0.22 0.37	ρ t t_{rel} setup

Table 5.6

Comparison of AMG01 performance with that of MGD1 for the convection-diffusion equation (5.1) with constant coefficients (5.2) as a function of the characteristic direction.

k	AMG01	MGD1	k	AMG01	MGD1	
0	0.00002 0.26 0.06 1.18	0.005 0.12 0.05 0.38	8	0.00002 0.26 0.06 1.18	0.005 0.12 0.05 0.37	ρ t t_{rel} setup
1	0.0003 0.45 0.13 3.28	0.001 0.12 0.04 0.37	9	0.00945 0.41 0.20 2.64	0.001 0.12 0.04 0.38	ρ t t_{rel} setup
2	0.00008 0.39 0.10 2.69	0.001 0.12 0.04 0.37	10	0.00006 0.39 0.09 2.78	0.001 0.12 0.04 0.38	ρ t t_{rel} setup
3	0.0007 0.40 0.13 2.81	0.0001 0.12 0.03 0.37	11	0.001 0.39 0.13 2.73	0.0001 0.12 0.03 0.37	ρ t t_{rel} setup
4	0.00002 0.27 0.06 1.18	0.00001 0.11 0.02 0.37	12	0.00002 0.27 0.06 1.16	0.00001 0.11 0.02 0.37	ρ t t_{rel} setup
5	0.0007 0.40 0.13 2.75	0.087 0.12 0.11 0.37	13	0.0005 0.40 0.12 2.80	0.085 0.12 0.11 0.37	ρ t t_{rel} setup
6	0.00005 0.39 0.09 2.68	0.249 0.12 0.20 0.37	14	0.00008 0.38 0.09 2.77	0.249 0.12 0.20 0.37	ρ t t_{rel} setup
7	0.0008 0.39 0.13 2.72	0.424 0.12 0.32 0.38	15	0.001 0.39 0.13 2.71	0.427 0.12 0.32 0.37	ρ t t_{rel} setup

In Table 5.5, the functions a and b are defined as follows:

(a) $a(x,y) = (2y-1)(1-x^2)$, $\quad b(x,y) = 2xy(y-1)$;

(b) $a(x,y) = 4x(x-1)(1-2y)$, $\quad b(x,y) = -4y(y-1)(1-2x)$;

(c) $a(x,y) = \begin{cases} (2y-1)(1-\bar{x}^2) & (\text{if } \bar{x} > 0) \\ 2y-1 & (\text{if } \bar{x} \leq 0), \end{cases} \quad b(x,y) = \begin{cases} 2\bar{x}y(y-1) & (\text{if } \bar{x} > 0) \\ 0 & (\text{if } \bar{x} \leq 0) \end{cases}$

where $\bar{x} := 1.2\,x - 0.2$.

The characteristic directions which correspond roughly to these coefficients, are shown in the table. We point out that there are stagnation points in case of examples (b) and (c) which require careful treatment in usual multigrid methods (cf. Börgers (1981)). AMG01 behaves very robustly for all cases and exhibits efficiency which is similar to, or even better than that observed before. This is in contrast to the behaviour of MGD1 which exhibits a slow-down of convergence if ε gets smaller. (Concerning an improvement of MGD1, see Hemker, Kettler, Wesseling and de Zeeuw (1983)).

Finally, Table 5.6 shows results corresponding to those in Table 5.5 for equation (5.1) with $\varepsilon := 10^{-5}$ and constants a, b defined by

$$a \equiv \cos\psi,\ b \equiv \sin\psi \text{ with } \psi = k\pi/8 \quad (k = 0,1,\dots,15). \qquad (5.2)$$

For certain of the characteristic directions (a,b), AMG01 convergence is especially fast. In particular, when a or b is zero, the problem becomes basically one-dimensional and AMG01 is almost a direct solver. (For $\varepsilon = 0$, it would be a direct solver.) This does not hold for other characteristic directions, but since most strong connections are used for interpolation, fast convergence is still obtained. Convergence is especially good for problems where the characteristic direction is accurately representable by the grid itself, making an optimal coarse grid point choice possible. MGD1, on the other hand, behaves quite unsatisfactorily for certain directions (cf. the corresponding remark above).

5.4 Application to complex-structured problems

In this section we want to demonstrate that the efficiency of AMG01 does not change when the underlying geometrical structure is more complex. For this we apply AMG01 to several finite element discretizations (using linear finite elements). For brevity, we confine ourselves to the Poisson equation (except for the last example) with Dirichlet boundary condition. As before, however, the performance of AMG01 is essentially the same for variable coefficient cases and/or other boundary conditions.

The examples treated in Table 5.7 are as follows:

(1) Poisson's equation on the unit square with a cut of length 1/2 (see Fig. 5.2). This example is to demonstrate that the convergence behaviour of AMG01 does not depend on the presence of singularities caused by the shape of the domain. This is in contrast to geometric multigrid methods where the speed of convergence slows down as a function of the strength of such singularities (cf., e.g., Stüben and Trottenberg (1981), Table 10.1); special techniques have to be employed in order to obtain the usual convergence behaviour (e.g., partial relaxation, see (Bai and Brandt (1983)).

(2) Poisson's equation on a triangle using a random triangulation (see Fig. 5.3). For such triangulations, the application of geometric multigrid is rather complicated as no simple and "natural" coarsening process exists.

(3) Poisson's equation on an L-shaped domain using the triangulation as shown in Fig. 5.4. This triangulation is "stretched" such that smaller triangles occur near the re-entrant corner (see Fig. 5.4).

(4) Poisson's equation on the same domain as before. Now, however, the triangulation is locally refined near the re-entrant corner (see Fig. 5.5).

(5) The diffusion equation $-\nabla(d\nabla u) = f$ on the domain displayed in Fig. 5.6 with diffusion coefficient d as defined by (a) $d \equiv 1$ and

$$\text{(b)}\; d := \begin{cases} 10000 & (\text{if } y \geq 0.8) \\ 1 & (\text{if } y < 0.8), \end{cases} \qquad \text{(c)}\; d := \begin{cases} 10000 & (\text{if } y \geq 0.8) \\ e^{5(x-y)} \cdot (\sin(10(x+y)) + 1.00001) & (\text{if } y < 0.8). \end{cases}$$

This example exhibits several difficulties as far as the application of geometric multigrid is concerned. First, there is no regular way to coarsen the grid in the upper part of the domain without losing too much information. Second, the domain contains square holes with a sidelength which is equal to that of the smallest triangles involved. These holes apparently cause even more difficulties in geometrically coarsening the grid. Third, there are several singularities due to the shape of the domain which would cause a slow-down of convergence in geometric multigrid. Finally, the diffusion coefficients (b) and (c) are strongly discontinuous.

The results in Table 5.7 show that the performance of AMG01 is hardly affected by all those difficulties which are inherent to the above problems. This confirms that the code is - in particular - suitable in case of (geometrically) complicated situations for which the application of geometric multigrid may be troublesome.

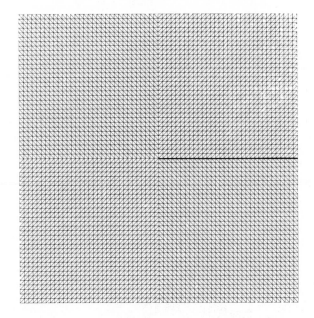

Fig. 5.2

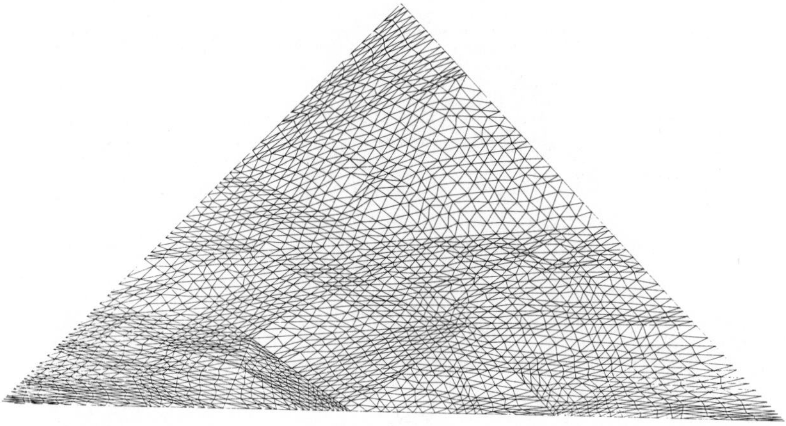

Fig. 5.3

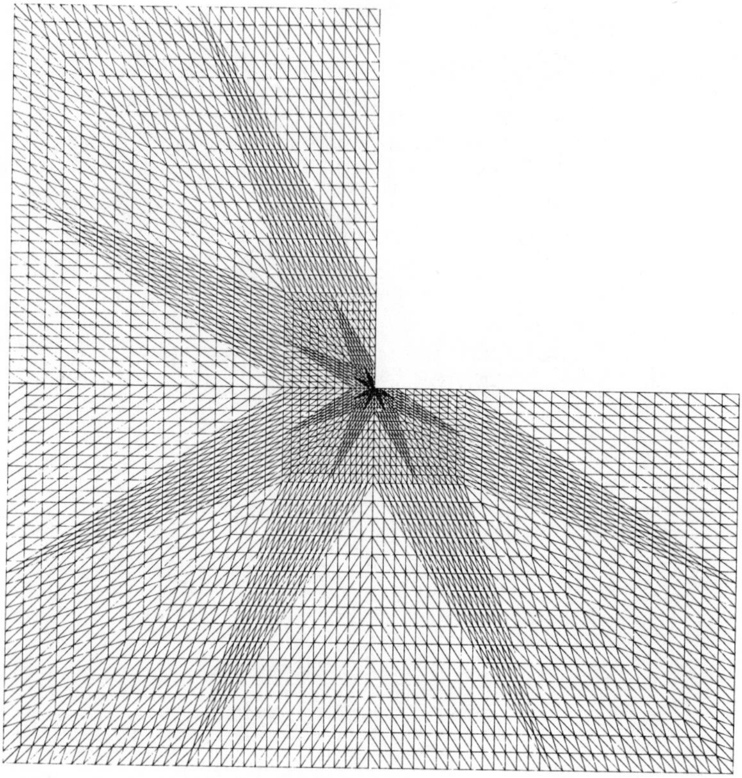

Fig. 5.4

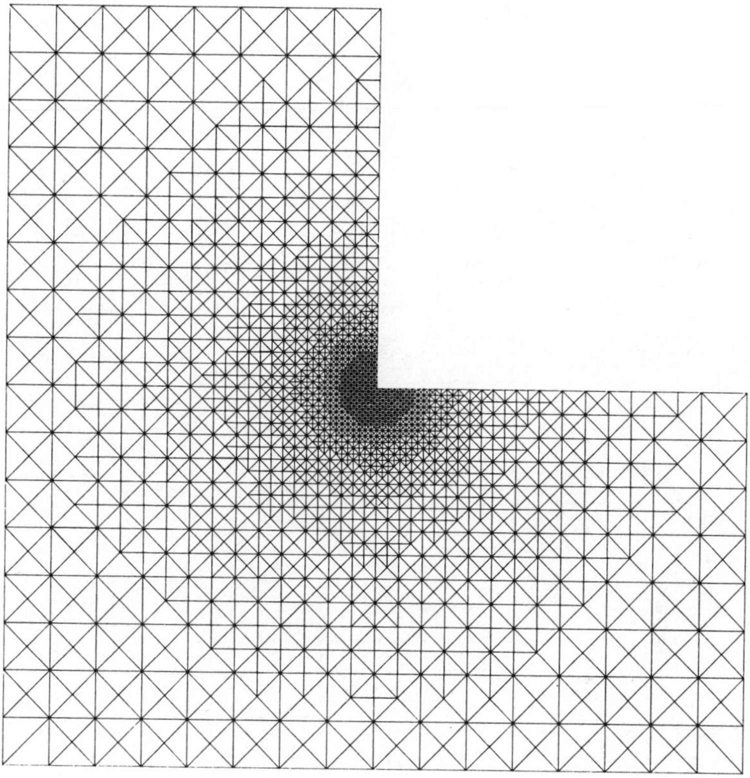

Fig. 5.5

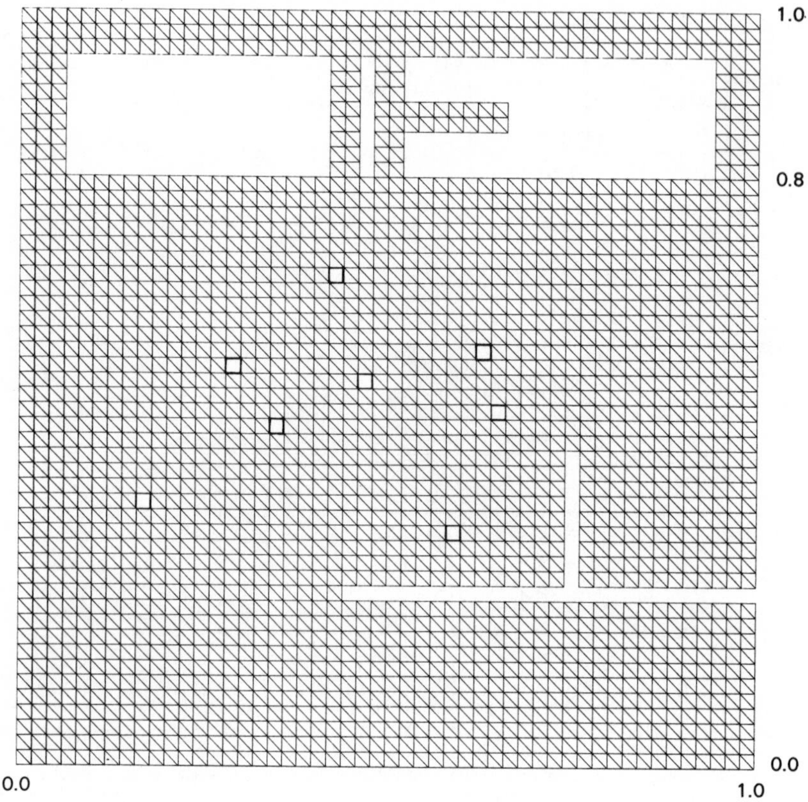

Fig. 5.6

Table 5.7

Performance of AMG01 for the examples (1) - (5).

example	convergence		CP-times			complexity		N
	V	V*	V	V*	setup	A	Ω	
(1)	0.063	0.017	0.31	0.35	2.30	2.27	1.69	3937
(2)	0.153	0.085	0.22	0.25	1.93	2.84	1.87	1953
(3)	0.118	0.058	0.32	0.37	2.70	2.91	1.89	2945
(4)	0.124	0.047	0.18	0.21	1.51	2.48	1.68	2232
(5a)	0.089	0.043	0.33	0.37	3.16	1.93	1.71	4064
(5b)	0.089	0.043	0.33	0.37	3.15	1.92	1.71	4064
(5c)	0.091	0.037	0.33	0.38	3.05	1.98	1.73	4064

Remark 5.3: The triangulations shown in Figs. 5.3 to 5.5 are actually constructed in a multigrid fashion (starting from some coarse first triangulation which is then refined recursively). Thus, geometric multigrid could be used here. However, this structure is not exploited by AMG01, and the triangulations shown may be regarded as representing more general situations for which really no natural coarsening process exists. In addition, in many applications the finest triangulation is given, and discovering the nesting of elements, or modifying code to give this information, is difficult. Thus, even though usual multigrid may be theoretically applicable, it is not practical.

Remark 5.4: We have already seen in Section 5.3 that AMG01 can work well on non-symmetric problems; so (1.2) is not a necessary condition for success. In addition, the sign conditions in (1.2) are not necessary either: Small positive off-diagonal elements occur in examples (2) and (3) from above but convergence is hardly affected. If, however, angles in the triangulation are allowed to get very large (giving a bad discretization not used in practice), the sign conditions in (1.2) are strongly violated and convergence of AMG01 deteriorates considerably. Another case where (1.2) is violated and good AMG01 convergence is obtained is with quadratic elements. Here, both positive off-diagonal entries and negative row sums are present. On the other hand, problems formulated using cubic elements (by including derivatives of the solution as unknowns) can cause problems. Not only do large positive off-diagonals result, but - without special care - the different types of unknowns can interfere with interpolation. Situations of this type are currently under study.

For all computations in this section we used the finite element package FEM2T which has been developed at the Universität Bonn (Institut für Angewandte Mathematik, Abteilung Angewandte Analysis). We thank the authors of this package for leaving us their code. In particular, we are indebted to H. Blum and A. Schüller for their help in preparing the finite element code in order to use it in connection with AMG01.

6. CONCLUSIONS AND FURTHER REMARKS

AMG01 has proved to be a robust and efficient solver for a variety of problems. In this paper, we have focused on finite difference and finite element discretizations of certain partial differential equations. For these problems, the present coarsening strategy and interpolation definition give solution efficiency which, measured by work required for some fixed reduction of the residual, approaches that for geometrically oriented robust multigrid methods. This makes AMG01 very attractive as a black-box solver, since all information needed in order to construct the coarser grids and associated operators is taken from the matrix itself. Of course, for this convenience the user must pay a price in terms of setup time and storage. Aside from convenience, the advantage of AMG01 over usual multigrid methods is its ability to solve more general problems, for example, problems with no regular geometric structure, such as those arising from more complex finite element discretizations (for examples, see Section 5.4).

There is still room for improvement in the present algorithm. The amount of storage required for the coarse-grid matrices and the interpolation weights can be quite high. This was briefly discussed, along with some possible remedies for special situations, in Section 4.3 (see, e.g., Remark 4.6). Also, throughout this paper, the parameters α and β were fixed. However, for some problems, better results can be obtained with different choices. This indicates that these parameters could be adapted automatically, perhaps even locally, to the problem at hand. One major change in the algorithm would be to allow interpolation from points which are not direct connections. This could reduce complexity considerably (especially in higher-dimensional problems, where storage is a primary concern), and thereby reduce both setup and solution time.

Present AMG research is primarily directed towards extending its range of applicability. In Chapter 5, it was shown that the assumptions in (1.2) need not hold strictly in order for AMG01 to work well. However, there are large classes of problems, not satisfying (1.2), on which the method will not perform satisfactorily, or to which it is not at all applicable in its present state.

As we have seen in the examples of convection-diffusion problems in Section 5.3, non-symmetry presents no problems when the sign conditions of (1.2) are satisfied. However, when central differencing is used for problems with large convection terms, resulting in positive off-diagonals much larger than the diagonal term, point-relaxation no longer smooths the error. A distributed relaxation scheme would be necessary. In addition, the residual equation for a point no longer gives a good interpolation formula.

The above problem belongs to a larger class, characterized by the loss of a "natural correspondence" between equations and unknowns. For matrices satisfying (1.2), this correspondence is clear. In the non-symmetric problem above, the i'th equation is no longer a good source of information for point i, and this correspondence is weakened. One of the main differences between AMG01 and geometric multigrid has been that AMG01 adapts the interpolation to "match" the point-relaxation. When there is no longer one equation associated with one point, both the relaxation scheme and the interpolation will require some generalization. There are many problems of interest in this category. For example, problems of this type arise when systems of differential equations are discretized in such a way that several unknowns are associated with each geometric point. Here, there is a correspondence between the set of equations and the set of unknowns associated with a particular point. In such a case, all the unknowns at that geometric point can be relaxed together, using the associated equations. Such a "block" process is a simple extension of the method presented in this paper.

REFERENCES

Alcouffe, R.E., Brandt, A., Dendy (Jr.), J.E. and Painter, J.M. (1981) The multigrid methods for the diffusion equation with strongly discontinuous coefficients, *SIAM J. Sci. Stat. Comput.*, **2**, pp. 430-454.

Bai, D. and Brandt, A. (1983) Local mesh refinement multilevel techniques, Report: Weizmann Institute of Science, Rehovot, Israel.

Börgers, C. (1981) Mehrgitterverfahren für eine Mehrstellendiskretisierung der Poissongleichung und für eine zweidimensionale singulär gestörte Aufgabe, Diplomarbeit, Institut für Angewandte Mathematik, Universität Bonn.

Brandt, A. (1982) Guide to multigrid development, in Multigrid Methods, (W. Hackbusch and U. Trottenberg, eds.) Lecture Notes in Mathematics, 960, Springer-Verlag, Berlin.

Brandt, A. (1983) Algebraic multigrid theory: the symmetric case, Preliminary Proceedings for International Multigrid Conference, April 6-8, 1983, Copper Mountain, Colorado (S.F. McCormick and U. Trottenberg, eds.).

Brandt, A., McCormick, S.F. and Ruge, J. (1982) Algebraic multigrid (AGM) for automatic algorithm design and problem solution. A preliminary report. Report, Inst. Comp. Studies, Colorado State Univ., Ft. Collins.

Brandt, A., McCormick, S.F. and Ruge, J. (1983) Algebraic multigrid (AGM) for geodetic survey problems. Preliminary Proceedings for International Multigrid Conference, April 6-8, 1983, Copper Mountain, Colorado (S.F. McCormick and U. Trottenberg, eds.).

Brandt, A., McCormick, S.F. and Ruge, J. (1984) Algebraic multigrid (AGM) for sparse matrix equations, in Sparsity and its applications (D.J. Evans, ed.), Cambridge Univ. Press.

Chatelin, F. and Miranker, W. (1982) Acceleration by aggregation of successive approximate methods, Linear Algebra Appl., **43**, pp. 17-47.

Dendy (Jr.), J.E. (1982) Black box multigrid, *J. Comp. Phys.*, **48**, pp. 366-386.

Dendy (Jr.), J.E. (1983) Black box multigrid for nonsymmetric problems, *Appl. Math. Comp.*, **13**, pp. 261-284.

Eisenstat, S.C., Gursky, M.C., Schultz, M.H. and Sherman A.H. (1977) Yale sparse matrix package, II. The nonsymmetric codes, Research report, 114, Dept. of Comp. Sci., Yale.

Foerster, H. and Witsch, K. (1982) Multigrid software for the solution of elliptic problems on rectangular domains: MGOO (Release 1), in Multigrid Methods (W. Hackbusch and U. Trottenberg, eds.), Lecture Notes in Mathematics, 960, Springer-Verlag, Berlin, pp. 427-460.

Hackbusch, W. and Trottenberg, U. (1982) Multigrid Methods, Proceedings of the Conference held at Köln-Porz, Nov. 23-27, 1981. Lecture Notes in Mathematics, 960, Springer-Verlag, Berlin.

Hemker, P.W., Kettler, R., Wesseling, P. and de Zeeuw, P.M. (1983) Multigrid Methods: development of fast solvers, *Appl. Math. Comp.*, **13**, pp. 331-326.

Stüben, K. and Trottenberg, U. (1982) Multigrid methods: Fundamental algorithms, model problem analysis and applications, in Multigrid Methods, Lecture Notes in Mathematics, 960, Springer-Verlag, Berlin, pp. 1-179.

Stüben, K., Trottenberg, U. and Witsch, K. (1983) Software development based on multigrid techniques, in Proceeding of the IFIP TC2 Working Conference on "PDE Software: Modules, Interfaces and Systems", Söderköping, Sweden (B. Engquist, ed.), Elsevier Science Pub. B.V. (North-Holland).

Stüben, K. (1983) Algebraic multigrid (AGM): experience and comparisons. *Appl. Math. Comp.*, **13**, pp. 419-452.

ALGEBRAIC FORMALISATION OF THE MULTIGRID METHOD IN THE SYMMETRIC AND POSITIVE DEFINITE CASE - A CONVERGENCE ESTIMATION FOR THE V-CYCLE

J.F. Maitre and F. Musy

(Ecole Centrale de Lyon, France)

ABSTRACT

To solve self-adjoint elliptic boundary value problems through a finite element discretization, we develop an algebraic formalisation of a family of multigrid iterations. The convergence is studied in the "energy-norm". We investigate here the case of the V-cycle (for the W-cycle see Maitre and Musy (to appear)). As an application, the behaviour of the convergence rate is studied for the S.O.R. smoothing with respect to the parameter ω.

1. INTRODUCTION

We consider the solution of the variational problem

$$(P)\quad \begin{cases} u^* \in H \\ a(u^*,v) = L(v), \ \forall v \in H, \end{cases}$$

where H is a finite dimensional Hilbert space endowed with the scalar product $a(.,.)$, and L is a linear functional defined on H.

We study the convergence of multigrid iterations in this variational framework, i.e. in the Hilbert space $(H,a(.,.))$, using tools such as orthogonal subspaces and projections.

In section 2, we introduce the algebraic formalisation of the multigrid iterations. In section 3, we give the expression of the operator associated with the multigrid iteration, a bound for its norm and the theorem of convergence for the V-cycle. We investigate in section 4 the behaviour of the convergence rate for the S.O.R. smoothing.

2. DEFINITION OF THE MULTIGRID ITERATION FAMILY

For the construction of the multigrid method, we use a sequence of increasing subspaces of H, U_k, $k = 0$ to m, satisfying

$$U_0 \subset U_1 \subset \ldots \subset U_m = H.$$

Let Θ_k be the "U_k-correction" mapping defined as follows

$$\Theta_k : H \longrightarrow U_k$$

$$u \longmapsto w = \Theta_k(u)$$

where w is the solution to the problem

$$(P_k) \begin{cases} w \in U_k \\ a(w,v) = a(u,v) - L(v) \quad \forall v \in U_k. \end{cases}$$

We denote by $\mathcal{F}_k$ the set of affine mappings f_k from H into H, with fixed point u*, defined by

$$f_k(u) = u - (I_k - F_k)\ \Theta_k(u)$$

with $F_k \in \mathcal{L}(U_k)$ (the space of linear mappings from U_k into U_k), and where I_k is the identity operator on U_k. We shall say that $f_k \in \mathcal{F}_k$ corresponds to $F_k \in \mathcal{L}(U_k)$ and denote by g_k the particular element of $\mathcal{F}_k$ corresponding to 0, the null mapping on U_k, i.e.

$$g_k(u) = u - \Theta_k(u).$$

Definition 1: A "(m+1)-level" multigrid iteration in H is defined by

$$u^0 \in H\ ;\ u^{r+1} = c_m(u^r),\ r \in \mathbb{N}, \tag{1}$$

where c_m is constructed recursively as follows

$$c_0 = g_0 \tag{2}$$

$$c_k = \psi_k \cdot c_{k-1}^{\mu_{k-1}} \cdot \phi_k,\ k = 1 \text{ to } m. \tag{3}$$

The integers μ_{k-1} are to be chosen in $\mathbb{N}^*$, and the "smoothers" ϕ_k, ψ_k in $\mathcal{F}_k$ (. denotes the composition product of mappings).

For more details on this formalisation of the multigrid method, namely the connection with the usual multigrid language, see Maitre and Musy (to appear).

3. CONVERGENCE OF THE MULTIGRID ITERATION

Within the framework of the Hilbert spaces $(U_k, a(.,.))$, we denote by

$U_k^{\perp}$ the subspace orthogonal in H to the subspace U_k;

P_V the projection in H onto any subspace V of H;

$P_V^{U_k}$ the projection in U_k onto the subspace $V \cap U_k$ of H;

i_k^l the injection from U_k into U_l $(l > k)$; and

$(i_k^l)^*$ the adjoint to i_k^l relatively to the Hilbert spaces $(U_k, a(.,.))$ and $(U_l, a(.,.))$.

By definition, we have

$$a(i_k^l u, v) = a(u, (i_k^l)^* v) \qquad \forall u \in U_k, \quad \forall v \in U_l.$$

Then, an expression for the operator associated with the multigrid iteration can be given (for more details see Maitre and Musy (to appear)).

<u>Proposition 1</u>: For k = 0 to m, the multigrid operator c_k defined by (2), (3) is in $\mathcal{F}_k$ and corresponds to the operator $C_k \in \mathcal{L}(U_k)$ defined recursively by

$$\left| \begin{array}{l} C_0 = 0 \\ C_k = N_k (i_{k-1}^k . C_{k-1}^{\mu_{k-1}} . (i_{k-1}^k)^* + P_{U_{k-1}^{\perp}}^{U_k}) M_k, \ k=1 \text{ to } m, \end{array} \right.$$

where N_k, M_k are the operators of $\mathcal{L}(U_k)$ corresponding respectively to ψ_k, $\phi_k \in \mathcal{F}_k$ of Definition 1 (3).

Within the framework of the solution of variational problem (P), it is natural and particularly appropriate to measure the error decrease by means of the "energy-norm", the norm of the Hilbert spaces $(U_k, a(.,.))$. This we denote by $\|.\|$, i.e.

$$\forall u \in U_k, \quad \|u\| = a(u,u)^{1/2}.$$

We define, as usual, the corresponding operator norms

$$\forall F_k \in L(U_k),\ \|F_k\| = \text{Sup}\left\{\|F_k u\|;\ u \in U_k,\ \|u_k\| = 1\right\}$$

and introduce the following notations:

$$\beta(M_k) = \rho((I-M_k^*M_k)^{-1}M_k^*P^{U_k}_{U^\perp_{k-1}}\ M_k), \tag{4}$$

$$\xi(M_k) = \left(\frac{\beta(M_k)}{1+\beta(M_k)}\right)^{1/2}, \tag{5}$$

where $M_k \in L(U_k)$ satisfies the requirement $\|M_k\| < 1$, M_k^* is the adjoint to M_k relative to the Hilbert space $(U_k, a(.,.))$ and ρ denotes the spectral radius. We emphasise that the number $\xi(M_k)$ is strictly less than one.

From Proposition 1, we derive $u^{r+1}-u^* = C_m(u^r-u^*)$. In order to give an upper bound for $\|C_m\|$, we prove the following lemma.

<u>Lemma 1</u>: Assume $\|M_k\| < 1$. Then

$$\|(i^k_{k-1}.C^{\mu_{k-1}}_{k-1}.(i^k_{k-1})^* + P^{U_k}_{U^\perp_{k-1}}\ )\ M_k\| \leqslant \text{Max}\left\{\|C^{\mu_{k-1}}_{k-1}\|,\ \xi(M_k)\right\} \tag{6}$$

with $\xi(M_k)$ defined by (4) and (5).

<u>Proof</u>: For simplicity, we prove (6) for the operator

$$(i.C.i^* + P_{U^\perp})\ M$$

with U subspace of H, $C \in L(U)$ and i the injection from U into H. Since $\|M\| < 1$, then $I-M^*P_UM$ is invertible. If we put $J = (I-M^*P_UM)^{-1}M^*P_{U^\perp}M$, the following inequality holds:

$$a(P_{U^\perp}Mu, P_{U^\perp}Mu) \leqslant \rho(J)\,[a(u,u)-a(P_UMu,P_UMu)],\quad \forall u \in H. \tag{7}$$

From the equality $i^*(Mu)=P_U Mu$ and Pythagoras' Theorem, one has

$$\|(i.C.i^*+P_{U^\perp})Mu\|^2 \leqslant \|C\|^2\|P_U Mu\|^2 + \|P_{U^\perp}Mu\|^2. \tag{8}$$

Equation (8) together with (7) imply

$$\|(i.C.i^*+P_{U^\perp})Mu\|^2 \leqslant (\|C\|^2 -\rho(J))\|P_U Mu\|^2 + \rho(J)\|u\|^2$$

and then, since $\|P_U Mu\|^2 < \|u\|^2$,

$$\|(i.C.i^*+P_{U^\perp})Mu\|^2 \leqslant \text{Max}\left\{\rho(J),\|C\|^2\right\}. \tag{9}$$

By using the equality

$$(I+(I-M^*M)^{-1}M^*P_{U^\perp}M)\ J = (I-M^*M)^{-1}M^*P_{U^\perp}M,$$

one easily proves

$$\rho(J) = \frac{\beta(M)}{1+\beta(M)} = \xi^2(M) \text{ which gives (6) from (9).} \ \square$$

Taking here an interest in the V-cycle case, we investigate the multigrid convergence with the assumption $\mu_k=1$, k=0 to m-1.

Theorem 1: If, for k=1 to m, the following inequalities are satisfied

$$\|N_k\| < 1,\ \|M_k\| < 1,$$

then the multigrid iteration sequence (u^r) constructed by (1), (2), (3) with $\mu_k=1$, k=0 to m-1, converges to the solution u^* to problem (P).

Furthermore, the error is strictly norm decreasing with

$$\|u^{r+1}-u^*\| \leqslant \xi_1\xi_2\ \|u^r-u^*\|,\quad \forall r \geqslant 0$$

where

$$\xi_1 = \max_{k=1 \text{ to } m}\left\{\xi(M_k)\right\} < 1,\ \xi_2 = \max_{k=1 \text{ to } m}\left\{\xi(N_k^*)\right\} < 1.$$

Proof: Since $\mu_k=1$, $k=0$ to $m-1$, the mapping c_m constructed from (2) and (3), can be also defined in the following manner:

$$c_m = c''_m \cdot c'_m \quad \text{with } c'_0 = c''_0 = g_0 \text{ and}$$
$$c'_k = c'_{k-1} \cdot \phi_k \,, \quad c''_k = \psi_k \cdot c''_{k-1} \,, \quad k = 1 \text{ to } m. \tag{10}$$

If we use Proposition 1 with $N_k=I_k$ and lemma 1, we obtain

$$\| c'_m(u) - u^* \| \leq \max_{k=1 \text{ to } m} \left\{ \xi(M_k) \right\} \| u-u^* \|, \quad \forall u \in H. \tag{11}$$

Thanks to the equalities

$$C^*_k = (i^k_{k-1} \cdot C^*_{k-1} \cdot (i^k_{k-1})^* + P^{U_k}_{U^\perp_{k-1}}) \, N^*_k,$$

$$\| C^*_k \| = \| C_k \|,$$

one proves in the same manner that

$$\| c''_m(u) - u^* \| \leq \max_{k=1 \text{ to } m} \left\{ \xi(N^*_k) \right\} \| u-u^* \|, \quad \forall u \in H. \tag{12}$$

Since c'_m and c''_m satisfy (10), Theorem 1 is proved from (11) and (12). □

We emphasise that, in Theorem 1, the only assumption about the smoothing is $\| M_k \| < 1$ which is satisfied in most practical cases. Nevertheless, we are left with giving estimates for $\xi(M_k)$. This is the purpose of the next section for a particular choice of M_k. A similar approach, in a matrix framework, can be found in McCormick (1984), but it is based on quantities different from ours. V-cycle convergence is also investigated by Bank and Douglas (1983), Braess and Hackbusch (1983), and Yserentant (1982).

4. CONVERGENCE RATE ESTIMATE IN THE CASE OF S.O.R. SMOOTHING

In this part, we give a bound for the "two-grid" quantity

$$\beta(M) = \rho((I-M^*M)^{-1} M^* P_{U^\perp} M) \tag{13}$$

with $M \in L(H)$, and U subspace of H, in order to obtain estimates of

$$\xi(M) = \left\{\frac{\beta(M)}{1+\beta(M)}\right\}^{1/2} \quad \text{(the mapping } \beta \longmapsto \frac{\beta}{1+\beta} \text{ is increasing).}$$

We restrict our investigations to the case of the S.O.R. method as smoothing procedure. More precisely, let w_i, i=1 to N be a basis of H and $\langle .,. \rangle$ be the scalar product defined by

$$\langle u,v \rangle = \sum_{i=1}^{N} x_i \bar{y}_i, \tag{14}$$

where $u,v \in H$ and $x,y \in \mathbb{C}^N$ are connected by the relations

$$u = \sum_{i=1}^{N} x_i w_i, \quad v = \sum_{i=1}^{N} y_i w_i.$$

We define the operator $A \in L(H)$ by

$$\langle Au,v \rangle = a(u,v) \quad \forall u,v \in H, \tag{15}$$

and the "stiffness" matrix K, supposed to be real, by

$$K_{i,j} = a(w_j,w_i) \quad i,j = 1 \text{ to } N. \tag{16}$$

From the splitting $K = D - E - E^t$ where D is a real diagonal and positive definite matrix, the S.O.R. iteration matrix is given by the expression

$$T = I - \left(\frac{1}{\omega} D - E\right)^{-1} K \quad \text{with } \omega \in \,]0,2[. \tag{17}$$

The associated operator $M \in L(H)$ is such that

$$M\left(\sum_{i=1}^{N} x_i w_i\right) = \sum_{i=1}^{N} (Tx)_i w_i. \tag{18}$$

Since K and D are symmetric and positive definite, the requirement $\|M\| < 1$ is satisfied.

<u>Proposition 2:</u> For the S.O.R. smoothing operator defined by (16), (17) and (18), the two-grid quantity $\beta(M)$ defined by (13) can be bounded as follows

$$\beta(M) \leqslant \rho(P_{U^\perp} A^{-1})\,\rho(H_\omega) \tag{19}$$

with $A \in L(H)$ defined by (14) and (15) and

$$H_\omega = \frac{\omega}{2-\omega}\,((1-\tfrac{1}{\omega})\,D-E^t)\,D^{-1}\,((1-\tfrac{1}{\omega})\,D-E). \qquad (20)$$

<u>Proof:</u> $\beta(M)=\rho((I-M^*M)^{-1}M^*P_{U^\perp}M)=\rho(P_{U^\perp}M(I-M^*M)^{-1}M^*)$. Hence

$$\beta(M)=\rho(A^{-1/2}P_{U^\perp}A^{-1/2}A^{1/2}M(I-M^*M)^{-1}M^*A^{1/2}). \qquad (21)$$

Since the operators $A^{-1/2}P_{U^\perp}A^{-1/2}$ and $A^{1/2}M(I-M^*M)^{-1}M^*A^{1/2}$ are self-adjoint, (21) yields

$$\beta(M) \leqslant \rho(A^{-1/2}P_{U^\perp}A^{-1/2})\,\rho(A^{1/2}M(I-M^*M)^{-1}M^*A^{1/2}). \qquad (22)$$

The equalities $\rho(A^{-1/2}P_{U^\perp}A^{-1/2})=\rho(P_{U^\perp}A^{-1})$ and $\rho(A^{1/2}M(I-M^*M)^{-1}M^*A^{1/2}) = \rho(A((I-MM^*)^{-1}-I))$ together with (22) show that

$$\beta(M) \leqslant \rho(P_{U^\perp}A^{-1})\,\rho(A[(I-MM^*)^{-1}-I]). \qquad (23)$$

The matrix T corresponds to the operator M through the relation (18). It is easy to show that K corresponds to A and that $K^{-1}T^tK = I-(\frac{D}{\omega}-E)^{-t}K$ corresponds to M^* through the same relation. Hence, the matrix

$$I-(I-(\frac{D}{\omega}-E)^{-1}K)(I-(\frac{D}{\omega}-E)^{-t}K)=(\frac{D}{\omega}-E)^{-1}(\frac{2-\omega}{\omega})D(\frac{D}{\omega}-E)^{-t}K$$

corresponds to $I-MM^*$. That implies

$$\rho(A((I-MM^*)^{-1}-I))=\rho(K(K^{-1}(\frac{D}{\omega}-E)^t(\frac{\omega}{2-\omega})D^{-1}(\frac{D}{\omega}-E)-I)).$$

Hence

$$\rho(A((I-MM^*)^{-1}-I))=\rho(H_\omega) \text{ with } H_\omega \text{ defined in (20).} \qquad (24)$$

(24) together with (23) proves (19). □

Remarks about Proposition 2

- In the case $\omega=1$, (19) yields

$$\beta(M) \leqslant \rho(P_{U^\perp}A^{-1})\rho(D)\rho(UL)$$

with the classical notations $L=D^{-1}E$, $U=D^{-1}E^t$.

- $\rho(D)$ and $\rho(LU)$ are often computable in practical cases, but this is not the case for $\rho(P_{U^\perp}A^{-1})$. Nevertheless, such a bound can be used for proving "h-independent" convergence of the multigrid iteration: see Bank and Douglas (1983), and Maitre and Musy (to appear).

The behaviour of $\rho(H_\omega)$ with respect to ω can be made precise. For that, as in the study of S.S.O.R. preconditioning (see for example Axelsson (1976)), we introduce the number

$$\delta = \sup_{x \in \mathbb{R}^N} \frac{x^t(E^tD^{-1}E - \frac{1}{4}D)x}{x^tKx} \tag{25}$$

which satisfies $\delta \geqslant -1/4$.

Proposition 3: The spectral radius of H_ω (see (20)) can be bounded as follows

$$\rho(H_\omega) \leqslant \rho(A)\phi(\omega) \tag{26}$$

$$\text{with } \phi(\omega) = \begin{cases} \dfrac{2-\omega}{4\omega} & \text{if } 0 < \omega \leqslant \dfrac{1}{1+\delta} \\[2ex] \dfrac{2-\omega}{4\omega} + \dfrac{(1+\delta)\omega-1}{2-\omega} & \text{if } \dfrac{1}{1+\delta} \leqslant \omega < 2 \end{cases} \tag{27}$$

where δ is defined by (25).

Moreover the values ω^* which minimise $\phi(\omega)$ over $\omega \in]0,2[$ and $\phi(\omega^*)$ are given by

$$\omega^* = \frac{1}{1+\delta}, \quad \phi(\omega^*) = \frac{1+2\delta}{4} \text{ if } \delta \leqslant 1/2$$

$$\omega^* = \frac{2}{1+\sqrt{2}\sqrt{1+2\delta}}, \quad \phi(\omega^*) = \frac{\sqrt{2}\sqrt{1+2\delta}-1}{2} \text{ if } \delta \geqslant 1/2.$$

Proof: By using the definitions of H_ω (20) and δ (25), we obtain the inequality

$$x^t H_\omega x \leqslant \frac{1}{\omega(2-\omega)} \left[\frac{(2-\omega)^2}{4} x^t Dx + ((1+\delta)\omega^2-\omega) x^t Kx \right].$$

Since H_ω is semi positive definite and K is positive definite, then

$$\rho(H_\omega) \leqslant \frac{2-\omega}{4\omega}\rho(D) \text{ if } \omega \leqslant \frac{1}{1+\delta}$$

$$\rho(H_\omega) \leqslant \frac{2-\omega}{4\omega}\rho(D) + \frac{(1+\delta)\omega-1}{2-\omega}\rho(K) \text{ if } \omega \geqslant \frac{1}{1+\delta}$$

and then (26), (27) follows from $\rho(D) \leqslant \rho(K) = \rho(A)$. □

According to the values of δ, the graph of $\phi(\omega)$ can be represented as follows:

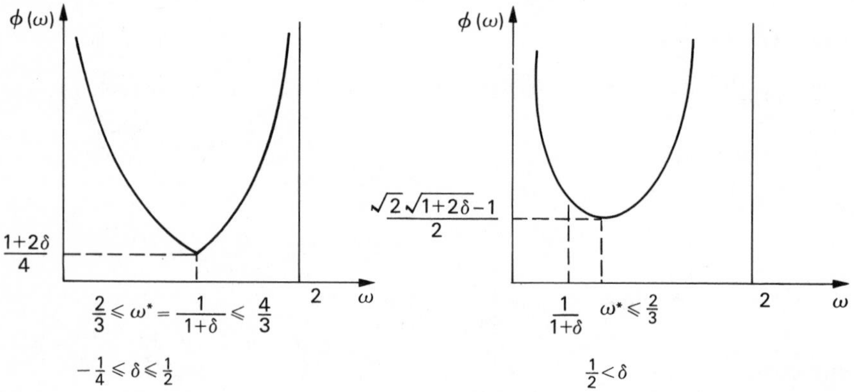

We remark that $\phi(\omega^*)$ satisfies $1/8 \leqslant \phi(\omega^*) \leqslant 1/2$ if $-1/4 \leqslant \delta \leqslant 1/2$ and $1/2 \leqslant \phi(\omega^*)$ if $\delta \geqslant 1/2$.

As an example, the classical discretization of Poissons equation on a square with Dirichlet conditions, using a regular mesh, verifies $\delta \leqslant 0$ (see Axelsson (1976)), which leads to $1 \leqslant \omega^* \leqslant 4/3$ with $\phi(\omega^*) \leqslant 1/4$.

REFERENCES

Axelsson, O. (1976) Solution of linear systems of equations: iterative methods - in sparse matrix techniques, Lecture Notes in Mathematics 572, Springer Verlag, 1-51.

Bank, R.E. and Douglas, C. (1983) Sharp estimates for multigrid rates of convergence with general smoothing and acceleration, Yale University, Computer Science Department Technical Report 277.

Braess, D. and Hackbusch, W. (1983) A new convergence proof for the multigrid method including the V-cycle. *SIAM J. Num. Anal.*, **20**, 967-975.

Maitre, J.F. and Musy, F. (To appear) Multigrid methods: convergence theory in a variational framework. *SIAM J. Num. Anal.*

McCormick, S. (1984) Multigrid methods for variational problems: further results. *SIAM J. Num. Anal.*, **21**, 255-263.

Yserentant, H. (1982) The convergence of multi-level methods for strongly nonuniform families of grids and any number of smoothing steps per level. R.W.T.H. Aachen-Institut für Geometrie und praktische Mathematik, Bericht Nr. 19.

A VARIATIONAL THEORY FOR MULTI-LEVEL ADAPTIVE TECHNIQUES (MLAT)

S. McCormick

(Department of Mathematics, Colorado State University)

ABSTRACT

The objective of this paper is to derive a theory for multi-level adaptive techniques (MLAT) for positive definite self-adjoint differential boundary value problems. This is done by developing a correction scheme version of MLAT and modifying the interpolation process at the interface between the coarse and fine grid regions. This results in a process that is equivalent to conventional multigrid where the fine level is the (extensive) composite grid.

1. INTRODUCTION

The objective of this paper is the development of a theory for multi-level adaptive techniques (most recently treated in Bai and Brandt (1983)) as they apply to positive definite self-adjoint differential boundary value problems. Theory is important in its own right, but the motive here is rather to give sound guidance for the implementation of MLAT on such problems.

In particular, Hackbusch offers an example that suggests difficulties in achieving the accuracy afforded by the local fine grid. Moreover, use of local grids can cause difficulties in conservation, for example, and though Bai and Brandt (1983) develop treatment for such difficulties, guidance for the general case is unclear. In short, a systematic and rigorous approach to the use of MLAT is needed.

A common trap in adaptive methods such as MLAT is that the user need not be aware of the discrete problem he is actually solving. It is often not enough to think pointwise in terms of the continuous problem; the inability to recognize the discrete problem that is actually being solved (if indeed one is) seems in conflict with rigor and often leads to procedures that are too ad hoc. This is embodied in the sometimes-held misconception that MLAT *must* be based on the full approximation scheme (FAS). On the contrary, it is possible to develop a correction scheme version of MLAT (which we henceforth refer to as CSMLAT) and this, in fact, is the subject of our paper. As we shall see, it leads naturally and very simply to a systematic development of MLAT techniques and its theory.

---

*This work was supported by the Air Force Office of Sponsored Research under grant no. FQ8671-83-01322, the Gesellschaft für Mathematik und Datenverarbeitung under a visiting research position, and the National Bureau of Standards under a consulting contract.

There are several issues that we do not discuss in the present paper, including the extension of the theory to the nonsymmetric and nonlinear cases, the relationship between CSMLAT and "FASMLAT", and various practical issues including grid adaption, accuracy estimation, and the use of more than two local grids. We choose instead to focus attention on the central issue and leave the implications to the reader and future papers. For this reason, we introduce CSMLAT with little comment and with little further reference to FASMLAT. However, the reader should not be deceived by the apparent simplicity of our development; some essential points are subtle and can easily be overlooked in practice.

Because of our development of CSMLAT, it may seem at first not to be of MLAT type. This stems from our focus on the "composite" (extensive) grid as opposed to the "fine" (nonextensive) grid. The proper interpretation can be gained, however, by noting that relaxation only involves operators defined on the coarse grid (as usual) and the fine part of composite grid.

We do not recommend for or against the use of CSMLAT in practice. We suggest rather that the use of MLAT in any form can be guided by the discrete problem. It is an advantage that CSMLAT forces this upon us, but other forms of MLAT may be more suitable for a specific application.

Finally, although we pose our theory in a variational setting, we do not mean to imply that CSMLAT should be implemented via finite elements. On the contrary, we see CSMLAT generally in terms of finite difference methods using finite elements possibly for guidance.

2. NOTATION AND ASSUMPTIONS

Suppose the continuous problem is given by

$$\mathcal{L}\mathcal{U} = f, \quad \mathcal{U} \in \mathcal{H}_1 \tag{2.1}$$

where $\mathcal{L}: \mathcal{H}_1 \to \mathcal{H}_2$ is linear, positive-definite and self-adjoint, $\mathcal{H}_1$, $\mathcal{H}_2$ are Hilbert spaces of functions defined on a region $\mathcal{D} \subset R^d$, $d \geq 1$, and $f \in \mathcal{H}_2$. We discretize (2.1) by

$$LU = f, \quad U \in H \tag{2.2}$$

where H is the composite grid space, $f \in H$, and $L: H \to H$ is linear, symmetric, and positive definite. Assume we are also given an equation

$$L^c U^c = f^c, \quad U^c \in H^c \tag{2.3}$$

where H^c is the coarse grid space, $f^c \in H^c$, and $L^c: H^c \to H^c$.

The interpolation or prolongation process is represented as a full rank matrix $I: H^c \to H$ with the restriction operator being its transpose $I^T: H \to H^c$. We assume the Galerkin condition

$$L^c \doteq I^T L I. \tag{2.4}$$

The assumptions on the grid structure are as follows. Let D and D^c be the grids in $\mathcal{D}$ corresponding to H and H^c, respectively. (We have in mind the picture in Figure 1.) We assume that D is the union of three disjoint sets D_1, D_2, and D_3 (and similarly for D^c) with the following properties:

i) the induced partitions of L and L^c are of the form

$$L = \begin{pmatrix} L_{11} & L_{12} & 0 \\ L_{21} & L_{22} & L_{23} \\ 0 & L_{32} & L_{33} \end{pmatrix} \text{ and } L^c = \begin{pmatrix} L^c_{11} & L^c_{12} & 0 \\ L^c_{21} & L^c_{22} & L^c_{23} \\ 0 & L^c_{32} & L^c_{33} \end{pmatrix};$$

ii) the induced partition of I is of the form

$$I = \begin{pmatrix} I & 0 & 0 \\ 0 & I & 0 \\ 0 & I_{32} & I_{33} \end{pmatrix}; \text{ and}$$

iii) the ratios of all mesh sizes of D and D^c in each coordinate direction are bounded. (In practice, if the mesh sizes in D_3 and D_3^c differ by much more than a factor of 2, intermediate local grids should be used. See McCormick (1984) for a treatment of the related FAC method with a theory that applies to the multi-level case.)

Note that these properties imply the following:

iv) $D_1^c = D_1$ and $D_2^c = D_2$; and

v) $L^c_{1j} = L_{1j}$, $j = 1, 2$.

We now suppose that $G: H \times H \to H$ is a relaxation process that applies to the grid points in $D_2 \cup D_3$ only. Let lower case denote approximations to the (upper case) solutions of (2.2) and (2.3) and let left arrow denote replacement. Then the relaxation

$$u \leftarrow G(u, f) \tag{2.5}$$

for solving (2.2) is assumed to satisfy

$$G_1\ (u,\ f) = u_1; \tag{2.6}$$

that is, G leaves alone the grid values that are associated with D_1.

3. CSMLAT

Given $u \in H$ as an approximation to the solution $U \in H$, then one two-level iteration of CSMLAT is as follows:

<u>Step 1</u> $u^c \leftarrow L^{c^{-1}} r^c$ where $r^c = I^T (f - Lu)$;

<u>Step 2</u> $u \leftarrow u + Iu^c$; and

<u>Step 3</u> $u \leftarrow G\ (u,\ f)$.

Note that Step 3 generally requires the relaxation of the equations (2.2) at points of D_2. We can actually implement Steps 2 and 3 by using "operator" interpolation (instead of I) on D_2 and restricting the basic relaxation process to D_3. This is equivalent to the above where G restricted to D_2 is Jacobi relaxation, but suggests an easier means of implementation. Note that relaxation on D_2 is unnecessary if $f - Lu$ is zero after interpolation in Step 2.

4. THEORY

The theory is an immediate consequence of the formulation in Section 3 and existing multigrid theory. The essence of this is given in the following.

<u>THEOREM 4.1.</u> Let $\hat{G}: H \times H \to H$ be any relaxation process that agrees with G on $D_2 \cup D_3$ (that is, $\hat{G}_i = G_i$ for $i = 2, 3$) and that satisfies $\hat{G}_1(u,\ f) = u_1$ whenever $(f - Lu)_1 = 0$. Then CSMLAT is unchanged if G is replaced by $\hat{G}$.

<u>Proof</u>: Condition v) shows that $(f - Lu)_1 = 0$ after Step 2 so that $\hat{G}(u,\ f) = G(u,\ f)$ in Step 3.

We may now apply any conventional multigrid theory to CSMLAT provided it applies to the discretizations L and L^c and admits relaxation schemes that extend G in the above way. This generally means that the theory must allow arbitrary ordering of the relaxation process and just one relaxation step per grid. For concreteness, we cite the following example. (More general relaxation schemes apply via a slight modification of the theory in McCormick (1983).)

THEOREM 4.2. Let G(u, f) be based on Richardson's method in $D_2 \cup D_3$ given by

$$G(u, f) = u + \frac{\omega}{\rho(L)} P(f - Lu)$$

where ρ denotes spectral radius, $0 < \omega < 2$, and $P = \begin{bmatrix} 0 & 0 & 0 \\ 0 & I & 0 \\ 0 & 0 & I \end{bmatrix}$ is the projection of D onto $D_2 \cup D_3$. Then the energy convergence factor of CSMLAT is at most $\alpha = \left(1 - \frac{\omega(2 - \omega)}{\delta}\right)^{1/2}$ where $\delta = \rho(L)\rho(L^{-1} - I L^{c^{-1}} I^T)$; that is,

$$\langle L\bar{e}, \bar{e}\rangle \leq \alpha^2 \langle Le, e\rangle$$

where $e = U - u$, $\bar{e} = U - G(u, f)$ and $\langle \cdot\ ,\cdot \rangle$ is the Euclidean inner-product. Finally, $\delta < \infty$ independently of the mesh size provided assumption A1 (with $a \geq 1$) of McCormick (to appear) is satisfied. (Note, however, that this bound generally depends on the ratio of the mesh sizes of D^c and D. See McCormick (1984) for the related FAC method with the established bound in fact independent of all mesh sizes.)

Proof: Let $\hat{G}(u, f) = u + \frac{\omega}{\rho(L)}(f - Lu)$ and apply Theorem 4.1 above and Theorem 3.1 of McCormick (1983).

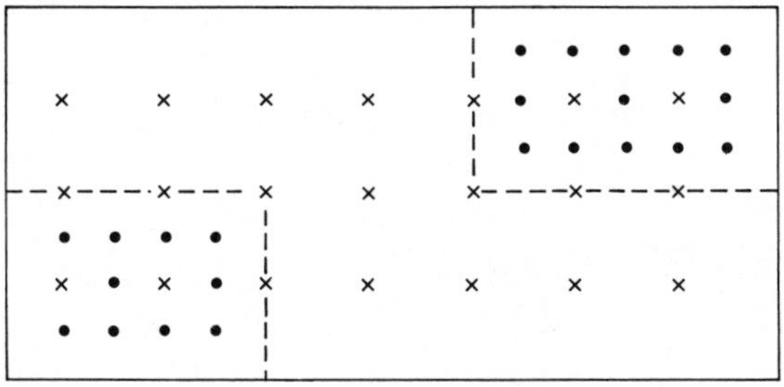

Figure 1

The coarse grid D^c is represented by x's, the composite grid D by x's and .'s., the actual boundary by solid lines, and the interface (D_2)

between coarse and the fine level by dashed lines.

REFERENCES

Bai, D. and Brandt, A. (1983) Local mesh refinement multilevel techniques, Res. rep., Dept. Appl. Math., Weizmann Inst. Sci., Rehovot, Israel.

Hackbusch, W., personal communication.

McCormick, S. (1983) Multigrid methods for variational problems: III, report, Inst. Comp. Studies, Colorado State Univ.

McCormick, S. (to appear) Multigrid methods for variational problems: further results, *SIAM J. Numer. Anal.*

McCormick, S. (1984) "Fast adaptive composite grid (FAC) methods: theory for the variational case", in Error Asymptotics and Defect Correction (K. Böhmer and H.J. Stetter, Eds.), a supplement to *Computing*.

SPECTRAL MULTIGRID METHODS FOR DIRICHLET PROBLEMS

T.N. Phillips

(Institute for Computer Applications in Science and Engineering)

T.A. Zang

(NASA Langley Research Center)

and

M.Y. Hussaini

(Institute for Computer Applications in Science and Engineering)

ABSTRACT

The systems of algebraic equations which arise from spectral discretizations of elliptic equations are full and direct solutions of them are rarely feasible. Iterative methods are an attractive alternative because Fourier transform techniques enable the discrete matrix-vector products to be computed with nearly the same efficiency as is possible for corresponding but sparse finite difference discretizations. For realistic Dirichlet problems preconditioning is essential for acceptable convergence rates. A brief description of Chebyshev spectral approximations and spectral multigrid methods for elliptic problems is given. A survey of preconditioners for Dirichlet problems based on second-order finite difference methods is made. New preconditioning techniques based on higher order finite differences and on the spectral matrix itself are presented. The preconditioners are analyzed in terms of their spectra and numerical examples are presented.

1. INTRODUCTION

Spectral methods involve representing the solution to a problem in terms of a truncated series of smooth global functions which are known as trial functions. Specifically, these functions are eigenfunctions of a singular Sturm-Liouville problem (Gottlieb and Orszag (1977)). This global character distinguishes spectral methods from finite difference and finite element methods. It is also responsible for their superior approximation properties, yielding accurate solutions with substantially fewer grid points than are required by finite difference methods.

Test functions are used to minimize the residual that results from the substitution of the series expansion of the solution into the differential equation. The choice of test functions distinguishes between the spectral Galerkin and spectral collocation methods. In the Galerkin approach the test functions are the same as the trial functions whereas in the spectral collocation method the test functions are shifted Dirac delta functions.

Research was supported by the National Aeronautics and Space Administration under NASA Contracts No. NAS1-17070 and No. NAS1-17130 while the first and third author were in residence at ICASE, NASA Langley Research Center, Hampton, VA 23665.

For many problems, especially nonlinear ones, the spectral collocation method is the easiest to implement, and the most efficient as well (Hussaini, Kopriva, Salas and Zang (1983)). The present discussion is confined to the spectral collocation method, and all future references to the spectral method will mean the spectral collocation method.

The matrices representing the discrete spectral collocation operator corresponding to elliptic problems (Zang, Wong, and Hussaini (1982)) are usually full. Even in the constant coefficient linear case, direct inversion of these matrices is usually expensive. Iterative schemes are a practical necessity. The condition number of the spectral matrices is large, and therefore effective preconditioning is necessary.

In recent years research in the field of elliptic equations has focused on multigrid methods. Basically, the multigrid method is a numerical strategy to solve partial differential equations by switching between finer and coarser levels of discretization. The characteristic feature of the method is the combination of a smoothing step and a coarse grid correction (Brandt (1977), Hackbusch (1980)). During the smoothing step the residuals are not necessarily decreased but smoothed. In the following correction step the discrete solution is improved by means of an auxiliary equation on a coarser grid. This results in an iterative method that is usually very fast and efficient.

2. FORMULATION OF THE PROBLEM

In this paper we consider the self-adjoint elliptic problem

$$\frac{\partial}{\partial x}\left(a(x,y)\,\frac{\partial u}{\partial x}\right) + \frac{\partial}{\partial y}\left(b(x,y)\,\frac{\partial u}{\partial y}\right) = f(x,y) \tag{2.1}$$

with Dirichlet boundary conditions in the domain $[-1,1] \times [-1,1]$. A proper representation of the solution to this Dirichlet problem employs Chebyshev polynomials. The details of implementing Chebyshev collocation methods for this type of problem have been given by Zang, Wong and Hussaini (1982) and by Hussaini, Salas and Zang (1983). This discretization of equation (2.1) may be written as

$$L_{sp}\, V = F, \tag{2.2}$$

where V is the vector of unknowns at the collocation points, F is the vector of the values of the right-hand side at the collocation points and L_{sp} is the vector-valued operator representing a spectral discretization of the left-hand side.

Iterative methods appear to be preferable to direct methods even for constant coefficient linear problems since the matrices representing L_{sp} are full. No fast direct methods are known for these systems. Iterative methods are appealing because the standard implementation of spectral discretizations employs Fast Fourier Transforms which reduces the cost of evaluating the left-hand side of (2.2) to $O(N \log N)$ operations, where N is the total number of unknowns. Iterative methods also have a clear advantage over direct methods in terms of storage.

Many iterative schemes can be described within the framework of defect correction. Let H be some approximation to L_{sp} and let V^n be the latest approximation to V. The simplest iterative scheme is the preconditioned Richardson's method,

$$V^{n+1} = V^n - \omega_n H^{-1} [L_{sp} V^n - F], \tag{2.3}$$

where ω_n is a relaxation parameter.

In practice the preconditioning matrix H is constructed to be an approximation to L_{sp} having the following properties:

(i) H has a sparse matrix structure;
(ii) H is easily invertible.

One choice for H considered here is an approximate LU decomposition of a finite difference approximation to (2.1), i.e.

$$H = LU$$

where L and U are lower and upper triangular matrices respectively.

3. AN HEURISTIC DISCUSSION OF SPECTRAL MULTIGRID

Of course, far better schemes are available than the simple Richardson's method. Multigrid methods have demonstrated their ability to accelerate many types of relaxation schemes (see, e.g., the conference proceedings edited by Hackbusch and Trottenberg (1982)). Multigrid methods have recently been developed for spectral discretizations of (2.1) by Zang, Wong, and Hussaini (1982, 1983). Streett, Zang, and Hussaini (1983) have shown that these techniques are extremely effective for the nonlinear potential flow problem of transonic aerodynamics. Preconditioning techniques play a crucial role in these spectral multigrid schemes. Our purpose here is to investigate additional choices for the preconditioning matrix.

We begin this discussion by looking at a one-dimensional model problem with periodic boundary conditions. For this problem we can obtain explicit expressions for the eigenvalues of the preconditioned system since the finite difference and spectral approximations have a common basis of eigenvectors. After describing the basic idea of multigrid we show how the spectral discretization for the Chebyshev model problem is constructed.

The fundamentals of spectral multigrid are perhaps easiest to grasp for the simple model problem

$$-\frac{d^2u}{dx^2} = f \tag{3.1}$$

on $[0,2\pi]$ with periodic boundary conditions. The Fourier approximation to the left-hand side of (3.1) at the collocation points $x_j = 2\pi j/N$ is

$$\sum_{p=-\frac{N}{2}+1}^{\frac{N}{2}-1} p^2 \hat{u}_p e^{ipx_j} , \tag{3.2}$$

where $\hat{u}_p$ are the Fourier coefficients of u. The eigenfunctions of this approximation are

$$\xi_j(p) = e^{2\pi ijp/N}, \tag{3.3}$$

with the corresponding eigenvalues

$$\lambda(p) = p^2, \tag{3.4}$$

where $j = 0,1,\ldots,N-1$ and $p = -\frac{N}{2} + 1,\ldots, \frac{N}{2} - 1$. The index p has a natural interpretation as the frequency of the eigenfunction.

Consider now the iterative process described by (2.3) with H taken to be the identity matrix, i.e., without any preconditioning. The iteration matrix, C, of this scheme is given by

$$C = I - \omega L_{sp}.$$

The iterative scheme is convergent if the eigenvalues, λ, of L_{sp} satisfy

$$|1 - \omega\lambda| < 1.$$

The best choice of ω is that for which

$$(1 - \omega\,\lambda_{max}) = -\,(1 - \omega\,\lambda_{min}),$$

where λ_{max} $(= N^2/4)$ and λ_{min} $(= 1)$ are the largest and smallest eigenvalues of L_{sp} respectively, for then the largest values of $\mu = 1 - \omega\lambda$ are equal in magnitude and have opposite sign (see Fox (1964)). One need not worry about the $p = 0$ eigenfunction since it corresponds to the mean level of the solution, which is at one's disposal for this problem. The optimal relaxation paramter for this single-grid procedure is

$$\omega_{SG} = \frac{2}{\lambda_{max} + \lambda_{min}}. \tag{3.5}$$

It produces the spectral radius

$$\mu_{SG} = \frac{\lambda_{max} - \lambda_{min}}{\lambda_{max} + \lambda_{min}}. \tag{3.6}$$

Unfortunately, $\mu_{SG} \simeq 1 - 8/N^2$, which implies that $O(N^2)$ iterations are required to achieve convergence.

This slow convergence is the outcome of balancing the damping of the lowest frequency eigenfunction with that of the highest frequency one in the minimax problem described above. The multigrid approach takes advantage of the fact that the low frequency modes ($|p| < N/4$) can be represented just as well on coarser grids. It settles for balancing the middle-frequency one ($|p| = N/4$) with the highest frequency one

($|p| = N/2$), and hence damps effectively only those modes which cannot be resolved on coarser grids. In (3.5) and (3.6), λ_{min} is replaced by $\lambda_{mid} = \lambda(N/4)$. The optimal relaxation parameter in this context is

$$\omega_{MG} = \frac{2}{\lambda_{max} + \lambda_{mid}} . \tag{3.7}$$

The multigrid smoothing factor

$$\mu_{MG} = \frac{\lambda_{max} - \lambda_{mid}}{\lambda_{max} + \lambda_{mid}} \tag{3.8}$$

measures the damping rate of the high-frequency modes. Alternatively, we may write

$$\mu_{MG} = \frac{\kappa_{MG} - 1}{\kappa_{MG} + 1}$$

where $\kappa_{MG} = \lambda_{max}/\lambda_{mid}$ is known as the multigrid condition number. In this example $\mu_{MG} = 0.60$, independent of N. The price of this effective damping of the high-frequency errors is that the low-frequency errors are hardly damped at all. Table 1 compares the single-grid and multigrid damping factors for N = 64. However, on a grid with N/2 collocation points, the modes for $|p| \in [N/8, N/4]$ are now the high-frequency ones. They get damped on this grid. Still coarser grids can be used until relaxations are so cheap that one can afford to damp all the remaining modes, or even to solve the discrete equations exactly. For the case illustrated in Table 1 the high-frequency error reduction in the multigrid context is roughly 250 times as fast as the single-grid reduction for N = 64.

Table 1

Damping Factors for N = 64

Frequency (p)	μ_{SG}	μ_{MG}
1	0.9980	0.9984
2	0.9922	0.9938
4	0.9688	0.9750
8	0.8751	0.9000
12	0.7190	0.7750
16	0.5005	0.6000
20	0.2195	0.3750
24	0.1239	0.1000
28	0.5298	0.2250
32	0.9980	0.6000

Orszag (1980) has proposed a preconditioning for spectral methods which amounts to using a low-order finite difference approximation for H.

Let $H^{(2)}$, $H^{(4)}$ and L_{sp} denote second-order, fourth-order and spectral discretizations of the operator $-d^2/dx^2$. The eigenvalues of these discretizations are given below:

$$\lambda_k^{(2)} = \frac{2[1 - \cos(k\Delta x)]}{(\Delta x)^2} ,$$

$$\lambda_k^{(4)} = \frac{\cos(2k\Delta x) - 16\cos(k\Delta x) + 15}{6(\Delta x)^2} ,$$

$$\lambda_k^{(\infty)} = k^2 .$$

The effective eigenvalues of the preconditioned iterations based on $\left(H^{(2)}\right)^{-1} L_{sp}$ and $\left(H^{(4)}\right)^{-1} L_{sp}$ are then given by

$$\Lambda_k^{(2)} = (k^2)\left(\lambda_k^{(2)}\right)^{-1}$$

$$= \frac{(k\Delta x)^2}{2[1 - \cos(k\Delta x)]} ,$$

$$\Lambda_k^{(4)} = (k^2)(\lambda_k^{(4)})^{-1}$$

$$= \frac{6(k\Delta x)^2}{\cos(2k\Delta x) - 16\cos(k\Delta x) + 15} .$$

Table 2

Properties of Finite Difference Preconditioning for the Model Problem

Finite Difference Order	Λ_{min}	Λ_{mid}	Λ_{max}	μ_{SG}	μ_{MG}
2	1.00	1.23	2.47	0.424	0.336
4	1.00	1.06	1.85	0.298	0.273
6	1.00	1.02	1.63	0.240	0.231

Similar results for higher even-ordered finite difference preconditionings are straightforward but tedious. The key properties of this class of preconditioning are given in Table 2. It will be seen in subsequent sections that these results on the spread of the eigenvalues of the preconditioned systems agree well with those obtained computationally in two dimensions for Dirichlet problems.

Unlike the original system, which has a condition number scaling as N^2, the preconditioned system has a condition number which is independent of N. The fourth-order finite difference operator offers around a 20% improvement in convergence rate over the second-order operator. This is partially offset by the additional cost of inverting the finite difference operator. The higher-order preconditionings are of doubtful utility.

The preconditionings are clearly most effective for the longer wavelength eigenfunctions, as reflected by how close Λ_{mid} is to 1. In fact, for the fourth-order version, Λ_{mid} is already so close to 1 that the multigrid convergence rate is only slightly faster than the single-grid rate. The advantage of multigrid for these preconditioned systems only shows up in two dimensions. Unlike the one-dimensional case, the inversion of the two dimensional operator is nontrival. It seems advisable in these situations to do only an approximate inversion of the finite difference operator, for example, by using an incomplete LU decomposition of H as the actual preconditioner. The outcome of this choice is that while Λ_{max} can be kept well under control, exhibiting a very slow growth with N, Λ_{min} plunges precipitously to zero. Fortunately Λ_{mid} remains virtually unchanged. Thus multigrid is attractive for these two-dimensional problems.

Recently, Haldenwang, Labrosse, Abboudi and Deville (1983) have obtained closed form expressions for the eigenvalues of $H^{-1}L_{sp}$ where L_{sp} is the Chebyshev representation of d^2/dx^2, while H is the corresponding finite difference approximation based on the Chebyshev collocation points. They show that the eigenvalues of $H^{-1}L_{sp}$ lie in the range $[1, \pi^2/4]$. Zang, Wong and Hussaini (1982) calculated the eigenvalues of the spectral approximation to the two-dimensional Laplace operator using the QR algorithm (Wilkinson (1965)). They found that the eigenvalues of interest exhibited the following behaviour:

$$\lambda_{max} = O(N^4) \;, \; \lambda_{mid} = O(N^2) \;, \; \lambda_{min} = O(1).$$

They also observed that all but the largest eigenvalues are good approximations to the eigenvalues of the continuous problem. It is this property which is responsible for the superior approximation nature of spectral methods. We note that the multigrid condition number is of $O(N^2)$ in this case and so preconditioning is essential for multigrid as well as single-grid iterations.

We describe the multigrid process by considering the interplay between two grids. The fine grid problem can be written in the form

$$L^f \, U^f = F^f.$$

The decision to switch to the coarse grid is made after the fine grid approximation V^f has been sufficiently smoothed by the relaxation process, i.e., after the high-frequency content of the error $V^f - U^f$ has been sufficiently reduced. The auxiliary equation on the coarse grid is·

$$L^c U^c = F^c,$$

where

$$F^c = R[F^f - L^f V^f].$$

The restriction operator R interpolates a function from the fine grid to the coarse grid. The coarse grid operator and correction are denoted by L^c and U^c, respectively. After an adequate approximation V^c to the coarse grid problem has been obtained, the fine grid approximation is updated using

$$V^f \leftarrow V^f + PV^c.$$

The prolongation operator P interpolates a function from the coarse grid to the fine grid.

We turn our attention to the model problem (3.1) but now defined on [-1, 1] with the Dirichlet boundary conditions. The expansion functions are the Chebyshev polynomials

$$T_n(x) = \cos\,(n \cos^{-1} x)$$

and the collocation points are

$$x_j = \cos\,(\pi j/N)\ ,\ j = 0,1,\ldots,\ N.$$

The discrete Chebyshev coefficients are given by

$$\hat{u}_n = \frac{2}{Nc_n} \sum_{j=0}^{N} c_j^{-1}\, u_j \cos\left(\frac{\pi j}{N}\right) \tag{3.9}$$

where

$$c_n = \begin{cases} 2 & \text{if} \quad n = 0 \text{ or } N, \\ 1 & \text{if} \quad 1 \leqslant n \leqslant N-1, \end{cases}$$

and $u_j = U(x_j)$. Thus the interpolating function is

$$\tilde{u}(x) = \sum_{n=0}^{N} \hat{u}_n T_n(x).$$

The analytic derivative of this function is given by

$$\frac{\partial \tilde{u}}{\partial x} = \sum_{n=0}^{N} \hat{u}_n^{(1)}\ T_n(x),$$

where the coefficients are computed recursively using

$$\hat{u}^{(1)}_{N+1} = \hat{u}^{(1)}_{N} = 0 ,$$

$$c_n \hat{u}^{(1)}_{n} = \hat{u}^{(1)}_{n+2} + 2(n+1)\, \hat{u}_{n+1} , \quad n = N-1, N-2, \ldots, 0.$$

The Chebyshev spectral derivatives at the collocation points are

$$\frac{\partial \tilde{u}}{\partial x}(x_j) = \sum_{n=0}^{N} \hat{u}^{(1)}_n \cos\left(\frac{\pi j n}{N}\right) . \tag{3.10}$$

The second derivative is calculated in a similar fashion. The sums in (3.9) and (3.10) may be efficiently evaluated using a version of the FFT in O(N log N) operations.

The natural prolongation operator in this application represents trigonometric interpolation. We describe this process in detail. On the coarse grid the discrete Chebyshev coefficients of the corrections u_j at the collocation points x_j are computed using

$$\hat{u}_n = \frac{2}{N_c \tilde{c}_n} \sum_{j=0}^{N_c} \tilde{c}_j^{-1} u_j \cos\left(\frac{\pi j n}{N_c}\right) , \quad n = 0,1,\ldots,N_c .$$

The fine grid approximation is then updated using

$$u_j \leftarrow u_j + \sum_{n=0}^{N_c} \hat{u}_n \cos\left(\frac{\pi j n}{N_f}\right) , \quad j = 0,1,\ldots,N_f ,$$

where $u_j = u(\tilde{x}_j)$ and $\tilde{x}_j$ are the fine grid collocation points. The restriction operator is chosen to be the adjoint of P.

4. A SURVEY OF SECOND-ORDER FINITE DIFFERENCE PRECONDITIONINGS

In this section we compare several types of preconditioning based on incomplete LU decompositions (see Meijerink and van der Vorst (1981)) of the matrix which represents the standard five-point second-order finite difference approximation to the differential equation (2.1). Two such preconditionings were discussed in Zang, Wong, and Hussaini (1983). The first, denoted by H_{LU}, has L identical to the lower triangular portion of H_{FD}, the finite difference matrix, and U chosen so that the two super diagonals of LU agree with those of H_{FD}. The second preconditioning H_{RS} has the diagonal elements of L altered from those of H_{FD} so as to ensure that the row sums of H_{RS} and H_{FD} are identical. We introduce a third type of preconditioning based on the strongly implicit method of Stone (1968).

The elements of L and U in all these instances can be easily computed by simple recursive formulae. The construction of the factors L and U are described in detail for Stone's method.

A five-point approximation to (2.1) at the pth mesh point can be written as

$$B_p u_{p-N} + D_p u_{p-1} + E_p u_p + F_p u_{p+1} + H_p u_{p+N} = q_p, \qquad (4.1)$$

or, in matrix form, as

$$H_{FD} \underset{\sim}{u} = \underset{\sim}{q} .$$

Stone's idea was to modify the matrix H_{FD} by a 'small' matrix M so that:

(i) the factorization of $\tilde{H} = H_{FD} + M$ into the product LU involves much less work than the standard LU decomposition of H_{FD};

(ii) the elements of L and U are easily calculated;

(iii) $\|M\| << \|H_{FD}\|$.

As a step to satisfying these criteria the factors L and U were chosen to have three non-zero diagonals, as shown in Fig. 1, corresponding to the diagonals b, d, and e, and e, f, and h respectively of H_{FD}.

The product LU has seven non-zero diagonals, the additional two being immediately interior to the B and H diagonals. The matrix M is taken to comprise these extra two diagonals. The elements of L and U can be computed from their relationships with the elements of H_{FD}.

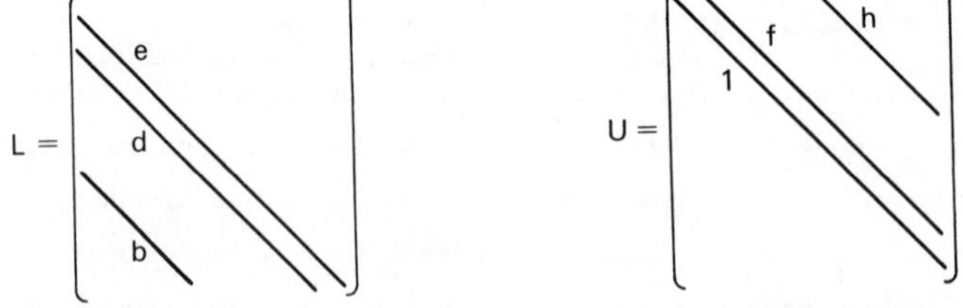

Fig. 1

Analogous with (4.1) the pth component of $M\underset{\sim}{u}$ can be written in the form

$$C_p u_{p-N+1} + G_p u_{p+N-1}. \tag{4.2}$$

Stone decided to diminish the magnitude of this term by subtracting from it a closely equivalent expression obtained by Taylor series expansions. Using these expansions it is easily shown that

$$u_{p-N+1} \simeq -u_p + u_{p+1} + u_{p-N}$$

and (4.3)

$$u_{p+N-1} \simeq -u_p + u_{p+N} + u_{p-1}.$$

Stone then introduced a parameter α, $0 < \alpha < 1$, and defined the pth component of $M\underset{\sim}{u}$ to be

$$C_p\{u_{p-N+1} - \alpha(-u_p + u_{p+1} + u_{p-N})\}$$

$$+ G_p\{u_{p+N-1} - \alpha(-u_p + u_{p+N} + u_{p-1})\}. \tag{4.4}$$

Hence the pth equation of $\tilde{H}\underset{\sim}{u} = \underset{\sim}{q}$ is, by (4.1) and (4.4),

$$(B_p - \alpha C_p)u_{p-N} + (D_p - \alpha G_p)u_{p-1} + \{E_p + \alpha(C_p + G_p)\}u_p + (F_p - \alpha C_p)u_{p+1}$$

$$+ (H_p - \alpha G_p)u_{p+N} + C_p u_{p-N+1} + G_p u_{p+N-1} = q_p . \tag{4.5}$$

The relationships between the elements of $\tilde{H} = H_{FD} + M$ and those of L and U are then given by

$$\begin{aligned}
b_p &= B_p/(1 + \alpha f_{p-N}) \\
d_p &= D_p/(1 + \alpha h_{p-1}) \\
e_p &= E_p + \alpha(b_p f_{p-N} + d_p h_{p-1}) - (b_p h_{p-N} + d_p f_{p-1}) \\
f_p &= (F_p - \alpha b_p f_{p-N})/e_p \\
h_p &= (H_p - \alpha d_p h_{p-1})/e_p ,
\end{aligned} \tag{4.6}$$

for $p = 1,\ldots,(N-1)^2$. Any terms with non-positive subscripts that occur in (4.6) are replaced by zero. For any fixed value of α, $0 < \alpha < 1$, (4.6) defines an incomplete LU-decomposition of H_{FD}. We denote this factorization by $H_{ST}(\alpha)$.

The eigenvalues of the iteration matrices $\tilde{H}^{-1} L_{sp}$ corresponding to these types of preconditioning have been computed numerically by the QR algorithm (see Wilkinson (1965)). The extreme eigenvalues for H_{FD}, H_{LU}, and H_{RS} are given in Zang, Wong, and Hussaini (1983), and those for H_{FD}, $H_{ST}(1.0)$ and $H_{ST}(0.9)$ in Table 3. A few of the eigenvalues at the lower end of the spectrum have small imaginary parts while the rest are completely real.

Table 3

Extreme Eigenvalues for Preconditioned Chebyshev Operator.

N	$H_{FD}^{-1} L_{sp}$		$H_{ST}^{-1}(1.0) L_{sp}$		$H_{ST}^{-1}(0.9) L_{sp}$	
	λ_{min}	λ_{max}	λ_{min}	λ_{max}	λ_{min}	λ_{max}
4	1.00	1.76	1.01	1.64	0.99	1.65
8	1.00	2.13	0.78	2.04	0.80	2.07
16	1.00	2.31	0.62	2.28	0.55	2.33
24	1.00	2.36	0.58	2.95	0.36	2.41

Table 4

Single-grid Condition Number.

N	$H_{LU}^{-1} L_{sp}$	$H_{RS}^{-1} L_{sp}$	$\bar{H}_{ST}^{-1}(1.0) L_{sp}$	$H_{ST}^{-1}(0.9) L_{sp}$
4	1.85	1.72	1.63	1.67
8	3.91	2.71	2.61	2.59
16	11.62	4.07	3.67	4.24
24	24.66	5.22	5.14	6.69

Table 5

Multigrid Condition Number.

N	$H_{LU}^{-1} L_{sp}$	$H_{RS}^{-1} L_{sp}$	$H_{ST}^{-1}(1.0)L_{sp}$	$H_{ST}^{-1}(0.9)L_{sp}$
8	1.79	2.07	1.70	1.71
16	2.12	2.92	2.09	2.08
24	2.26	3.79	2.81	2.15

In order to examine the effectiveness of these preconditionings from the multigrid point of view, we need to know the smallest high-frequency eigenvalue. The numerical results indicate that this is 1.22 for H_{LU} and 1.45 for H_{RS}, independent of N. For H_{ST} this value lies between 1.10 and 1.21. Tables 4 and 5 contain the single-grid and multigrid condition numbers respectively for H_{LU}, H_{RS}, $H_{ST}(1.0)$ and $H_{ST}(0.9)$. Here we see that $H_{ST}(0.9)$ is more effective as a preconditioner for multigrid iterations. The multigrid condition number for $H_{ST}(1.0)$ lies between those for H_{LU} and H_{RS}.

In Table 3 we see that the maximum eigenvalue of $H_{ST}^{-1}(0.9)L_{sp}$ grows more slowly than that of $H_{ST}^{-1}(1.0)L_{sp}$ with increasing N. For values of N greater than 24 the eigenvalue calculation using the QR algorithm becomes too expensive. Using the power method the maximum eigenvalue of $H_{ST}^{-1}(0.9)L_{sp}$ for N = 32 was calculated and found to be 2.45. Thus the maximum eigenvalue of $H_{ST}^{-1}(0.9)L_{sp}$ exhibits slow growth with N. This means that the multigrid condition number does not increase drastically with N. However, it is doubtful whether the choice of α in Stone's algorithm is problem independent and this may detract from the robustness of this preconditioning.

To complete this section we look at one other type of incomplete LU-decomposition which is due to Wesseling (1982a). In this decomposition the sparsity of the factors L and U differ from those in Fig. 1 by the addition of extra non-zero diagonals c and g which are located immediately interior to the b and h diagonals respectively. The main diagonal of U is specified to be unity. The elements of L and U are computed from those of H_{FD} (see (4.1)) recursively as follows:

$$b_p = B_p$$

$$c_p = -b_p f_{p-N}$$

$$d_p = D_p - b_p g_{p-N}$$

$$e_p = E_p - b_p h_{p-N} - c_p g_{p-N+1} - d_p f_{p-1} \qquad (4.7)$$

$$f_p = (F_p - h_{p-N+1} c_p)/e_p$$

$$g_p = -d_p h_{p-1}/e_p$$

$$h_p = H_p/e_p,$$

for $p = 1,\ldots,(N-1)^2$. Quantities that are not defined are replaced by zero. This preconditioning will be known as the seven-diagonal preconditioning and we denote it by H_{SD}. The error matrix M contains two non-zero diagonals. It can be shown that the norm of the error matrix of the seven-diagonal preconditioning is smaller than those corresponding to H_{LU} and H_{RS}.

In Table 6 we give details of the seven-diagonal preconditioning for different values of N. Included in this table are the extreme eigenvalues of $H_{SD}^{-1} L_{sp}$ and the single-grid and multigrid condition numbers. The smallest high frequency eigenvalue was found to be 1.23 independent of N. The results indicate that H_{SD} is the most effective preconditioner considered to date.

Table 6

Details of Seven-Diagonal Preconditioning.

N	λ_{min}	λ_{max}	κ_{SG}	κ_{MG}
4	1.00	1.76	1.77	-
8	0.85	2.16	2.54	1.76
16	0.46	2.38	5.22	1.93
24	0.25	2.47	9.81	2.01

5. HIGHER-ORDER FINITE DIFFERENCE PRECONDITIONING

Here we consider the possibility of choosing the preconditioning matrix $\tilde{H}$ to be a fourth-order finite difference representation of (2.1).

For a general non-uniform grid a compact nine-point finite difference approximation cannot be constructed since the set of equations for the coefficients of the scheme is inconsistent. Instead an approximation was constructed based on fourth-order finite difference formulae for each of the second derivatives separately. The coefficients of the function values at the grid points are functions of the mesh lengths which define the non-uniform mesh and are given in the Appendix. At internal points the approximation to the second derivative reduces to the following in the case when the mesh is uniform:

$$\frac{d^2u}{dx^2} = \frac{1}{12h^2}\{-u_{i-2} + 16u_{i-1} - 30u_i + 16u_{i+1} - u_{i+2}\} + O(h^4). \qquad (5.1)$$

The approximation at points one mesh length from the boundary is constructed using the same number of points as interior equations thus maintaining the same order of accuracy. However, in doing this the symmetry of the sparsity pattern of the coefficient matrix is destroyed. As before, it is an incomplete LU decomposition of this finite difference matrix that is used to precondition (2.3). An algorithm due to Wesseling (1982b) was used to perform this decomposition. The factors L and U have the same sparsity pattern as the lower and upper triangular portions of H_{FD4} respectively and the corresponding diagonal elements of L and U are equal. We let H_W denote this preconditioning.

Table 7

Extreme Eigenvalues for Preconditioned Chebyshev Operator.

N	$H_{FD4}^{-1} L_{sp}$		$H_W^{-1} L_{sp}$	
	λ_{min}	λ_{max}	λ_{min}	λ_{max}
8	1.00	1.59	0.51	1.75
16	1.00	1.79	0.19	2.03
24	1.00	1.83	0.09	2.13

Table 8

Condition Numbers for $H_W^{-1}L_{sp}$.

N	Single-grid	Multigrid
8	3.43	1.73
16	10.68	2.01
24	23.67	2.11

Table 7 contains the largest and smallest eigenvalues of $H_{FD4}^{-1} L_{sp}$ and $H_W^{-1} L_{sp}$. Table 8 presents the single-grid and multigrid condition numbers for the matrix $H_W^{-1} L_{sp}$. We note that the multigrid condition numbers presented here compare favourably with those obtained using second-order finite difference preconditioning. The smallest high frequency eigenvalue was found to be 1.01, independent of the value of N. A disadvantage of this preconditioning is that more work is needed to perform the decomposition.

6. SPECTRAL PRECONDITIONING

The final type of preconditioning we investigate is that based on the spectral matrix, L_{sp}, itself. As stated earlier the spectral matrix is full and hence costly to invert. Let H_5 be the matrix containing five diagonals of L_{sp}, the positions of which correspond to the non-zero diagonals of the finite difference matrix based on the five point formula. Let the matrix H_9, illustrated in Fig. 2, be the corresponding matrix with nine non-zero diagonals.

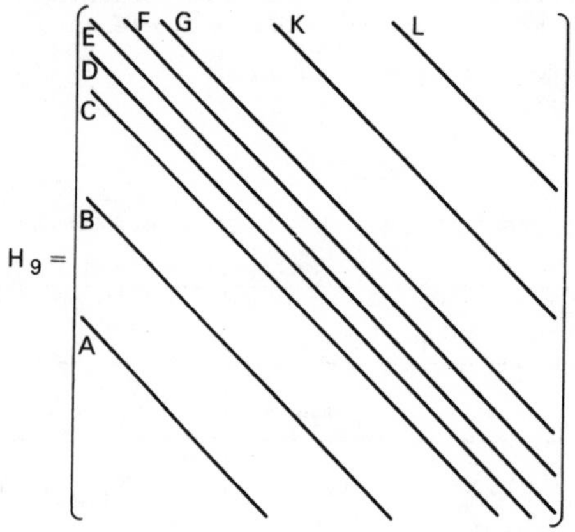

Fig. 2

Experimentally, we found that several eigenvalues of $H_5^{-1} L_{sp}$ were negative, thus showing that this matrix is not positive definite. The eigenvalues of $H_9^{-1} L_{sp}$ were also computed and found to be positive. Therefore, H_5 was discarded as a potential preconditioner and the usefulness of H_9 examined further.

An approximate LU-decomposition of H_9 is used to precondition (2.2), i.e., H is taken to be the product of a lower triangular matrix L and an upper triangular matrix U. We perform this decomposition according to the criterion described in Section 5 for constructing H_{LU}, i.e., the factors L and U are chosen so that L is identical to the lower triangular portion of H_9 and U is such that the four super diagonals of LU agree with those of H_9. The sparsity pattern of the factors L and U is shown in Fig. 3. We denote this preconditioning by H_{sp}.

The elements of the matrices L and U can be computed recursively as follows:

$$
\begin{aligned}
a_p &= A_p \\
b_p &= B_p - a_p k_{p-2N} \\
c_p &= C_p \\
d_p &= D_p - c_p f_{p-1} \\
e_p &= E_p - d_p f_{p-1} - c_p g_{p-2} - b_p k_{p-N} - a_p \ell_{p-2N} \qquad (6.1) \\
f_p &= (F_p - d_p g_{p-1})/e_p \\
g_p &= G_p/e_p \\
k_p &= (K_p - b_p \ell_{p-N})/e_p \\
\ell_p &= L_p/e_p,
\end{aligned}
$$

for $p = 1,\ldots,(N-1)^2$. Any terms with non-positive subscripts occurring in (6.1) are replaced by zero.

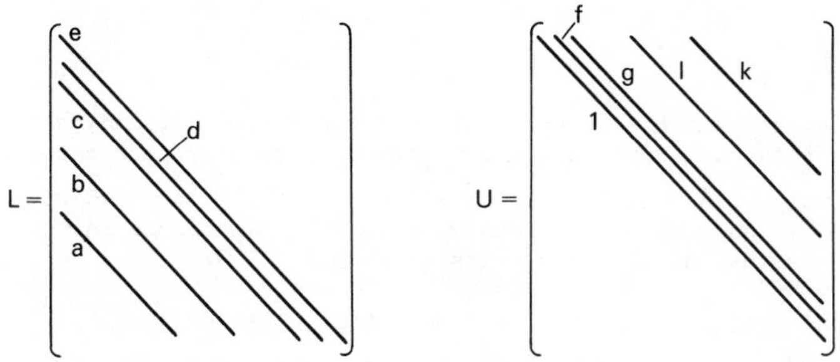

Fig. 3.

Table 9 contains the largest and smallest eigenvalues of $H_9^{-1} L_{sp}$ and $H_{sp}^{-1} L_{sp}$. Table 10 presents the single-grid and multigrid condition numbers for the matrix $H_{sp}^{-1} L_{sp}$. The smallest high frequency eigenvalue was found to be 0.90, independent of the value of N. For N = 32 we were able to calculate that the maximum eigenvalue of $H_{sp}^{-1} L_{sp}$ was 1.52. We conclude from these results that H_{sp} is likely to be an effective preconditioner in multigrid applications since the observed behaviour of the condition number compares favourably with that of the other preconditioners considered in this paper.

Table 9

Extreme Eigenvalues for the Preconditioned Chebyshev Operator.

N	$H_9^{-1} L_{sp}$		$H_{sp}^{-1} L_{sp}$	
	λ_{min}	λ_{max}	λ_{min}	λ_{max}
8	0.63	1.19	0.39	1.31
16	0.23	1.33	0.12	1.47
24	0.11	1.37	0.05	1.51

Table 10

Condition Numbers for $H_{sp}^{-1} L_{sp}$.

N	Single-grid	Multigrid
8	3.40	1.44
16	12.53	1.63
24	27.87	1.69

7. NUMERICAL RESULTS

Here we investigate the performance of various preconditioning ideas developed in this paper within the SMG method. The operation count for a relaxation sweep of SMG is $O(N \log N)$ where $N = (N-1)^2$ compared with $O(N)$ for finite difference methods. Thus we make our comparisons in terms of machine time instead of the work units of Brandt (1977).

The measure used is the equivalent smoothing rate, defined by μ_e. This was introduced in earlier work on SMG by Zang, Wong, and Hussaini (1983) and is defined as follows. In some preliminary calculations, the average time τ_0 required for a single fine-grid relaxation is determined. For an actual multigrid calculation let r_1 and r_2 be the residuals after the first and last fine-grid relaxations respectively, and let τ be the total CPU time. Then we define μ_e by

$$\mu_e = \left(\frac{r_2}{r_1}\right)^{1/((\tau/\tau_0) - 1)} .$$

There are many variants of the multigrid method. In the one used here we first solve the problem on the coarsest grid, that solution is then interpolated to the next finer level to serve as the initial approximation for a multigrid iteration involving these two levels, etc. The sizes of the grids on the coarsest and finest levels are 4×4 and 32×32 respectively. Internal checks based on the anticipated smoothing rates were used in governing decisions to switch levels. Non-stationary Richardson iteration employing three distinct parameters was used for relaxation. We used the correction scheme of Brandt (1977) with random numbers for the initial guess. On lower levels the right-hand sides were obtained by applying the appropriate restriction operator to the finest level right-hand side.

The test problems are specified by

$$a(x,y) = b(x,y) = 1 + \varepsilon \exp(\cos(\beta\pi(x+y))),$$

$$u(x,y) = \sin(\sigma\pi x + \pi/4) \sin(\sigma\pi y + \pi/4).$$

The parameters of the test problem are given in Table 11. Problem 1 has constant coefficients and is also well-resolved by the Chebyshev collocation method. On the other hand, for Problem 3 the coefficients of the equation oscillate so rapidly that the finest grid cannot resolve them. The converged solution of the collocation equations on the finest grid has an error of order 1. This is a test of whether the Chebyshev SMG method is robust enough to converge on such a problem. Table 12 contains the values of the equivalent smoothing rates for various preconditioners.

Table 11

Parameters of the Test Problems

Problem	ε	σ	β
1	0.00	1	1
2	0.20	2	2
3	1.00	5	10

Table 12

Equivalent Smoothing Rates

Problem	H_{LU}	$H_{ST}(0.9)$	H_{SD}	H_{sp}
1	0.26	0.16	0.12	0.24
2	0.60	0.65	0.47	0.54
3	0.76	0.84	0.76	0.82

The results show the seven-diagonal finite difference preconditioning and the spectral preconditioning are comparable in terms of efficiency and give an improvement over the other preconditioners except on Problem 3 where the performance of all the preconditioners is about the same. However, since in our computations to date the spectral matrix is never stored, the elements of H_{sp} are more expensive to compute than H_{SD}. A closer examination revealed that the factorization accounted for around 1.5% of the total CPU time when a finite difference preconditioner was used. The figure for the spectral preconditioning was around 15%. Notice the superb performance of Stone's preconditioning with $\alpha = 0.9$ on the constant coefficient problem. This performance is not maintained on the more difficult problems which demonstrates that the choice of the parameter α is problem dependent. The seven-diagonal preconditioning also performs extremely well on the constant coefficient problem. We did not experiment with the fourth-order finite difference preconditioning since to compute the approximation in the non-constant coefficient case was thought to be too laborious.

8. CONCLUSIONS

In this paper we have developed efficient iterative techniques for solving the algebraic equations which arise from the discretization of a self-adjoint elliptic equation using the spectral collocation method. An advantage of using spectral methods is that they possess superior approximation properties compared with finite difference and finite element methods. In practice, this means one can obtain the same accuracy with fewer mesh points than are needed for finite difference or finite element methods. Acceleration of the basic iterative scheme has been enhanced by employing the multigrid method. The need for preconditioning has been demonstrated and efficient preconditioners for multigrid iterations presented.

APPENDIX

Higher-order Finite Differences on Non-uniform Grids

Suppose that we require an approximation to the second derivative of u at point C in Fig. 4. We assume that the approximation has the form

$$\frac{d^2u}{dx^2} \simeq au_A + bu_B + cu_C + du_D + eu_E, \tag{A.1}$$

where the coefficients a, b, c, d and e are to be determined.

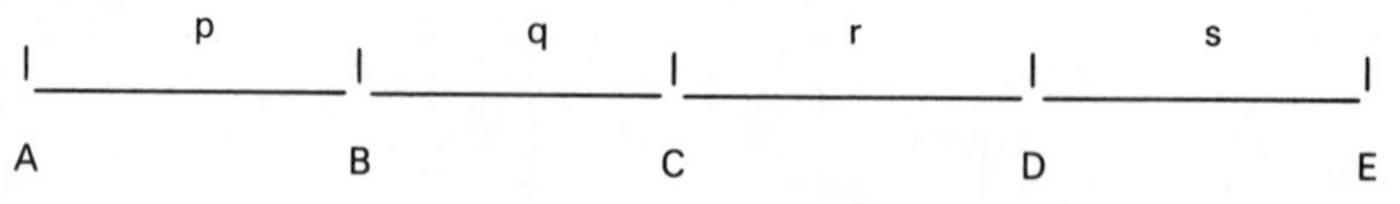

Fig. 4

The right-hand side of (A.1) is expanded in a Taylor's series about the point C. The coefficient of the second derivative is set to unity while those of u and the first, third, and fourth derivatives are set to zero. This results in a system of five linear equations for the coefficients appearing in (A.1). The accuracy of the approximation is $O(h^3)$ where h is a typical mesh length.

If we define $G = p + q$ and $H = r + s$, then the coefficients are given by

$$a = \frac{-2q(H + r) + 2Hr}{[Gp(G + H)(r + G)]},$$

$$b = \frac{2G(H + r) - 2Hr}{[pq(q + r)(H + q)]},$$

$$d = \frac{2H(G + q) - 2Gq}{[rs(q + r)(G + r)]},$$

$$e = \frac{-2r(G + q) + 2Gq}{[Hs(H + q)(H + G)]},$$

$$c = -a - b - d - e.$$

REFERENCES

Brandt, A. (1977) Multi-level adaptive solutions to boundary-value problems, *Math. Comp.* **31**, pp 330-390.

Fox, L. (1964) Numerical Linear Algebra. Oxford, Clarendon Press.

Gottlieb, D. and Orszag, S. A. (1977) Numerical Analysis of Spectral Methods: Theory and Applications. CBMS-NSF Regional Conference Series in Applied Mathematics No. 26, Philadelphia: SIAM.

Hackbusch, W. (1980) Convergence of multigrid iterations applied to difference equations, *Math. Comp.*, **34**, pp 425-440.

Hackbusch, W. and Trottenberg, U. (1982) (eds.) Multigrid Methods, Proceedings of the Conference held at Köln-Porz, West Germany, Lecture Notes in Mathematics, 960, Springer-Verlag, Berlin.

Haldenwang, P., Labrosse, G., Abboudi, S. and Deville, M. (1983) Chebyshev 3-D spectral and 2-D pseudospectral solvers for Helmholtz equation. Submitted to *J. Comput. Phys.*

Hussaini, M. Y., Kopriva, D. A., Salas, M. D., and Zang, T. A. (1983) Spectral Methods for the Euler Equations, AIAA 6th Computational Fluid Dynamics Conference, Danvers, Massachusetts, USA, Paper No. 83-1942-CP.

Hussaini, M. Y., Salas, M. D. and Zang, T. A. (1983) Spectral methods for inviscid, compressible flows. In "Advances in Computational Transonics" (W. G. Habashi, ed), Pineridge Press, Swansea.

Meijerink, J. A. and Van Der Vorst, H. A. (1981) Guidelines for the usage of incomplete decompositions in solving sets of linear equations as they occur in practical problems, *J. Comput. Phys.*, **44**, pp 134-155.

Orszag, S. A. (1980) Spectral methods for problems in complex geometries, *J. Comput. Phys.*, **37**, pp 70-92.

Stone, H. L. (1968) Iterative solution of implicit approximations of multi-dimensional partial differential equations, *SIAM J. Numer. Anal.*, **5**, pp 530-558.

Streett, C. L., Zang, T. A. and Hussaini, M. Y. (1983) Spectral multigrid methods with applications to transonic potential flow. ICASE Report 83-11, NASA Langley Research Center, Virginia, USA.

Wesseling, P. (1982a) A robust and efficient multigrid method. In Multigrid Methods (W. Hackbusch and U. Trottenberg, eds.), Lecture Notes in Mathematics, 960, Springer-Verlag, Berlin.

Wesseling, P. (1982b) Theoretical and practical aspects of a multigrid method, *SIAM J. Sci. Statis. Comput.*, **3**, pp 387-407.

Wilkinson, J. H. (1965) The Algebraic Eigenvalue Problem. Oxford, Clarendon Press.

Zang, T. A., Wong, Y. S. and Hussaini, M. Y. (1982) Spectral multigrid methods for elliptic equations, *J. Comput. Phys.*, **48**, pp 485-501.

Zang, T. A., Wong, Y. S. and Hussaini, M. Y. (1983) Spectral multigrid methods for elliptic equations II, ICASE Report 83-12, NASA Langley Research Center, Virginia, USA.

MG TECHNIQUES FOR STAGGERED DIFFERENCES*

B. Favini and G. Guj

(Dipartimento di Meccanica e Aeronautica, Università di Roma 'La Sapienza', Italy)

SUMMARY

Numerical solutions for elliptic equations, approximated on a staggered grid, are evaluated by a multigrid technique. Several aspects of the method are analysed, with particular attention to the definition of transfer operators and the enforcement of Dirichlet boundary conditions. The solution accuracy on the coarse grids is improved by a suitable definition of the restriction operator. Furthermore the same accuracy is obtained by either an implicit or an explicit treatment of the boundary conditions, providing in this case the boundary conditions are assigned at every sweep. Some numerical test cases are performed to assess the accuracy and the efficiency of the proposed multigrid technique.

1. INTRODUCTION

The numerical modelling of the heat transfer and fluid motion phenomena has become more and more sophisticated in order to better represent real physical processes and to improve the accuracy and the stability of the discrete approximations. The location of variables plays a peculiar role: in fact the discrete formulation is able to represent principal properties of the original continuum problem only if the variables are placed in a suitable way. In the solution of Navier-Stokes equations the major difficulties connected with the enforcement of the incompressibility constraint are overcome by adopting the MAC scheme (Harlow and Welch (1965)). In this scheme the velocity components are defined at the cell midsides, while the pressure is displaced to the centre of the cell. Moreover Brandt emphasizes (Brandt and Dinar (1979)) that the discrete approximation obtained with the MAC scheme for an elliptic differential operator has a good discrete ellipticity measure, thus it is stable. Analogously in heat transfer problems, or other scalar transport phenomena, the integral conservation law can be satisfied exactly for each individual cell by defining the temperature at the centre of each cell (Zedan and Schneider (1982)).

For Cauchy-Riemann equations, on the other hand, the only possibility of transforming the discrete formulation of the system of two equations into a single equation of higher order is to use a staggered grid. Then, as for Navier-Stokes equations, this scheme shows a good ellipticity (Brandt and Dinar (1979)).

---

* This research has been supported by the 'Nucleo di Meccanica Numerica' Università di Roma 'La Sapienza' under contract M.P.I. 08/01/04.

The implementation of a multigrid solver for staggered differenced problems should incorporate specialised smoothing procedures, suitable transfer operators and boundary condition representation. Up to now the efforts have been focused primarily on relaxation techniques either developing multigrid-orientated methods such as the Convective Successive Line Relaxation (Dinar (1979)) and the Distributive Gauss-Seidel (Brandt and Dinar (1979)), or adapting special matrix solvers to the multigrid structure (Hackbusch and Trottenberg (1982)).

The aim of this study is directed to analyze the role of transfer operators and of the enforcement of boundary conditions on staggered grids. For linear problems solved by a standard multigrid technique with globally defined grids, the accuracy of the solution on the finest grid is defined only by the truncation error associated with the discrete operator, while the coarse grid problems (CGP) depend on the restriction operator of the solution and of the residual. Therefore, in the adaptive technique, using subgrids defined only in regions of higher gradients, the solution on the finer globally defined level would depend on the transfer operators. Nonlinear problems would behave in the same way as the linear ones, but the analysis cannot be performed theoretically and only numerical experiments could enlighten the question. Several authors (Brandt and Dinar (1979), Hackbusch and Trottenberg (1982), Fuchs (1981)) have defined the restriction of the solution so as to satisfy some properties associated with the original differential problems: e.g. in the solution of Navier-Stokes equations the restriction is chosen so as to verify the conservation of the mass also on the coarser levels. However, better accuracy can be obtained with the 'local inversion method' (LIM) (Hyman (1977)) which uses the discrete formulation of the problem to compute the solution at new locations, simply rotating the reference system.

The enforcement of Dirichlet boundary conditions on staggered grids leads to the same difficulties associated with the treatment of Neumann boundary conditions on non-staggered grids. The relevant aspect is the presence of a residual associated to the boundary equations: this residual has to be projected onto the coarse grid. However, by modifying the multigrid cycle we can avoid the transfer of any value at the boundary; it is sufficient to enforce the boundary conditions after every relaxation sweep. Otherwise, we can assign directly the boundary conditions modifying the discrete operator near the boundary.

The analysis of the properties of the restriction is developed in the next section, together with the problem of the enforcement of the boundary conditions. In the third section some numerical test cases are presented in order to verify the accuracy of the proposed multigrid technique. In order to emphasize the features of staggered differences only linear problems are considered.

2. SOLUTION AND RESIDUAL TRANSFERS

We can define, following the notation proposed by Brandt in Hackbusch and Trottenberg (1982), an algebraic problem on a level k

$$L^k u^k = f^k \tag{1}$$

where L^k, u^k and f^k are respectively the discrete operator obtained by discretising the original continuum operator, the vector of the unknowns and the right hand side vector. The coarse grid problem gives:

$$L^{k-1}u^{k-1} = \hat{f}^{k-1} \tag{2}$$

where the expression for the modified source term is:

$$\hat{f}^{k-1} = L^{k-1}\hat{I}_k^{k-1}u^k + I_k^{k-1}r^k . \tag{3}$$

In Zedan and Schneider (1982) , $\hat{I}_k^{k-1}$ and I_k^{k-1} are the restriction operators respectively of the solution and of the residual, u^k is a smooth approximation of the solution on level k, and r^k the residual computed on level k, i.e.

$$r^k = f^k - L^k u^k . \tag{4}$$

Substituting this expression in (3) and reordering the terms gives:

$$\hat{f}^{k-1} = I_k^{k-1}f^k + \tau_k^{k-1} \tag{5}$$

where τ_k^{k-1}, the local truncation error, is given by

$$\tau_k^{k-1} = L^{k-1}\hat{I}_k^{k-1}u^k - I_k^{k-1}L^k u^k . \tag{6}$$

Equation (6) can be interpreted as the approximation of the difference between the truncation errors of levels k and k-1 (Hackbusch and Trottenberg (1982)).

In order to simplify the two level analysis, the algebraic problem on the coarse grid is solved by a direct elimination. A new approximation of the CGP solution results from (2) and (3):

$$u^{k-1} = (L^{k-1})^{-1}(L^{k-1}\hat{I}_k^{k-1}u^k + I_k^{k-1}r^k), \tag{7}$$

which for linear problems results in

$$u^{k-1} = \hat{I}_k^{k-1}u^k + (L^{k-1})^{-1}I_k^{k-1}r^k . \tag{8}$$

From the new approximation we can compute the correction which will be interpolated to the finer grid:

$$v^{k-1} = u^{k-1} - \hat{I}_k^{k-1}u^k . \tag{9}$$

Substituting equation (8) in (9) gives

$$v^{k-1} = (L^{k-1})^{-1} I_k^{k-1} r^k \ . \tag{10}$$

Therefore the correction does not depend on the projection of the actual finest solution u^k on level k-1, but only on the restriction of the residual. Consequently, also the solution of the finest level will not depend upon the restriction operator $\hat{I}_k^{k-1}$.

The role of $\hat{I}_k^{k-1}$ may be analysed by considering a linear problem that is solved by a multigrid scheme designed for nonlinear problems, i.e. a FAS scheme (Brandt (1977)) and not by a correction scheme. In fact, at convergence the residual on level k equals zero, and, from (8), the solution on the coarse level k-1 is given by

$$u^{k-1} = \hat{I}_k^{k-1} u^k \ . \tag{11}$$

Therefore, the restriction operator can spoil the accuracy of the solution on the coarse grid. If a double discretisation (Brandt (1981)) is adopted the residual does not tend to zero at convergence, that is the solution on the coarse grid depends also on the restriction operator I_k^{k-1}.

For non-staggered grids a natural definition of the restriction is the 'injection', i.e. to assign to a node at the coarse level the value of the corresponding node of the successive finer level (Fig. 1).

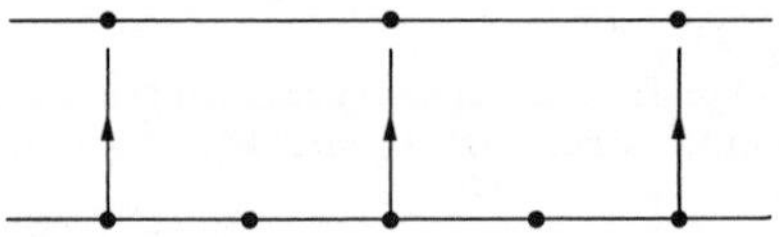

Fig. 1 Non-staggered grids: the solution restriction is evaluated by injection

For staggered grids, since there is no correspondence among points on the two levels, an interpolation operator has to be used to transfer both the solution and the residual (Fig. 2). Therefore the order of restriction operator should be no less than the order of the truncation error on the finer level.

A really simple way to perform a sufficiently accurate restriction is to compute, by means of the finite difference approximation of the continuum model, a new value at a fictitious node on the fine level, located in correspondence to a node on the coarse level (Hyman (1977)). Let us consider now the following one dimensional problem:

$$u_{xx} = b \quad \text{on }]0,1[$$
$$u(o) = u(1) = 0 \, , \tag{12}$$

discretised by a centred second order scheme:

$$\frac{u_{i+1} - 2u_i + u_{i-1}}{\lambda^2} = b_i \, . \tag{13}$$

The projection of the solution from the fine to the coarse grid, (Fig. 2), is computed by means of (14):

$$u_{i+\frac{1}{2}}^{k-1} = (u_i^k + u_{i+1}^k - \lambda^2 b_i)/2 \, . \tag{14}$$

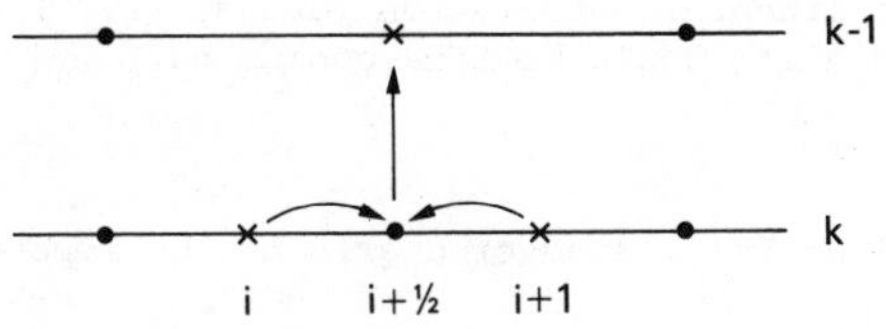

Fig. 2 Staggered grids: the coarse grid value i + ½ is computed at level k and then transferred to level k - 1.

With minor modification, simply rotating the reference system, we can apply the same scheme to a two and three dimensional problem (Fig. 3).

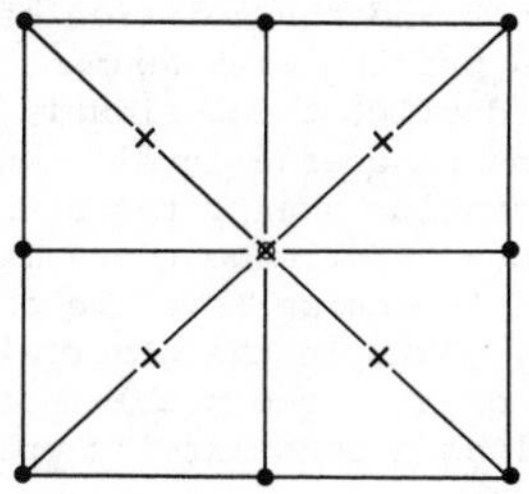

Fig. 3 Staggered grids: local inversion method for two dimensional problems

If the level k is the actual finest level then the right hand side is known, while if the level is a coarser one we can compute it by interpolation. Unless the problem is characterised by rapidly varying coefficients or a rapidly varying right hand side, the residual restriction may be evaluated by a bilinear interpolation (Fig. 4). Otherwise, better results may be obtained by means of an averaging operator, such as the optimal weighting or the full weighting (Brandt (1977)).

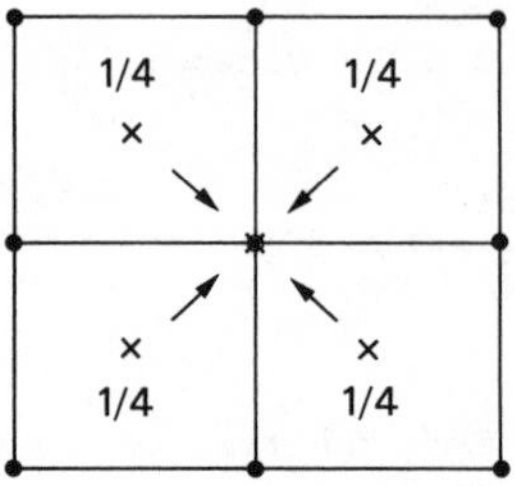

Fig. 4 Staggered grids: residual restriction by bilinear interpolation

For the prolongation of the correction it can be sufficient to use a constant or piecewise linear interpolation (Brandt (1977)). Moreover a higher order interpolation would be time consuming.

3. BOUNDARY CONDITIONS

The boundary conditions for a staggered grid may be imposed in two ways:

(a) defining a virtual cell external to the domain of the problem and assigning the boundary condition by means of an interpolation;

(b) without introducing external cells, modifying the discrete operator near the boundaries.

In the first case it is easier to program the relaxation procedure but the transfer of the residuals and the solution near the boundaries is quite troublesome. As for a problem with Neumann boundary conditions defined on a non-staggered grid, the residuals at the boundaries have to be transferred from fine to coarse level. An alternative can be to reassign, after each relaxation sweep, the boundary conditions. In fact, in this way, the residual at the boundary is equal zero. Besides, while moving from fine to coarse level we can avoid the need to project the external values. But, in the other direction, the presence of the external cells creates the problem of defining the correction at the boundary, unless a constant interpolation for boundary cells is used.

Moreover, if it is possible to use the same discrete operator of the internal cells, the truncation error near the boundaries depends on the order of the interpolation. Let us consider again the one dimensional problem (12). The boundary condition may be enforced by means of a linear interpolation (Fig. 5) which has second order accuracy.

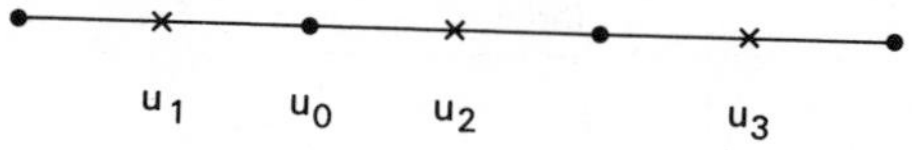

Fig. 5 Enforcement of boundary conditions on staggered grids

The external value u_1 is computed by:

$$u_1 = 2u_0 - u_2 + O(\lambda)^2 . \qquad (15)$$

Substituting this expression in (13), a formula with zero order of accuracy results:

$$\frac{u_3 - 2u_2 + u_1}{\lambda^2} = \frac{u_3 - 3u_2 + 2u_0}{\lambda^2} + O(\lambda)^0 . \qquad (16)$$

The accuracy will be of the first order by means of a quadratic interpolation:

$$u_1 = \frac{u_3 - 6u_2 + 8u_0}{3} + O(\lambda)^3, \qquad (17)$$

which, on substituting in (10), gives:

$$\frac{u_3 - 2u_2 + u_1}{\lambda^2} = \frac{4}{3}\,\frac{u_3 - 3u_2 + 2u_0}{\lambda^2} + O(\lambda) . \qquad (18)$$

Without the external cells we can avoid all these problems by paying once the price of the greater computational work of the relaxation procedure.

4. NUMERICAL EXAMPLES

The first test case analysed is a two dimensional heat conduction problem. The boundary conditions have been chosen so that we can obtain an exact solution which is a function of one variable only. The analytical solution with the boundary conditions shown in Fig. 6 is the parabola

$$\theta = c/2 \times (1 - x) . \qquad (19)$$

This problem has been solved by three different schemes. For all schemes the Laplacian is discretised by the usual five point stencil, which yields second order accuracy, and the prolongation is a constant interpolation on each individual cell.

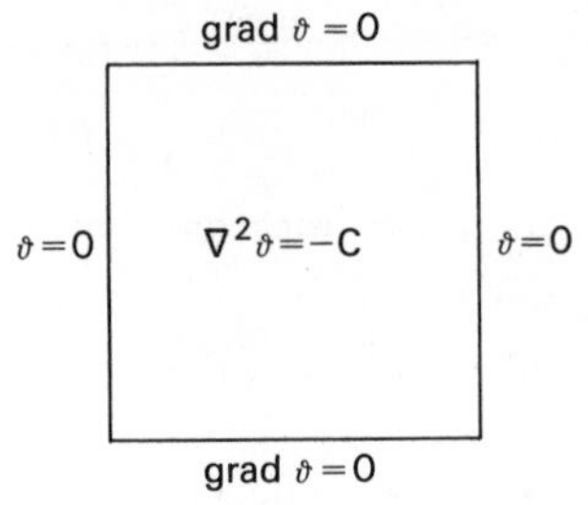

Fig. 6 One dimensional test case

The numerical experiments have been obtained by means of the same multi-grid cycle of the 'sawtooth' type with three sweeps on each grid (Guj and Favini (1982). In scheme A the same restriction operator for both residual and solution is assumed and the boundary conditions are enforced by means of a linear interpolation. For each grid the accuracy of this scheme is of the first order, but the results obtained in a multigrid cycle with three levels exhibit an anomalous behaviour (Table 1); better accuracy is obtained on the intermediate grid rather than on the finest level. This result may be explained examining the related effects of the enforcement of boundary conditions and of the restriction. The error induced by the linear interpolation at the boundary is balanced by the modified right hand side. This behaviour depends on the particular shape of the solution (19). In fact the second order error of the bilinear restriction equals the error of (15), because the second derivative of the parabola (19) is a constant. In the scheme B the boundary conditions are imposed by a quadratic interpolation. Therefore, it shows a second order accuracy. The improved enforcement of boundary conditions permits us to isolate the effect of the restriction. The numerical results on the finest level agree very well with the analytical solution, while the accuracy of the solution on the coarser level is spoiled by the restriction operator (11).

The third scheme C uses the LIM technique, defined in the previous section, to perform the transfer of the solution from fine to coarse grid. The projection of the residual and the enforcement of boundary conditions are defined as in the scheme B. The solution shows the same accuracy on each level. The slight differences between the error norms depend on the definition of the local truncation error (6), which is only an approximation of the differences between the real truncation errors on different grids.

The L_2 norm of the residual at convergence is the same for all three schemes, and also the smoothing factor (Brandt (1981)) is always about 0.5.

The implicit treatment of boundary conditions does not improve the convergence rate nor the accuracy of the solution with respect to model C. Therefore, results for this scheme are not presented. Let us consider now more severe test cases which satisfy Laplace's equation, as the previous one, but with different boundary conditions and source term. This class of problems has the general solution

$$\theta(x,y) = \sin(m\pi x^p)\ \sin(n\pi y^q) \qquad (20)$$

Table 2 shows the numerical results obtained by schemes B and C, setting m = n = q = 1 and p = 2.

Table 1

L_2 Norm of the differential error of problem (19).

A - Linear interpolation of boundary conditions and restriction

B - Second order interpolation of b.c and first order restriction.

C - Second order interpolation of b.c and LIM restriction.

CASE / LEVEL	A	B	C
4 x 4	.31 - 02	.39 - 02	.22 - 06
8 x 8	.16 - 06	.48 - 03	.15 - 06
16 x 16	.27 - 03	.98 - 07	.83 - 07

The results confirm substantially the theoretical analysis. The differences among the three schemes are not so relevant as found for the first test case. For equation (20) the boundary condition enforcement is more accurate with a linear interpolation, because the function (20) exhibits a point of inflexion at each boundary.

Table 2

L_2 Norm of the differential error of problem (20).

A, B, C - as in Table 1.

CASE / LEVEL	A	B	C
4 x 4	.469 - 02	.534 - 02	.968 - 04
8 x 8	.699 - 03	.748 - 03	.455 - 04
16 x 16	.392 - 04	.652 - 04	.248 - 04
32 x 32	.268 - 04	.133 - 04	.133 - 04

REFERENCES

Brandt, A. (1981) Multi-Grid Solvers for Non-Elliptic and Singular-Perturbation Steady-State Problems. Weizmann Institute of Science, Rehovot, Israel.

Brandt, A. and Dinar, N. (1979) Multi-Grid Solutions to Elliptic Flow Problems. ICASE Report, 79-15.

Brandt, A. (1977) Multi-Level Adaptive Solutions to Boundary-Value Problems. *Mathematics of Computation,* **31**, 138, p. 333.

Dinar, N. (1979) Fast Methods for the Numerical Solutions of Boundary Value Problems. PhD Thesis, Weizmann Institute of Science, Rehovot, Israel.

Fuchs, L. (1981) Multi-Grid Solution of the Navier-Stokes Equations on Non-Uniform Grids, in 'Multigrid Methods' (H. Lomax Ed.), NASA Report, CP-2202.

Guj, G. and Favini, B. (1982) Multi-Grid Technique in General Curvilinear Coordinates. In Proceedings of the International Symposium on Refined Modelling Flows, Paris.

Hackbusch, W. and Trottenberg, U. (1982) Multi-Grid Methods. Proceedings of the Conference held at Köln-Porz, November 23-27, 1981. Lecture Notes in Mathematics, **960**, Springer-Verlag, Berlin.

Harlow, F.H. and Welch, J.E. (1965) Numerical Calculations of Time-Dependent Viscous Incompressible Flow of Fluid with Free Surface. *The Physics of Fluids,* **9**, pp. 21-82.

Hyman, J.M. (1977) Mesh Refinement and Local Inversion of Elliptic Partial Differential Equations. *Journal of Computational Physics,* **23**, p. 124.

Zedan, M. and Schneider, G.E. (1982) A Physical Approach to the Finite-Difference Solution of the Conduction Equation in Generalized Coordinates, *Numerical Heat Transfer,* **5**, p. 1.

MULTIGRID ANALYSIS OF LINEAR ELASTIC STRESS PROBLEMS

K.E. Barrett, D.M. Butterfield, S.E. Ellis, C.J. Judd and J.H. Tabor

(Department of Mathematics, Coventry (Lanchester) Polytechnic, Coventry)

1. INTRODUCTION

Much of the work described so far in the conference has concentrated on the detailed analysis of a variety of multigrid techniques, emphasis being placed on such aspects as their rate of convergence, their range of application and their computational implementation. However, sight must not be lost of the overall design tool that is typically required in many engineering problems. It is not sufficient to simply have efficient solution procedures, it is also necessary to have robust user friendly methods which permit the transition between the mathematical model of the real life situation and the discretised system to be made easily, especially at the pre- and post-processing stages. An engineer is not really interested in solving the equation system for a 1000 mode finite element stress problem in a second or so, if it takes him a day or more to set up the mesh and loading data, and an equal time to plot the resultant stress contours. What he needs is a complete package, which enables him to set up and analyse such a size of problem, using a small data set derived from an engineering drawing, in the order of an hour of real time.

Typically, the computational tools available can be assumed to consist of a time shared mainframe system of reasonable power, allowing about 5 minutes CPU time/hour connect time with about a megabyte of store, or to consist of a desk top machine of comparable capacity. Additionally it may be expected that the analysis would be performed on a medium resolution graphics terminal with screen dump or high resolution hard copy plotting facilities available in the immediate working environment.

These were the constraints imposed on the project described in this report, the industrial sponsors Lucas CAV Ltd. requiring that the computational package should be capable of being used by engineering technicians who had a sound knowledge of stressing problems but little knowledge of finite element or multigrid solution procedures. The initial aim involved the development of a suite of programs which could analyse the stresses developed in a fuel injection nozzle similar to that shown in Fig. 1.1, under the assumption that an axisymmetric analysis would be adequate. Other applications included an analysis of the pressures developed in the lubrication films in a range of fuel injection pumps, involving complex multiply connected regions, as shown in Fig. 1.2.

2. FINITE ELEMENT PACKAGE DESIGN

The aim of the project described here was to provide the engineer with an efficient readily usable design tool able to cope with a limited range of two dimensional or axisymmetric problems, involving linear elastic stress analysis or linear steady state diffusion problems, on

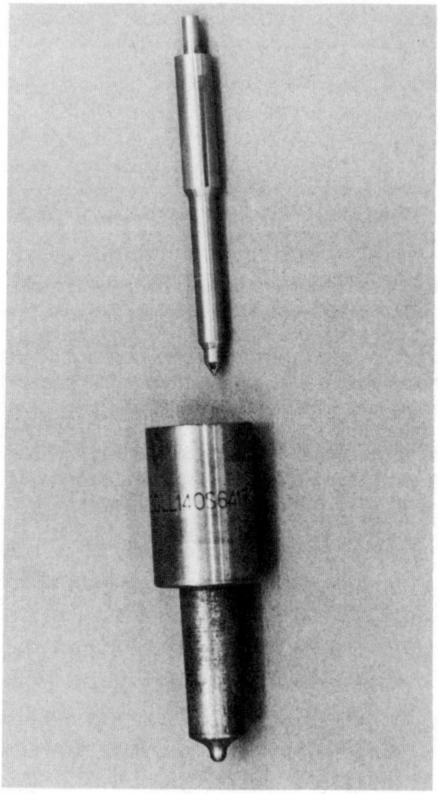

Fig. 1.1 Fuel injection nozzle

both simply connected and multiply connected regions. Fig. 2.1 shows a simple plain stress test problem in which a sloping boundary is fixed and a uniform normal load is imposed within a circular cutout.

In the finite element analysis of such a problem there are a number of identifiable stages:

(a) the boundary of the region has to be defined and a finite element mesh created on the region,

(b) the problem type has to be described, which involves the specification of the governing equations and boundary conditions,

(c) the discrete finite element equations need to be solved and

(d) post processing is required to plot stress contours, displacements, isobars for lubrication problems and isotherms for temperature problems.

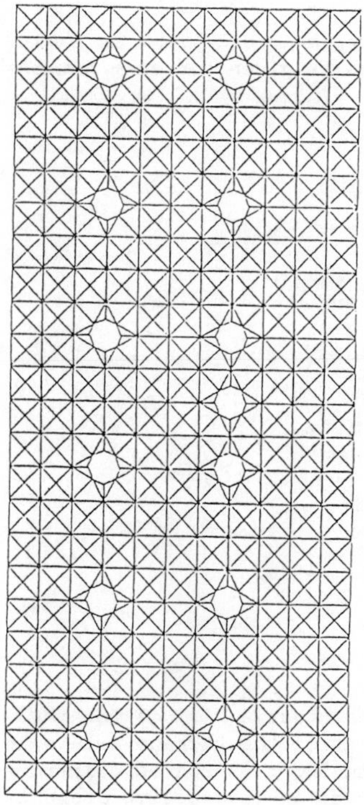

Fig. 1.2 Lubrication film finite element mesh

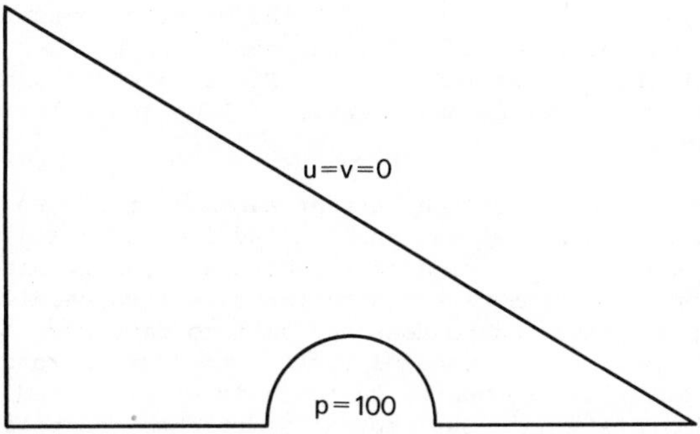

Fig. 2.1 Standard test problem

Table 2.1

Sample execution times for a 16 x 32 mesh on the test problem (order of a process is given for an m x m mesh).

Process	Order	Connect	Minutes CPU	Ratio
Set up boundary	1	14	1	14
Calculate 3 meshes using multigrid	m^2			
Impose boundary conditions	m	4	0.2	20
Solve discrete equations using multigrid	m^2	5	1	5
Generate two stress contour plots	m^2	7	0.5	14
TOTAL		30	2.7	11

Table 2.1 gives representative timings for the various stages in this process, in the actual run of the package with a 16 x 32 mesh for the test problem, on a Harris 500 time shared mainframe system with a BBC 'B' Microcomputer acting as an interactive graphics terminal (more details on comparative speeds are presented later). For a problem on an m x m mesh, the maximal orders of the various processes are given, where it is being assumed that a multigrid solution procedure, based on Wesseling's saw tooth method (Wesseling (1982)), is used to solve the equations for the internal mesh points and later those for the displacements. The figures include the time needed by the user to react to the terminal prompts. As the solution for the displacements is the least interactive phase the ratio of 5 for the connect/CPU times is a measure of the load on the time shared system. For this particular example a problem of 1122 unknowns has been easily set up and solved well within the target time.

Prior to the creation of the current package, the most time consuming parts of solving a plain stress problem involved the data preparation for the mesh and the imposition of the boundary conditions. Although there are mesh generation packages on the market to resolve the first of these problems, the second does not seem to have been given enough emphasis. The most time consuming part of the current package is now the equation solution procedure, so there is still a need for further development in this area. Our approaches to each stage of the complete process will now be outlined, starting with the mechanism for setting the initial boundary and creating the internal mesh.

3. BOUNDARY AND MESH GENERATION

In many engineering devices the outline contours are described in terms of straight line and circular arc segments, so the mesh generation program was constructed to accept information of that character. Fig. 3.1 shows the construction of the boundary for the test problem, prompts being given at appropriate stages for information such as the co-ordinates for successive points, whether or not a circular arc is required, its radius and sense. Allowance is made for error correction at any stage and a graphical display is created as the boundary construction proceeds. In the case of multiply connected regions, simply connected regions are meshed first and then added together to form a composite mesh.

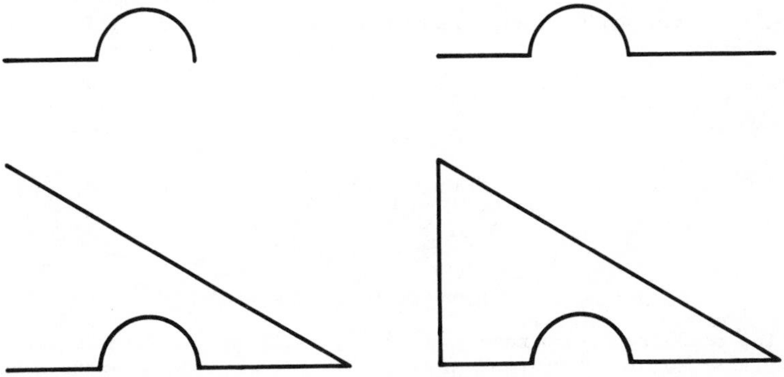

Fig. 3.1 Four stages in definition of region to be meshed

In the simply connected case, after the construction of a complete boundary, the internal mesh points are computed by using the Thompson boundary fitted co-ordinate method. In this technique the region is mapped onto a square, as illustrated in Fig. 3.2, and the points in the interior found by solving the 'inverse' problem for the coupled elliptic system:

$$\xi_{xx} + \xi_{yy} = P(\xi,\eta) \ , \ \eta_{xx} + \eta_{yy} = Q(\xi,\eta), \tag{3.1}$$

where P and Q are prescribed functions. The inverse equations are

$$\alpha \, x_{\xi\xi} - 2\beta x_{\xi\eta} + \gamma x_{\eta\eta} = -J^2(x_\xi P + x_\eta Q), \tag{3.2}$$

$$\alpha \, y_{\xi\xi} - 2\beta y_{\eta\eta} + \gamma y_{\eta\eta} = -J^2(y_\xi P + y_\eta Q), \tag{3.3}$$

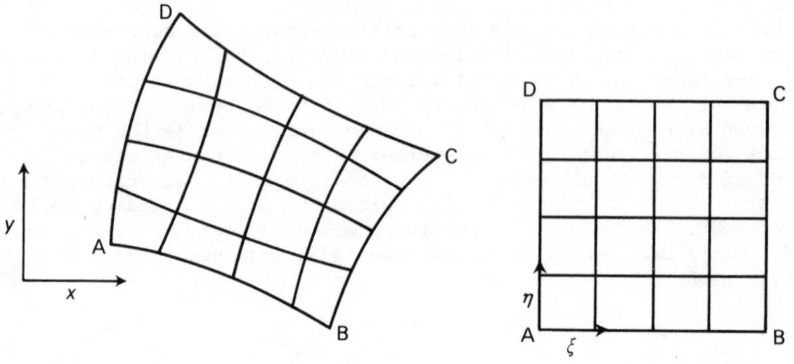

Fig. 3.2 Mapping from x-y co-ordinates to unit square in ξ-η plane

where

$$\alpha = x_\eta^2 + y_\eta^2 , \quad \beta = x_\xi x_\eta + y_\xi y_\eta , \quad \gamma = x_\xi^2 + y_\xi^2$$

and $J = x_\xi y_\eta - x_\eta y_\xi$.

As the boundary co-ordinates are known, (3.2) and (3.3) give rise to a non-linear coupled Dirchlet problem. The presence of non-zero P and Q allows the mesh to be attracted to particular points or lines; in practice P and Q can usually be taken to be zero.

A seven point discretisation was employed for (3.2) and (3.3) at each internal point of a regular square grid in the (ξ,η) plane so that the ILU multigrid procedure of Wesseling (1982) could be used. An initial interpolation was used to estimate the coefficients α,β,γ and J and their first derivatives, these were then frozen while a single multigrid sweep was used on each of the equations to estimate new values for x and y. This process was repeated until convergence was achieved. The multigrid code, based on the saw tooth procedure, has been enhanced at the detailed programming level so that it runs faster than that available as described in Hemker et al. (1983). The enhancements are independent of the particular problem being solved as they only depend on the precise restriction and prolongation strategies being employed.

In the Wesseling saw tooth method for a linear system

$$A^\ell \underset{\sim}{U}^\ell = \underset{\sim}{f}^\ell$$

on ℓ grids, where ℓ denotes the finest grid, the first step is to perform an ILU decomposition on A^ℓ, i.e. to write $A^{\ell}\,\underset{\sim}{U}^{\ell} = \underset{\sim}{f}^{\ell}$ as $L^{\ell}\,U^{\ell}\,\underset{\sim}{U}^{\ell} = f^{\ell}$, (3.4)

where the entries in L^{ℓ} and U^{ℓ} are in precisely the same locations as those in the lower and upper triangular parts of A^{ℓ}. When A^{ℓ} corresponds to a seven (nine) point discretisation the correction matrix has precisely two (four) non-zero diagonals in the locations shown in Fig. 3.3. (The nine point C matrix is incorrectly described in Table 1, p. 13 of Kettler and Meijerink (1982).) The multigrid method involves the propagation of error terms from the finest to the coarsest grids and vice-versa, so restriction and prolongation operators are needed. For the seven point discretisation these were taken to be those proposed by Wesseling (1982) as illustrated in Fig. 3.4. In addition the Galerkin method was used to calculate the system matrices on the coarser grids

i.e. $$A^{k-1} = R^{k}A^{k}P^{k}, \qquad k = \ell(-1)2, \tag{3.5}$$

where R^{k} is the restriction operator from grid k to grid k-1 and P^{k} is the prolongation operator from grid k-1 to grid k.

The complete saw tooth cycle is given by

$$\underset{\sim}{f}^{\ell-1} = R^{\ell}C^{\ell}(\underset{\sim}{U}^{\ell} - \bar{\underset{\sim}{U}}^{\ell})$$

<u>for</u> $k = \ell - 1(-1)2$ <u>do</u> $\underset{\sim}{f}^{k-1} = R^{k}\underset{\sim}{f}^{k}$

$$\underset{\sim}{U}^{1} = (L^{1}U^{1})^{-1}\,\underset{\sim}{f}^{1}$$

<u>for</u> $k = 2(1)\ell - 1$ <u>do</u> $\underline{U}^{k} = L^{k}U^{k})^{-1}\,(C^{k}P^{k}\underset{\sim}{U}^{k-1} = \underset{\sim}{f}^{k})$

$$\bar{\underset{\sim}{U}}^{\ell} = P^{\ell}\underset{\sim}{U}^{\ell-1} + \underset{\sim}{U}^{\ell}$$

$$\underset{\sim}{U}^{\ell} = (L^{\ell}U^{\ell})^{-1}\,(C^{\ell}\bar{\underset{\sim}{U}}^{\ell} + \underset{\sim}{f}^{\ell}). \tag{3.6}$$

It might have been advantageous to replace this process by a full approximation storage scheme as proposed by Brandt (1977) for general use with non-linear problems. This was not done. Table 3.1 shows the total time in microseconds, in a non-optimised code for 10 iterations per fine grid point for the case of a linear Poisson problem solved on a 65 x 65 grid on the Harris 500 system, alongside the individual times in the various substages. For the mesh generation problem the fine grid system matrix is updated after every two steps so that the total time for one cycle is

$$(1133 + 350 + 2*(4172/10))*4225 = 9.8 \text{ seconds.}$$

		A	L	U	C
(a)	7 point	xx	..	xx	x..
		xxx	xx.	.xx	...
		xx	xx	..	..x
(b)	9 point	xxx	...	xxx	x...
		xxx	xx.	.xx	x...x
		xxx	xxx	...	...x

Fig. 3.3 Entries in the L U and C matrices for 7 and 9 point ILU decompositions of A

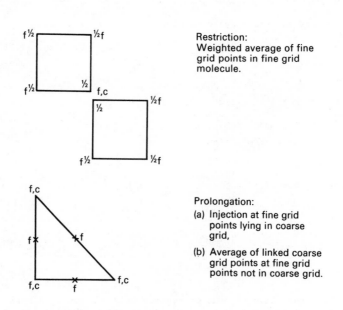

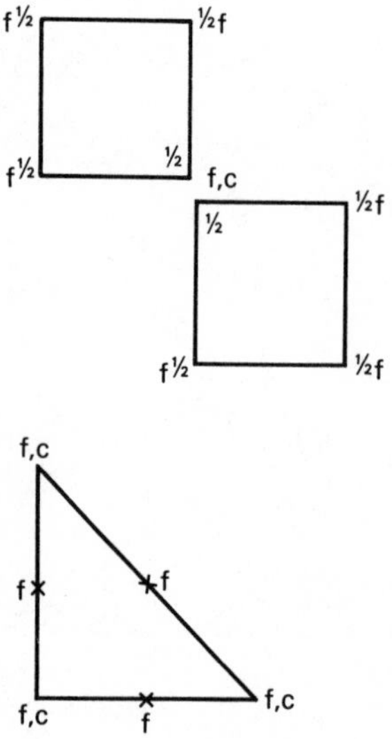

Restriction:
Weighted average of fine grid points in fine grid molecule.

Prolongation:
(a) Injection at fine grid points lying in coarse grid,
(b) Average of linked coarse grid points at fine grid points not in coarse grid.

Fig. 3.4 Seven point restriction and prolongation operators

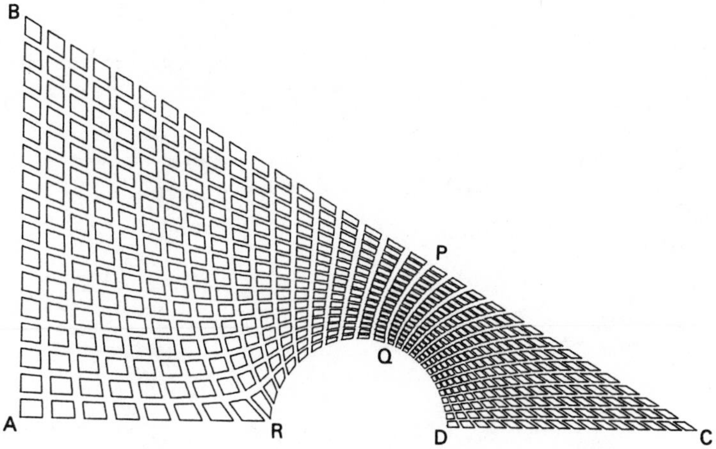

Fig. 3.5 4 noded quadrilateral mesh

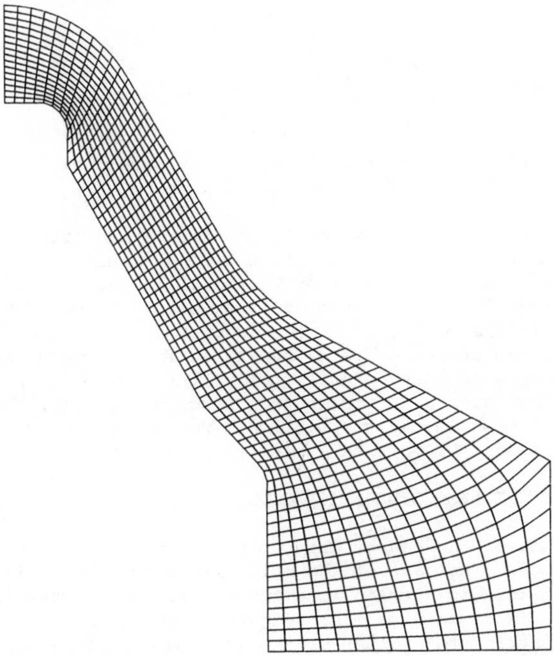

Fig. 3.6 Cross section of tip of fuel injection nozzle

Table 3.1

Times for multigrid code for 65 x 65 Poisson problem on Harris 5000

Time in microseconds/gridpoint (figures in brackets are Cyber 205 times)

Process	Non-optimised N	Optimised 2D O2D	Optimised 1D O1D	Wesseling W	N/O1D	W/O1D
Generate Coarse grid matrices	1133	118	80	611(21)	14.0	7.6
Perform ILU decomposition	350	197	133	171(7.3)	2.6	1.3
Solution	1189	1111	813	1286(35.5)	2.3	1.6
$C(\underset{\sim}{u} - \underset{\sim}{\bar{u}})$	513	273	237	539(15.6)	2.2	2.3
$C\underset{\sim}{u} + \underset{\sim}{f}$	760	374	318	711(21)	2.4	2.2
Prolongation	180	107	107	174(7.1)	1.7	1.6
Restriction	499	115	115	107(4.0)	4.3	0.9
10 Iterations	4172	2083	1687	3013(124)	2.5	1.8
Set up and perform 5 iterations	3569	1356	1056	2289	3.4	2.2
Set up and perform 2 iterations	2317	732	550	1385	4.2	2.5

The coarse grid matrix generation plus the ILU decomposition thus accounts for 49% + 15% (=64%) of the total time.

Fig. 3.5 shows the final mesh obtained after three attempts at placing the boundary points. Sides AB and CD are uniformly divided into 16 sections each whereas sides BC and AD are subdivided further into two and three sections respectively by points P, Q and R so that there are 18 sections on BP, 14 on PC, 14 on DQ, 10 on QR and 8 on RA. Subdivisions of the boundary are a usually sufficient mechanism for obtaining a reasonable mesh. Additionally, internal point positions can be changed by using a cursor control process for final mesh adjustment. The mesh in the computational region is then non-uniform. Fig. 3.6 shows a 16 x 64 mesh on a fuel injection nozzle.

4. OPTIMISATION OF MESH GENERATION CODE

The generation of the coarse grid operators has been shown to be the most time consuming part of the multigrid algorithm so it is clearly of benefit to make this section more efficient. Each entry in the coarse grid matrix is made up of a weighted combination of entries in the fine grid matrix, the values of the entries generally not being known explicitly. Using the notation described by Wesseling (1982) where a Greek letter α denotes a 2-tuple (α_1,α_2) for the index of a grid point, the entries required for the computation of $A^{k-1}_{\alpha\delta}$ are

$A^k_{2\alpha+\beta,2\delta+\gamma-\beta}$, where $2\alpha+\beta$ corresponds to one of the seven fine grid points surrounding the coarse grid point indicated by α, β, γ and δ each take the values (0,-1), (1,-1) (-1,0), (0,0), (1,0), (-1,1) and (0,1) corresponding to the points in the seven point molecule, and contribute to the result provided $2\delta+\gamma-\beta$ is within the molecule. It can be shown that this occurs 85 times out of 343 possible combinations of β, γ and δ. Thus, if the coding involves the statement

$$A^{k-1}_{\alpha\delta} = A^{k-1}_{\alpha\delta} + \mu\, A^k_{2\alpha+\beta,2\delta+\gamma-\beta},$$

this will be executed 85 times for each internal coarse grid point. As the weights μ and the contributing nodes are the same at all internal points,each element of the coarse grid should only be accessed once. The code shown below prints out the required information for a general coarse grid point.

```
For Iδ = 1(1)7 Do
   Output Iδ
   δ = molecule(Iδ)
   For weight ∈ W Do
      Output weight
      For Iβ = 1(1)7 Do
         β=molecule(Iβ)
         For Iγ = 1(1)7 Do
            γ=molecule(Iγ)
            μ = μβμγ
            If μ = weight Then
               If 2δ+γ-β ∈ molecule Then
                  find K such that 2δ+γ-β = molecule(K)
                  Output Iβ,K
               End If
            End If
         End For
      End For
   End For
End For
```

The function "molecule(J)" returns the 2-tuple corresponding to the J'th diagonal of the system matrix, for example, molecule(4)=(0,0).

"W" is the set of all possible weights, i.e. [1,2,4]

For each of the seven non-zero diagonals of the coarse grid matrix, denoted by I_δ, the algorithm outputs the elements of the fine grid matrix which contribute to the coarse grid matrix element on that diagonal.

For each fine grid matrix element, two numbers between 1 and 7 are output.

1) I_β denotes the number of the fine grid point. β (=molecule(I_β)) is the position of the point relative to the fine grid point which coincides with the coarse grid point.

2) K is the number of the non-zero diagonals of the fine grid matrix.

The contributions with the same weight will be grouped together.

The output can then be used to set up a FORTRAN routine which will calculate the coarse grid matrices. The contributions may be typed in by hand in the order in which they are given by the program, or suitable formatting statements could be placed in the program to enable it to produce the FORTRAN code directly.

A program written in BBC BASIC on the BBC micro-computer was used. It produces FORTRAN code which can be transferred to the Harris computer and incorporated into the multigrid routine. It can be used to generate code with one-dimensional arrays or a two-dimensional array, and for a seven or nine point molecule.

An example of the code (using a two-dimensional array) produced by the program is shown overleaf. The variables used are as follows:

```
C
C    Coarse grid element for diagonal number 1
C
      A(KCORSE,1)=
     . 0.2500*(A(KA,2)+A(KA,3)+A(KA,4)+A(KB,1)+A(KB,3)+
     . A(KC,1)+A(KC,2))+
     . 0.5000*(A(KA,1)+A(KD,1))
C
C    Coarse grid element for diagonal number 2
C
      A(KCORSE,2)=
     . 0.2500*(A(KA,2)+A(KA,5)+A(KB,1)+A(KB,4)+A(KB,5)+
     . A(KE,1)+A(KE,2))+
     . 0.5000*(A(KB,2)+A(KD,2))
```

```
C
C     Coarse grid element for diagonal number 3
C
       A(KCORSE,3)=
     . 0.2500*(A(KA,3)+A(KA,6)+A(KC,1)+A(KC,4)+A(KC,6)+
     . A(KF,1)+A(KF,3))+
     . 0.5000*(A(KC,3)+A(KD,3))
C
C     Coarse grid element for diagonal number 4
C
       A(KCORSE,4)=
     . 0.2500*(A(KA,4)+A(KA,5)+A(KA,6)+A(KB,3)+A(KB,4)+
     . A(KB,7)+A(KC,2)+A(KC,4)+A(KC,7)+A(KE,1)+A(KE,4)+
     . A(KE,6)+A(KF,1)+A(KF,4)+A(KF,5)+A(KG,2)+A(KG,3)+
     . A(KG,4))+
     . 0.5000*(A(KA,7)+A(KB,6)+A(KC,5)+A(KD,1)+A(KD,2)+
     . A(KD,3)+A(KD,5)+A(KD,6)+A(KD,7)+A(KE,3)+A(KF,2)+
     . A(KG,1))+
     . 1.0000*(A(KD,4))
C
C     Coarse grid element for diagonal number 5
C
       A(KCORSE,5)=
     . 0.2500*(A(KB,5)+A(KB,7)+A(KE,2)+A(KE,4)+A(KE,7)+
     . A(KG,2)+A(KG,5))+
     . 0.5000*(A(KD,5)+A(KE,5))
C
C     Coarse grid element for diagonal number 6
C
       A(KCORSE,6)=
     . 0.2500*(A(KC,6)+A(KC,7)+A(KF,3)+A(KF,4)+A(KF,7)+
     . A(KG,3)+A(KG,6))+
     . 0.5000*(A(KD,6)+A(KF,6))
C
C     Coarse grid element for diagonal number 7
C
       A(KCORSE,7)=
     . 0.2500*(A(KE,6)+A(KE,7)+A(KF,5)+A(KF,7)+A(KG,4)+
     . A(KG,5)+A(KG,6))+
     . 0.5000*(A(KD,7)+A(KG,7))
```

A is the system matrix on all grids, an index being kept for each grid. A(I,1) is the entry for grid point I on the first diagonal, A(I,2) is the second diagonal, etc.

KCORSE is the position in the array A of the coarse grid point, this will range over all the grid points in each coarse grid.

KD is the position in the array of the corresponding fine grid point.

KA, KB, KC, KE, KF, KG are the positions of the surrounding fine grid points i.e.

KA corresponds to the 2-tuple (0,-1),
KB corresponds to the 2-tuple (+1,-1),
KC corresponds to the 2-tuple (-1,0),
KE corresponds to the 2-tuple (+1,0),
KF corresponds to the 2-tuple (-1,+1),
KG corresponds to the 2-tuple (0,+1).

These points correspond to the β 2-tuple in the algorithm, thus, if $I_\beta = 1$, the program prints out "KA", if $I_\beta = 2$, "KB" is printed, etc.

The diagonal is determined by the value of K in the algorithm, if K = 1, the program prints out "A(K?,1)", if K = 2, it prints out "A(K?,2)", etc.

In a similar manner, I_δ determines the coarse grid diagonal.

The code produced by the program should be more efficient for a number of reasons.

1) For each coarse grid point, the coarse grid array is accessed seven times rather than 170 times.

2) For each coarse grid array element, the contributions with the same weight are grouped together. This reduces the number of multiplications from 85 to 15, per coarse grid point.

3) Treatment of the boundary points is more efficient. Each fine grid point must be tested to see if it is outside the region. Previously, this test was performed 85 times per coarse grid point. The generated code requires the test to be performed once only per coarse grid point.

4) The three loops involving δ, β and γ are completely eliminated.

A further advantage of the code generator is that code using seven one-dimensional arrays can be produced by modifying a couple of statements in the generator.

The generator code was then timed and compared with the previous code. Note that, as the fine grid contributions with the same weight have been grouped together, the number of operations has reduced from 170 per coarse grid point, to 93. The results for a routine using one two-dimensional array are given in table 3.1.

Thus the new method has reduced the time for the generation of the coarse grid operators by a factor of 9.6. As this stage was the most time-consuming stage, the overall time will be significantly reduced. In particular, the time for one cycle of the non-linear solver is reduced from 9.8 seconds, to 5.5 seconds.

The section of code concerned with the generation of the coarse grid operators has been optimised and is now one of the least time-consuming stages. Attention is thus turned to the other stages of the algorithm in an effort to further improve the overall performance.

Most of the stages in the algorithm require special treatment for the boundary points (and the points next to the boundary). For example, the ILU decomposition at a point refers to the points surrounding that point. If that point is a boundary point, some of these points will not exist, the entries at these points must be taken as zero. There are two ways to deal with these points.

1) Loop over all points, testing each one for proximity to each boundary.

2) Use a separate loop for each special case, e.g. top row, right hand row, internal points, corner points, etc.

The first method produces more compact code and is probably easier to write and maintain. However, a large number of "IF" statements must be executed at every point, for example, the coding for the ILU decomposition requires 17 at each point. If the second method is used, no "IF" statements are needed thereby saving on execution time.

The sections of code performing the ILU decomposition, the solution step, the multiplication by the error matrix (C), and the prolongation, were originally programmed using the first method. They were then rewritten using the second method, resulting in significant improvements in execution time. The times are also given in table 3.1 in column O2D.

The routine was originally written in a general manner, so that it could be used for other molecules as well, for example, a nine point molecule. However, to allow this, an array of 2-tuples were required. This array gives the positions of the required fine grid points, relative to the fine grid point which coincides with the coarse grid point at which a value is required. Partly because of the need to deal differently with the boundary points, this array was accessed six times for each of the seven molecule entries, that is, 42 times per coarse grid point. This routine was thus rewritten. Separate loops were used for each type of boundary point, and the relative positions of the fine grid points were entered directly into the code, thus removing the need for the 2-tuple array.

5. POISSON PROBLEMS

It has been seen that the meshes are derived by mapping from the physical (x,y) plane to the computational (ξ,η) plane. As for some problems it became necessary to carry out minor local adjustments of the mesh leading to a non-uniform mesh in the computational plane, it was decided to discretise the governing equations for any physical problem in the physical plane, and to use finite element techniques rather than transform the equation(s) to (ξ,η) co-ordinates.

In the case of Poisson equations such as

$$\nabla^2 U = f, \tag{5.1}$$

linear elements on triangles lead directly to 7 point molecules and bilinear quadrilateral elements to 9 point molecules, so the application of Wesseling's saw tooth multigrid process is straight-forward and many examples have been described elsewhere. Better accuracy is of course obtained with higher order elements such as 6 noded quadratic triangles or 8 or 9 noded quadrilaterals. In these cases the computational molecules are much larger so ILU decomposition becomes elaborate. In the case of the 6 noded triangle there is a very special connection between the element stiffness matrices on two successive grid levels. Fig. 5.1 shows a 6 noded quadratic triangle consisting of 4 linear sub-triangles. If A B C are coarse grid points on grid level k-1 and DEF are mid points, which are fine grid points on grid level k, then the quadratic stiffness matrix for the Laplace operator may be calculated on the coarse grid as Q^{k-1} and the corresponding linear stiffness matrices may be calculated on levels k-1 and k as A^{k-1} and A^k. The relationship

$$Q^{k-1} = 4/3\ A^k - 1/3\ \tilde{A}^{k-1}$$

then holds i.e. the quadratic stiffness matrix is just a linear combination of two linear stiffness matrices on two successive grids. This forms the basis for an iterative process for quadratic elements driven by the ILU decomposition for the linear stiffness matrices. If the quadratic finite element discretisation of (5.1) is

$$Q^\ell\ \underset{\sim}{U}^\ell = \underset{\sim}{g}^\ell, \tag{5.2}$$

then this may be written as the iteration

$$A^\ell\ \underset{\sim}{U}^\ell_{new} = \beta(\underset{\sim}{g}^\ell - Q^\ell\ \underset{\sim}{U}^\ell) + A^\ell\ \underset{\sim}{U}^\ell_{old}, \tag{5.3}$$

where β is a free parameter. The choice $\beta = \frac{3}{4}$ gives the iteration scheme

$$L^\ell\ U^\ell\ \underset{\sim}{U}^\ell_{new} = \tfrac{3}{4}\ \underset{\sim}{g}^\ell + \tfrac{1}{4}\ \tilde{A}^{k-1}\ \underset{\sim}{U}^\ell_{old} + C^\ell\ \underset{\sim}{U}^\ell_{old},$$

where $\tilde{A}^{k-1}$ is the linear stiffness matrix for the coarse grid, expanded with zero entries in the rows and columns corresponding to fine grid points, so both the right side matrices are sparse. This scheme is generally found to converge to the quadratic results in about double the number of iterations required for the linear case.

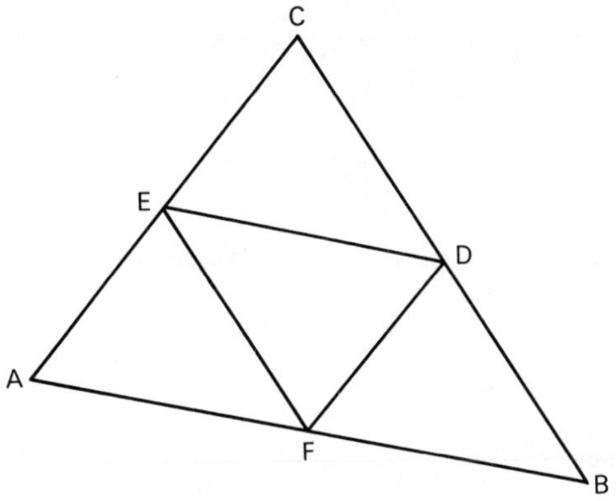

Fig. 5.1 Linear and quadratic triangles

6. LINEAR ELASTIC PROBLEMS

The major application of the package is to linear elastic stress problems for plain stress or strain or axisymmetric stress. The governing equations for the displacements (u,v) in the x and y directions in the case of plane strain are

$$G\left\{\nabla^2 u + \frac{1}{1-2\nu}\frac{\partial}{\partial x}\left(\frac{\partial u}{\partial x} + \frac{\partial v}{\partial y}\right)\right\} + F_x = 0, \tag{6.1}$$

$$G\left\{\nabla^2 v + \frac{1}{1-2\nu}\frac{\partial}{\partial y}\left(\frac{\partial u}{\partial x} + \frac{\partial v}{\partial y}\right)\right\} + F_y = 0, \tag{6.2}$$

where $G = \frac{E}{2(1+\nu)}$, E is Young's modulus, ν is Poisson's ratio and F_x and F_y are the Cartesian components of the imposed body force. Problems associated with (6.1) and (6.2) may be cast into the form of a variational principle in which it is required to minimise the energy

$$\pi(u,v) = \frac{1}{2}\int_{\Omega} \underset{\sim}{\varepsilon}^T \underset{\sim}{\sigma}\, da - \int_{\Omega} \underset{\sim}{F}^T \underset{\sim}{\delta}\, da + \int_{\partial\Omega_1} pds\ , \tag{6.3}$$

where the strain $\underset{\sim}{\varepsilon}^T$ is $(u_x, v_y, u_y + v_x)^T$, the stress $\underset{\sim}{\sigma}$ is

$$G \begin{pmatrix} \alpha & \beta & 0 \\ \beta & \alpha & 0 \\ 0 & 0 & 1 \end{pmatrix} \underset{\sim}{\varepsilon}, \text{ the displacement } \underset{\sim}{\delta}^T \text{ is } (u,v)^T$$

and p is a normal surface load,

$$\text{with} \quad \alpha = \frac{2(1-\nu)}{1 - 2\nu} \quad \text{and} \quad \beta = \frac{2\nu}{1 - 2\nu},$$

subject to the conditions that $\underset{\sim}{\delta}$ may be given on part of the boundary of the region Ω and the load p may be given on another part $\partial\Omega_1$.

A standard finite element procedure can be used with (6.3) to produce the set of coupled equations for the nodal displacements u and v for a range of element choices. When linear interpolation on triangles was used for u and v it was found that asymmetric solutions were found for problems with symmetric boundary conditions so the basic elements were from that time chosen to be bilinear quadrilaterals, it being anticipated that the solution for higher order element schemes could be handled in a way similar to that used in the Poisson case.

Table 6.1

Sample stress calculations

Geometry	Type	Fix	Load	Mesh NX*NY	Iterations /Sig fig	Solution Time secs MG	Front
Unit square $0 \leqslant x \leqslant 1$ $0 \leqslant y \leqslant 1$	Plane stress	Fixed y=0	Load y=1	17x17	7/6	3.1	8.1
				33x33	6/6	10.8	84.0
Triangle with cut out Figure	Plane stress	Fixed on slope	load in cut out	17x9	10/6	2.1	2.2
				9x17	13/6	2.5	2.2
				17x33	17/6	12.0	16.5
Nozzle figure	Axisymmetric stress	None	Internal pressure + balancing plane load	9x33	16/6	5.9	4.4
				33x9	18/6	6.6	4.4
				17x65	18/6	24.5	34.6
				65x17	19/6	25.7	34.6
Hook	Plane stress	x=0	Inner face	9x33	426/4	131	18.8
				33x9	324/4	102	18.8
	Laplace	T=0, 1 top/bottom	-	9x33	9/6	0.7	1.7
		T=0, 1 at ends	-	9x33 (triangles)	15/6	1.1	

Table 6.1 summarises a series of numerical experiments, comparing the multigrid with a frontal solution procedure, for a range of problems with various geometries and loading. In the unit square case with one edge fixed and the other loaded, the frontal method (which depends essentially on the number of unknowns times the front width squared, i.e. m^3 for an m x m element problem) is slower than the multigrid method certainly for m = 16 or larger. In the standard problem of the loaded triangle, the multigrid method loses to the frontal procedure if the mesh is numbered in the wrong fashion and the total number of variables is sufficiently small, a break-even mesh being about 8 x 16. The problem that generated this research concerns a fuel injection nozzle in which an internal uniform pressure load is matched by a balancing load to remove any rigid body motion. No difficulties have been found in obtaining convergence for the multigrid method. Compared to the frontal procedure the breakdown point seems to be a 9 x 33 mesh. The final problem concerns a loaded hook, as illustrated in Fig. 6.1, in which the left hand face is held fixed and the right hand inner face is subject to a uniform load in the positive x-direction. The frontal method gives solutions in a reasonable time comparable with those found for the nozzle, whereas the multigrid procedure is barely convergent. No problems were found in calculating the temperatures within the hook with the 'top' and 'lower' curved boundaries maintained at differing constant temperatures of 0 and 1. In the case where the temperatures at the left hand face and the end of the hook were given as 0 and 1, no difficulties were found in obtaining solutions with linear triangles, whereas bilinear quadrilaterals barely converged at all. It is not clear at the present time whether or not the strong difference in performance is a matrix conditioning problem or a feature of the multigrid algorithm.

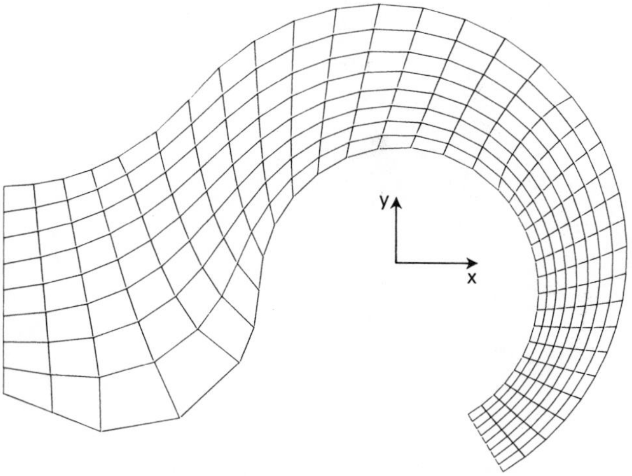

Fig. 6.1 Finite element mesh for a hook

7. CONCLUSIONS

It has been demonstrated that multigrid techniques can certainly be used for generating smooth meshes on simply connected regions via the Thompson approach. Here accuracy is not the foremost criterion, so the 7-point ILU decomposition method for Wesseling may be employed. If accurate results are needed then finite element procedures based on triangulations, which lead to 7-point computational molecules, are to be avoided and bilinear quadrilateral elements are to be preferred, especially for stress problems. On the other hand in some mixed boundary value problems with Neumann conditions given on 'long' boundaries the matrices based on triangular finite elements allow convergent algorithms to be developed whereas those based on bilinear quadrilaterals do not. The precise reasons why this occurs need further investigation.

REFERENCES

Brandt, A. (1977) Multi-level adaptive solution to boundary value problems. *Maths. of Comp.*, **31**, pp. 333-390.

Hemker, P.W., Wesseling, P. and Zeeuw, P.M. de (1983) A portable vector code for autonomous multigrid modules. Dept. of Numerical Maths. NW154/83 Maths. Centrum, Amsterdam.

Kettler, R. and Meijerink, J.A. (1982) A multigrid method and a combined multigrid conjugate gradient method for elliptic problems with strongly discontinuous coefficients in general domains. Pub. 604, University of Delft.

Wesseling, P. (1982) A robust and efficient multigrid method. Report 82-05 Dept. of Maths. and Informatics, University of Delft.

MULTIGRID METHODS FOR OIL RESERVOIR SIMULATION IN THREE DIMENSIONS

T. Scott

(United Kingdom Atomic Energy Authority, Winfrith)

ABSTRACT

One of the essential components of an oil reservoir simulator is an efficient technique to solve the systems of algebraic equations arising from discretisation of the governing partial differential equations. This paper describes the development of a robust and efficient multigrid solution algorithm and its implementation for solving typical problems encountered in the modelling of petroleum reservoirs. The method is used to compute results for two and three-dimensional cases, involving in the region of up to 5000 grid-blocks. A comparison in terms of CRAY CPU time is provided between different multigrid smoothing strategies employed by the algorithm.

1. INTRODUCTION

The purpose of this paper is to describe the development and application of a selected multigrid method to solve the non-separable diffusion equation defined in up to three dimensions. This type of equation plays a central role in the theory of flow through porous media and provides a basis for setting up the governing partial differential equations of oil reservoir simulation. Allowing for discretisation in time, the diffusion equation at each time step may be written compactly as a self-adjoint boundary value problem,

$$\nabla . \underline{K} \nabla u = f. \tag{1.1}$$

Concerning notation, $\underline{K}$ is proportional to the permeability tensor, u is the dependent variable, in this case pressure, and f a functional term evaluated at the previous time step. The problem is to solve for u implicitly throughout the reservoir domain Ω subject to prescribed boundary conditions (usually no-flow Neumann conditions) on the boundary $\partial\Omega$.

The approach to deriving a solution to equation (1.1) is to discretise the system using finite differences. Hence the continuous problem is reduced to a discrete system of linear equations to be solved for u. It is common practice in oil reservoir simulation to adopt a block centred grid structure whereby the region Ω is divided up into blocks and the pressure solved for at the centre of each block.

Given this block centred scheme, the objective has been to develop and apply an efficient and robust multigrid algorithm to solve the approximating difference equations. By robust is meant a technique which can be successfully applied to a wide variety of domains with properties defined according to the second order tensor $\underline{K}$. Thus $\underline{K}$ may involve sharp discontinuities due, for example, to faults in the rock matrix, or exhibit large anisotropic variations from one region of the reservoir to another.

Hence the task required the identification of a reliable multigrid algorithm which could quickly and efficiently solve upwards of 700 linear equations involving the pressure distribution. Further, it was necessary to derive a method for obtaining the coarse-grid difference equations from the previous fine-grid difference equations, since only the coefficient matrix is readily available rather than the partial differential equation itself. Earlier multigrid work (Brandt 1977a) relied on knowledge of the differential equation itself being at hand, so that coarse-grid representations of the finite difference equations could be formulated by returning to and discretising this equation as necessary.

In response to these demands a multigrid solution package has been developed which can be added to an existing oil reservoir simulator in order to provide an alternative iterative solution procedure. This solution package is particularly suitable for field studies involving up to several thousand grid blocks.

Section 2 begins with a brief description of the multigrid strategy used in the present work, before going on to discuss the finer mathematical points and details of the computational aspects. Application and results of the method to representative examples form the material of Section 3 and concluding remarks are presented in Section 4.

2. MULTIGRID ALGORITHM

In order to establish notation and for completeness, a brief description is first given of the multigrid strategy used to solve equation (1.1). It is assumed throughout that the time dependence of the diffusion equation has been discretised and that one is now concerned with an elliptic boundary value problem to be solved at each time step. Rectangular Cartesian coordinates are used as a basis for spatial discretisation.

The algorithm is written with the option of starting iteration with either some user supplied initial approximation to the solution on the finest level or alternatively at the coarsest level of discretisation with a direct solution method. Thus the algorithm incorporates a cyclic option or the full multigrid procedure (Brandt 1982). Both work in a fixed mode, determined after a number of trial runs on representative example problems. The so-called automatic prescription developed by Nicolaides (1977) within the context of general finite element systems is employed to compute the various grid communication operators and coarse level difference equations.

2.1 Brief outline

Let $G^1, \ldots, G^M$ denote a sequence of discretisation grids such that $G^k \subset G^{k+1}$ with G^k a typical member of this grid family. G^M represents the finest grid and G^1 the coarsest. The centres of the grid blocks referred to in Section 1 are located at the nodes of G^k. Suppose the corresponding mesh sizes are $h_1 > \ldots > h_M$ where for simplicity a square uniform grid is assumed; standard coarsening is used to define the step lengths, namely $h_k = 2h_{k+1}$.

The discretised form of equation (1.1) is expressed as

$$L^M U^M = f^M \tag{2.1}$$

on G^M, the grid on which the problem is presented to the multigrid package. L^M represents the finite difference form of the differential operator appearing in equation (1.1) and U^M the exact solution of this system of linear equations. It is assumed that iteration is to begin on G^M with some initial approximation u_0^M to U^M; the extension to full multigrid iteration is relatively straightforward.

The starting point is to write equation (2.1) in residual form on G^k, that is

$$L^k v^k = f^k, \tag{2.2}$$

where $v^k = U^k - u^k$ and f^k is the residual function; u^k is some approximate solution to U^k so that v^k is interpreted as a correction term. As is well known from the multigrid literature, it is the correction v^k which is smoothed by relaxation on the auxiliary grids comprising the grid hierarchy. That is, those error components with Fourier wavelength of the order of h_k are most efficiently eliminated by just one or two relaxation sweeps on G^k.

Having smoothed v^k in this way, the problem is then transferred to G^{k-1} according to the technique

$$L^{k-1} v^{k-1} = f^{k-1} \tag{2.3}$$

where the residual term f^{k-1} is defined to be given by

$$f^{k-1} = I_k^{k-1} (f^k - L^k v^k), \tag{2.4}$$

with I_k^{k-1} a suitable residual transfer operator. The usual smoothing procedure is again repeated, this time on G^{k-1}, whereafter v^{k-1} may be interpreted as a coarse-grid correction to v^k. The initial approximation required to begin relaxation on equation (2.3) is taken to be the null solution. Having solved the problem on G^{k-1} sufficiently well, the function v^k is corrected as follows:

$$v^k \to v^k + I_{k-1}^k v^{k-1}, \tag{2.5}$$

where I_{k-1}^{k} denotes interpolation from G^{k-1} to G^{k}. The choice of interpolation and residual transfer operator is discussed in the following section. Hence the correction v^k has been modified according to line (2.5) and should be a better value for the next step in the algorithm. This new value is further smoothed by relaxation before being used as a correction to v^{k+1} on the next finer grid G^{k+1}. On the other hand if $k = M$, this value is treated as the new approximate solution to the problem on G^M (equation (2.1)) and checked for convergence.

The above sequence of steps is repeated for all $k \leq M$, cycling down to the coarsest level G^1 and returning to the finest level G^M. Convergence is investigated by comparing the L_2 norm of the residual with some prescribed tolerance ε, usually $\varepsilon = 0.01$. In the current program of work, the coarsest grid equations are solved directly by Gaussian elimination rather than by relaxation. This is relatively cheap compared to calculations on other grids due to the reduction in the number of grid points at this level of discretisation.

The algorithm described above is essentially similar to the Correction Scheme introduced by Brandt (1977a). Computations begin on G^M by setting $v^M = u_0^M$ rather than zero and applying relaxation before transferring the problem to G^{M-1}. Moreover, it is essential to this scheme that the partial differential equation be linear in order that a meaningful equation can be written down for v^k.

Alternatively, u_0^M may be calculated by the package by starting on G^1 with a direct solution method and interpolating to G^M by a process of multigrid cycling. This relieves the user of having to supply a suitable u_0^M and is particularly useful at the first time step where a good initial starting value is not always available. Hence full multigrid iteration is applied at the first time step. At subsequent time steps, because the pressure distribution is generally a slowly varying function within the overall solution procedure, the previous distribution at the last time step can be used as an initial starting approximation on the finest grid for the new time step. This is referred to as a purely cyclic algorithm.

The multigrid strategy outlined above defines the basic framework of the current algorithm developed for simulation studies. Some further details are covered in the next section.

2.2 Mathematical operators

Having established the principal features of the multigrid algorithm, it is now appropriate to specify in more detail the choice of the various mathematical operations constituting the scheme, such as those concerned with communication between grids and smoothing.

As already mentioned in Section 2.1, at the first time step the basic procedure is the full multigrid approach whereby in order to obtain a good initial approximate solution u_0^M, the algorithm begins on G^1 where the system of linear equations is solved exactly by a non-iterative scheme. This solution is interpolated to G^2, refined by a multigrid cycle and then interpolated to G^3. This is repeated until the algorithm reaches G^M and hence provides a good starting solution at this level. Ideally only a few multigrid sweeps on this approximation are necessary to yield an acceptable solution to equation (2.1). It is important to note that this technique eliminates the extra work which would be incurred by an unfortunate starting solution on G^M in the cyclic multigrid approach (Brandt 1977a).

In addition, a fixed mode of operation is chosen as opposed to the accommodative mode. That is, instead of inserting switching criteria to determine how many relaxation sweeps need to be performed on a given grid G^k ($k > 1$) before the algorithm should switch to another grid, the number of sweeps per grid is prescribed in advance. In other words the number is fixed at the outset, taking into account the nature of the reservoir problem and the effectiveness of the selected relaxation procedure at smoothing the error correction term. Moreover this mode of operation thus eliminates the additional code required to implement switching criteria and hence a saving on execution time.

The mathematical operators are described below in turn.

Relaxation:

Three options have been considered,

(i) Point successive relaxation,

(ii) Line successive relaxation,

(iii) Alternating line relaxation.

The first is straightforward application of Gauss-Seidel iteration where the solution at each grid point is up-dated in turn using the latest values available at neighbouring points. Standard ordering is adopted.

The second option up-dates a line of point values simultaneously, the direction of the line being parallel to either one of the three co-ordinate axes (x, y or z). This is particularly useful for solving problems involving a discontinuity in the coefficient terms $\underline{K}$ of equation (1.1) where the line of relaxation should be optimised with respect to the line of discontinuity.

The third choice combines line relaxation of option (ii) in such a way that in three dimensions a single sweep is defined to incorporate taking lines parallel to first the x-axis, then the y-axis and thirdly the z-axis. For problems which exhibit severe anisotropy, alternating line relaxation is particularly appropriate. It can also be used to treat efficiently the difficulties associated with discontinuities.

None of the above relaxation procedures requires an iteration parameter, as for example in the more conventional method of line successive over-relaxation.

Interpolation:

This is concerned with transfer of the solution vector from G^{k-1} to G^k and in Section 2.1 is denoted by the symbol I^k_{k-1}. For fine-grid points which coincide with coarse-grid points, the identity operator is applied. Otherwise use is made of the discretisation equation about the particular fine-grid point labelled (i, j, k) for which one wants to interpolate a value:-

$$L^q_{i,j,k} \; U^q_{i,j,k} = f^q_{i,j,k}, \tag{2.6}$$

where $1 < q \lesssim M$. The inclusion of the right hand side $f^q_{i,j,k}$ is optional. In the derivations to be described in this section it is however ignored in order to retain a higher degree of clarity. The multigrid results reported in Tables 1-3 have been obtained using an interpolation operator which excludes $f^q_{i,j,k}$. Generally, exclusion of this right hand side function was found to lead to a more efficient interpolation scheme. Referring to Fig. 1, there are in three dimensions 3 types of points to which equation (2.6) is applied.

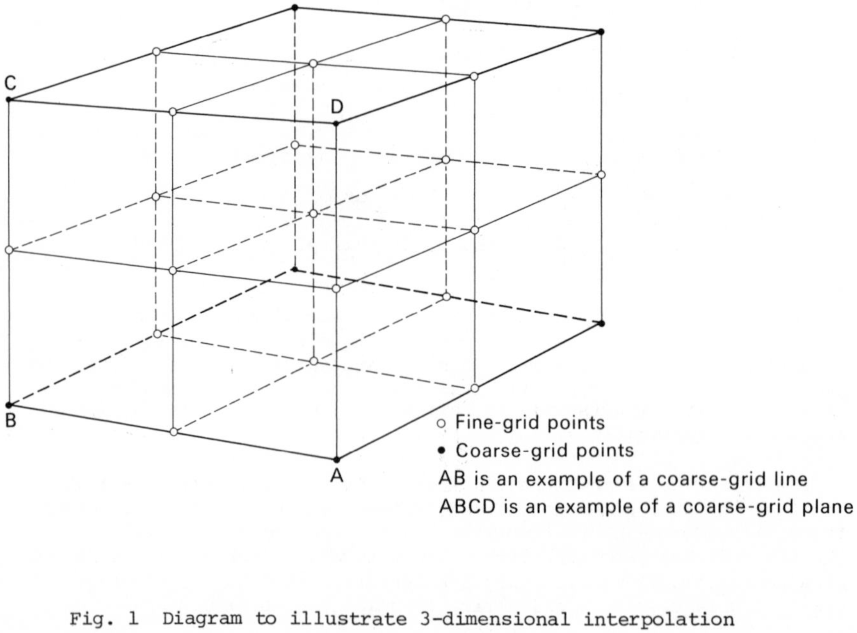

Fig. 1 Diagram to illustrate 3-dimensional interpolation

The first type consists of those points which lie on coarse-grid lines but do not coincide with coarse-grid points. For such points equation (2.6) is averaged in the two directions perpendicular to the coarse-grid line in question (one direction if the problem is two-dimensional). This simply amounts to adding the coefficients of $L^q_{i,j,k}$ in the respective direction(s). The result is an equation in the three fine-grid points which lie on the coarse-grid line. The two end-points coincide with coarse-grid points (by definition of the grid structure) and are therefore known. On substitution for these points, the equation is then solved for the unknown (mid-point) value.

The second type of point consists of those on coarse-grid planes but not lying on coarse-grid lines. In two dimensions, using the above techniques to obtain values at the fine-grid points within the plane which coincide with coarse-grid points and coarse-grid lines, equation (2.6) is solved for the fine-grid value in question. For three-dimensional cases, equation (2.6) is averaged, employing a similar technique to that described above, in the direction perpendicular to the plane. The resulting equation (defined in a plane) is then treated as if it were a two-dimensional example.

The final type of point only occurs in three-dimensional problems and does not lie on a coarse-grid plane (rather the centre point of the cube illustrated in Fig. 1). However the 26 neighbouring (nearest) fine-grid point values are all known from application of the above averaging processes (8 points coincide with coarse-grid points, 12 points lie at the mid-point of coarse-grid lines, 6 points lie at the centre of coarse-grid planes) and hence no averaging is needed. Equation (2.6) is solved for the centre point value using the 26 previously interpolated values.

It is noted that because the central difference equation (2.6) is used to derive the interpolation formulae, maximum use is made of the information supplied by the discrete problem. As a further note, this approach to interpolation preserves continuity of $\underline{K}\nabla u$ (equation (1.1)) which is more accurate physically than ordinary linear interpolation which only preserves continuity of ∇u. The scheme adopted here is more appropriate to reservoir calculations which frequently involve jump discontinuities of several orders of magnitude in $\underline{K}$. Ordinary linear interpolation is only reasonably accurate for problems involving changes in $\underline{K}$ of no more than a single order of magnitude.

Residual Transfer:

This refers to the definition of the operator I^{k-1}_k appearing in equation (2.4) which is used to transfer the residual from G^k to G^{k-1}. Preliminary applications of multigrid merely used the fine-grid values which happened to coincide with the required coarse-grid points. This technique, sometimes referred to as injection, has however been shown by Brandt (1977b) to lead to a 40% degradation in the rate of convergence for Neumann problems compared to the corresponding Dirichlet problem. Such difficulty is overcome by transferring weighted averages of neighbouring residuals instead. Indeed if L^k possesses highly varying coefficients, the residuals after relaxation are also highly varying so that it becomes necessary to use some residual weighting scheme to correctly represent them on G^{k-1}. The approach adopted in the present study follows along these lines and is more robust and consistent with the interpolation process.

In fact the residual transfer operator is defined to be the transpose of the interpolation operator. That is, in three dimensions for example, one writes down the interpolation formula for the third type of point discussed in the above section on interpolation. Each of the 26 neighbouring point values are weighted in some way, namely the interpolation weights. These weight coefficients are transposed with respect to opposite points in the discretisation molecule so that the residual at the coarse-grid point in question is obtained as a weighted sum of the residuals at the 26 nearest fine-grid points.

Coarse-Grid Difference Operator:

Given the fine-grid difference operator L^k, the interpolation operator (I^k_{k-1}) and hence the residual transfer operator $(I^k_{k-1})^T$, then following Nicolaides (1977), an automatic way to construct the coarse-grid operator is by the following:-

$$L^{k-1} = (I^k_{k-1})^T L^k I^k_{k-1}. \qquad (2.7)$$

The action of the product of operators on the right hand side can be elucidated by considering its effect on some coarse-grid function $u_{I,J,K}$. Upper case subscripts are employed to denote coarse-grid values and in what follows lower case to denote fine-grid values. The interpolation operator (I^k_{k-1}) transfers from the coarse grid G^{k-1} to the fine grid G^k whence the fine-grid difference operator L^k acts on each of the $u_{i,j,k}$ variables. The residual transfer (or restriction) operator $(I^k_{k-1})^T$ then returns the problem to the coarse-grid, thus providing the coefficients of each of the $u_{I,J,K}$ comprising the discretisation molecule on G^{k-1}.

The derivation of the coarse-grid difference operator coefficients is a tedious task and involves lengthy algebraic formalism. Even if L^k is based on a 7-point discretisation molecule (5-point in two dimensions), L^{k-1} will involve 27 points (9 in two dimensions). On subsequent coarser grids the L^{k-1} discretisation molecule will again comprise 27 points in three dimensions, 9 in two dimensions. The final form of the coarse-grid difference operators is a long and complicated expression, particularly in three dimensions. Various cross-checks on the algebra are maintained by taking advantage of the symmetry of the problem. A similar approach to the coarse-grid difference operator, summarised by equation (2.7), has been utilised by Alcouffe et al (1981) and Dendy (1982) who both confined their studies to two-dimensional problems.

2.3 Computational Details

Having outlined the mathematical operators to be employed, consideration is now given to some of the other details of the multigrid algorithm. The grid hierarchy is selected so that for points within the domain of interest, coarse-grid lines and planes coincide with fine-grid lines and planes. In order to ensure that the true boundary of the problem is included on each grid, a fictitious set of boundary points is added outside the domain of interest, the distance of these points from the true boundary depending on the step-length of the grid to which the points belong. This means that the boundary of each coarse grid is not identical with the boundary of any other grid. Essentially an aid to programming (Dendy 1982), the foregoing may be summarised by saying that if a typical grid point is denoted by (i,j,k) then the grid itself is defined to be the set $\{(i,j,k): 1 \leqslant i \leqslant i_m, 1 \leqslant j \leqslant j_m, 1 \leqslant k \leqslant k_m\}$ with i_m, j_m, k_m the maximum number of points in the x, y and z directions respectively. The fictitious boundary points are given by the sets:-

$$\{(1,j,k), (i_m,j,k): 1 \leqslant j \leqslant j_m, 1 \leqslant k \leqslant k_m\}$$

$$\{(i,1,k), (i,j_m,k): 1 \leqslant i \leqslant i_m, 1 \leqslant k \leqslant k_m\}$$

$$\{(i,j,1), (i,j,k_m): 1 \leqslant i \leqslant i_m, 1 \leqslant j \leqslant j_m\}.$$

As remarked earlier, on the coarsest level of discretisation the resultant system of difference equations is solved exactly by a direct (Gauss elimination) method. Because the coefficient matrix is unchanged, the upper and lower triangular matrices of the decomposition are stored for solution at this level during subsequent multigrid cycles. It is recognised that relaxation could be used relatively inexpensively at this stage but solving directly eliminates the additional work necessary if the first approximation used to start the iteration is considerably in error.

The number of relaxation sweeps to be performed on each grid apart from the coarsest is fixed to be one before transfer to another grid, whether up or down the grid hierarchy. The only other exception occurs when the algorithm reaches the finest grid, whereupon two sweeps are carried out. Other combinations of relaxation sweeps have been tried such as one sweep when the previous grid was lower down the hierarchy (a coarser grid), two sweeps when the previous grid was higher up in the hierarchy (a finer grid) and three sweeps at the finest level. However, the first mentioned strategy was found to be generally the most efficient in terms of computational time for the examples to be discussed in Section 3. Depending on whether or not the prescribed tolerance criterion for solution to the problem has been satisfied at the finest level, the algorithm will either terminate or begin another multigrid cycle, cycling through all the coarser levels to the coarsest G^1 and returning through the same sequence in reverse order to G^M. Hence the algorithm utilises the so-called V-cycle. Moreover, the algorithm will only move to the grid immediately finer or coarser than the current grid; it will never skip a grid.

On the issue of storage it is necessary to store for each grid G^k, $k = 1, \ldots, M-1$, the interpolation weights and the coefficients of the difference operator L^k. For two-dimensional problems, assuming a 5-point differencing scheme for L^M, this amounts to approximately 10n locations where n is the number of unknowns on G^M. In addition, storage has been included for the solution itself and the right hand side for each of the computational grids. On the other hand for three-dimensional cases with a 7-point discretisation molecule on the finest grid, approximately 12n locations are required where again n is the number of unknowns on the finest grid. In order to conserve on the number of arithmetic operations, additional storage is utilised to retain a number of auxiliary arrays. This increases the allocation to approximately 14n and 17n locations respectively for two and three dimensions. It is perhaps worth saying at this point that a significant proportion of the solution time may be spent in computing the interpolation weights and especially the L^k coefficients. This is found to be particularly true for three-dimensional cases and will be reflected in the results recorded in Tables 1-3.

The multigrid algorithm described in this section has been applied to a number of example problems, selected to include the effects of severe anisotropy and sharp discontinuity in the coefficient terms of $\underline{K}$, typical of those encountered in reservoir simulation studies. It is these test problems that are discussed in the following section.

3. REPRESENTATIVE EXAMPLES

In reservoir simulation, one is frequently concerned with solving a partial differential equation which describes a time evolution process. As already stated in Section 1, the usual strategy is to consider the problem at successive time steps so that it becomes necessary to solve a boundary value problem at each such step. The current implementation of the multigrid package is within a simulator which assumes incompressible flow. Various representative examples have been considered in two and three dimensions which involve anisotropic and discontinuous heterogeneities and allow for injector and producer wells via a relatively simple model.

Returning to the general equation (1.1), this is expressed in rectangular Cartesian co-ordinates in the form

$$\frac{\partial}{\partial x} K_x \frac{\partial u}{\partial x} + \frac{\partial}{\partial y} K_y \frac{\partial u}{\partial y} + \frac{\partial}{\partial z} K_z \frac{\partial u}{\partial z} = f , \qquad (3.1)$$

where $f = f(x,y,z)$ and $K_\alpha = K_\alpha(x,y,z)$. The analogous equation in two dimensions is defined in the x-y plane. In writing down equation (3.1), the common assumption is made that the co-ordinate axes of the reference system are aligned along the principal axes of the tensor $\underline{K}$. Different functional forms are considered for (K_x, K_y, K_z) and the right hand side function f.

Applying second order central differences to reduce equation (3.1) to a discrete system on the finest grid G^M, one readily obtains a matrix system of linear equations equivalent to equation (2.1). In the notation of oil reservoir simulation, the transmissibility terms are defined to be the K_α. The right hand side function provides the possibility of including source and sink terms at appropriate points of interests.

The finite difference equations associated with each representative example are solved on the CRAY-1 and a comparison is provided in Tables 1-3 of the times taken with different smoothing operators. The time required for solution by Gaussian elimination with D4 ordering is also included. One could also compare with other iterative methods apart from multigrid. However, these are generally sensitive to the initial approximate solution and may converge quickly if this choice is close to the exact solution or on the other hand grind away for some time if the approximation differs considerably from the exact result. Full multigrid iteration as used in the present study, however, determines its own good initial starting solution right at the beginning by solving the reduced system of equations at the coarsest level and interpolating this solution to the finest grid using multigrid cycling. Iteration is terminated when the L_2 norm of the residuals on G^M becomes less than ε where $\varepsilon = 0.01$.

The domain of the representative examples is taken to be square in two dimensions and cubic in three dimensions; the code however is not restricted to these configurations and can be readily applied to rectangular regions. Both types of domain are divided up into block-centred grids (a single layer of grid blocks in two dimensions) with u to be calculated at the centre of each block. $\underline{K}$ denotes the transmissibility between adjacent grid-blocks. The examples have been selected on the basis of variation in $\underline{K}$ and three cases studied on each of the two domains. The two-dimensional examples are discussed first.

3.1 Two-dimensional examples

As remarked earlier the domain is set in the x-y plane; both 33 x 33 grid-block (N = 33) and 65 x 65 grid-block (N = 65) configurations are considered in turn to define the finest level of discretisation. The step-length is taken to be scaled to unity in both coordinate directions at this level. No-flow (Neumann) conditions are imposed along all boundaries. A source with strength proportional to N is located in the corner grid-block labelled (1,1) and a sink of similar strength in the opposite corner (N,N). Within the context of reservoir engineering these correspond to a simple description of injection and production wells respectively.

The degree of continuity and isotropy of the medium is governed by $\underline{K}$. Below are listed the properties of $\underline{K}$ for each of the three problems.

Example 1:

Continuous and anisotropic everywhere with $K_x = 1$ and $K_y = 10^2$ for N = 33 and N = 65.

Example 2:

Discontinuous and isotropic such that for

$$i_1 \leqslant i \leqslant i_2 \text{ and } j_1 \leqslant j \leqslant j_2,\ K_x = K_y = 10^{-3}$$

and elsewhere $K_x = K_y = 1$.

For N = 33: $i_1 = j_1 = 12$ and $i_2 = j_2 = 22$.

For N = 65: $i_1 = j_1 = 22$ and $i_2 = j_2 = 44$.

Example 3:

Discontinuous and anisotropic such that $K_x = 1$ everywhere and $K_y = 1$ except for $j_1 \leqslant j \leqslant j_2$ where $K_y = 10^2$ and for $j_3 \leqslant j \leqslant j_4$ where $K_y = 10$.

For N = 33: $j_1 = 16$, $j_2 = 21$, $j_3 = 22$, $j_4 = 27$.

For N = 65: $j_1 = 30$, $j_2 = 41$, $j_3 = 42$, $j_4 = 53$.

Using multigrid, the number of auxiliary coarser meshes set up in addition to the finest grid was 4 for the examples with N = 33 and 5 for those with N = 65.

All problems were solved on a CRAY-1 machine. The amount of CPU time in seconds needed to obtain a convergent solution is recorded in Table 1. Apart from the left hand column headed GD4, which lists the time required to solve by Gaussian elimination preceded by D4 ordering, the table provides a summary of the times taken when each of the smoothing procedures discussed in Section 2.2 is employed within the multigrid scheme. Hence the column under PSR denotes results obtained using point successive relaxation; LSR(α) corresponds to use of line successive relaxation by lines parallel to the direction(s) indicated by α. Results in the column furthest to the right were derived from using alternating line successive relaxation. For each multigrid run, the time needed to perform the preliminary calculations to set up the various interpolation weights and coarse-grid difference operators is also included in the figure given. This set-up time is also tabulated separately in Table 3 together with the time required to perform a complete multigrid V-cycle using the respective smoothing operator. The number of complete multigrid cycles used by the algorithm in order to determine the solution is quoted in parentheses below each of the figures in Table 1. A dash indicates that an excess of multigrid cycles (greater than 10) was needed to achieve convergence.

From Table 1 it is immediately observed that in each case, solution of these problems using multigrid iteration is faster than using the non-iterative approach. Indeed for such problems which involve 1089 and 4225 unknowns at the finest level of discretisation, one would expect that an iterative method should prove superior over a direct elimination

Table 1

CRAY CPU time in seconds required to solve Two-Dimensional Examples by Multigrid Iteration.

33 x 33 Grid-Blocks

Example	GD4	PSR	LSR(x)	LSR(y)	LSR(x,y)
1	0.231	-	-	0.075 (4)	0.110 (4)
2	0.231	0.089 (7)	0.090 (6)	0.099 (6)	0.128 (5)
3	0.231	-	-	0.099 (6)	0.128 (5)

65 x 65 Grid-Blocks

Example	GD4	PSR	LSR(x)	LSR(y)	LSR(x,y)
1	2.770	-	-	0.322 (6)	0.502 (6)
2	2.773	0.337 (9)	0.392 (9)	0.439 (9)	0.502 (6)
3	2.772	-	-	0.440 (9)	0.566 (7)

Number of Multigrid V-Cycles given in parentheses

procedure. This is true not only in terms of computational time but also concerning storage demanded by the algorithm; the amount of store needed by the Gauss D4 algorithm becomes prohibitively excessive for larger domain problems due to the size of the bandwidth.

The speed of convergence within the multigrid scheme depends crucially on the choice of relaxation. In the case of anisotropic problems it is clearly desirable to choose the optimum direction for relaxation, namely the direction in which K_α is greatest. A similar observation is well known when using line successive over-relaxation (LSOR) to solve the finite difference equations at the finest level of discretisation alone (Aziz and Settari, 1979). On the other hand, example 2 which includes discontinuity but no anisotropy converges quickly for each choice of relaxation, a result which could be anticipated from the symmetry of the problem.

Generally, given the optimum relaxation method, it is seen from Table 1 that less than 0.1 second is needed to solve examples on a 33 x 33 grid and 0.3-0.4 second to solve problems on a 65 x 65 grid configuration. Use of alternating line relaxation, though requiring fewer relaxation sweeps, is however not to be recommended for these examples since for each comparison, more time was necessary in order to achieve similar accuracy to using single line relaxation or indeed point successive relaxation in some cases. On the other hand, if time is not of prime importance, smoothing by alternating line relaxation is noted to produce a convergent solution in all cases.

3.2 Three-dimensional examples

As in section 3.1 above, two levels of finest discretisation are considered, the first comprising 9 x 9 x 9 grid-blocks ($N = 9$) and the second 17 x 17 x 17 grid-blocks ($N = 17$), each grid-block being scaled so as to have edges of length unity. No-flow (Neumann) conditions are imposed over all boundaries of the domain. Source and sink terms consist of a source in the grid-blocks labelled $(1,1,k)$ with strength proportional to N^2 and a sink diagonally opposite in the grid-blocks with index coordinates (N,N,k) of similar strength, $k = 1, 2, \ldots, N$. In oil reservoir terminology, taking the z-axis to be in the vertical direction, this corresponds to simple modelling of an injection and production well completed in all layers of the reservoir.

These examples are essentially three-dimensional geometric analogues of the two-dimensional problems.

Example 1:

Continuous and anisotropic everywhere with $K_x = 1$, $K_y = 10$ and $K_z = 10^{-1}$ for $N = 9$ and $N = 17$.

Example 2:

Discontinuous and isotropic such that for

$$i_1 \leqslant i \leqslant i_2,\ j_1 \leqslant j \leqslant j_2 \text{ and } k_1 \leqslant k \leqslant k_2,\ K_x = K_y = K_z = 10^{-1}$$

and elsewhere $K_x = K_y = K_z = 1$.

For $N = 9$: $i_1 = j_1 = k_1 = 4$ and $i_2 = j_2 = k_2 = 6$.

For $N = 17$: $i_1 = j_1 = k_1 = 7$ and $i_2 = j_2 = k_2 = 11$.

Example 3:

Discontinuous and anisotropic such that $K_x = K_y = 1$ everywhere and $K_z = 1$ except for $k_1 \leqslant k \leqslant k_2$ where $K_z = 10^2$ and for $k_3 \leqslant k \leqslant k_4$ where $K_z = 10$.

For $N = 9$: $k_1 = 5$, $k_2 = 5$, $k_3 = 6$, $k_4 = 7$.

For $N = 17$: $k_1 = 9$, $k_2 = 11$, $k_3 = 12$, $k_4 = 14$.

Apart from the finest grid on which the problems are discretised by the simulator, multigrid solution of these examples used 2 additional coarser grids when N = 9 and 3 additional grids when N = 17.

Table 2

CRAY CPU time in seconds required to solve Three-Dimensional Examples by Multigrid Iteration.

9 x 9 x 9 Grid-Blocks

Example	GD4	PSR	LSR(x)	LSR(y)	LSR(z)	LSR(x,y,z)
1	0.573	-	-	0.083 (4)	-	0.126 (3)
2	0.573	0.082 (5)	0.076 (4)	0.083 (4)	0.084 (4)	0.101 (2)
3	0.573	-	-	-	0.084 (4)	0.126 (3)

17 x 17 x 17 Grid-Blocks

Example	GD4	PSR	LSR(x)	LSR(y)	LSR(z)	LSR(x,y,z)
1	-	-	-	0.511 (5)	-	0.896 (4)
2	-	0.475 (6)	0.512 (6)	0.575 (6)	0.517 (5)	0.746 (3)
3	-	-	-	-	0.517 (5)	0.896 (4)

Number of Multigrid V-Cycles given in parentheses.

Again all three-dimensional examples were solved on a CRAY-1 machine. The associated CPU times are tabulated in Table 2 where the notation follows that adopted in Table 1. A full description of the symbols and headings has already been given in Section 3.1 above. The times needed to perform the preliminary calculations for setting up the grid transfer operators and the coarse-grid difference equations are presented in Table 3. Also included in this table are the respective times for performing complete multigrid V-cycles starting at the finest grid and cycling down grid by grid to the coarsest level before returning similarly to the finest level. In the case of Gaussian elimination with D4 ordering, the routine employed in this study was unable to handle the greater bandwidth encountered on the larger (17 x 17 x 17) grid configuration. Otherwise a dash in the columns referring to multigrid calculations indicates that more than 10 multigrid iterations (complete cycles) were required in order to satisfy the convergence criterion.

Inspection of results in Table 2 reveals a similar behaviour to those reported for the two-dimensional runs in Table 1. It is anisotropy which causes the greater difficulty compared to discontinuity; example 2 which involves only discontinuity is readily amenable to solution using any of the smoothing procedures considered. On the other hand a degree of sensitivity is demonstrated by examples 1 and 3, which both include anisotropic effects, to the choice of relaxation procedure. Again for these problems, the line of relaxation should be taken in the direction of greatest K_α in order to achieve maximum efficiency.

On comparing with the two-dimensional results, the increased bandwidth of the present coefficient matrices causes a degradation in execution time as far as the Gauss D4 method is concerned. Multigrid however is able to solve the examples on the 9 x 9 x 9 grid (729 points) in a comparatively short time. Indeed, for grids involving in the region of 700 unknowns in the three-dimensional geometry of this study, multigrid is able to derive a convergent solution in less than 0.1 second. Turning to the larger grid, possessing 4913 unknowns, multigrid iteration is able to solve the problems in less than 0.5 second using the optimum relaxation technique. Solution of the three-dimensional problem is generally more time consuming than the two-dimensional case with a similar number of unknowns. This is due in part to the increase in the amount of preliminary computations (cf. Table 3) and also the more complicated nature of the 3-D operators, especially the discretisation operator discussed in Section 2.2.

Table 3

Average CRAY CPU time in seconds required for Multigrid Preliminary Calculations (MGPC) and for each Multigrid V-Cycle using Point Successive Relaxation MGCY(PSR), using Line Successive Relaxation MGCY(LSR) and using Alternating Line Relaxation MGCY(ALR).

Two-Dimensional Examples

	33 x 33 Grid	65 x 65 Grid
MGPC	0.010	0.036
MGCY(PSR)	0.011	0.035
MGCY(LSR)	0.014	0.046
MGCY(ALR)	0.023	0.082

Three-Dimensional Examples

	9 x 9 x 9 Grid	17 x 17 x 17 Grid
MGPC	0.022	0.118
MGCY(PSR)	0.011	0.061
MGCY(LSR)	0.013	0.072
MGCY(ALR)	0.031	0.193

As for the two-dimensional cases, alternating line relaxation is able in each example to produce a convergent solution using less than the current upper limit of 10 iterations. However, the scheme is overall more demanding on time compared to the multigrid algorithm incorporating the most efficient smoother available.

4. CONCLUDING REMARKS

A state-of-the-art multigrid algorithm capable of solving elliptic partial differential equations in two and three dimensions has been developed and applied to a number of example problems based on situations occurring in oil reservoir simulation. These calculations have been selected to test the robustness of this particular choice of algorithm for handling problems exhibiting strong anisotropic properties and sharp jump discontinuities in the coefficient function $\underline{K}$ of the differential equation (1.1).

Given a convergent relaxation procedure which has the appropriate smoothing property, multigrid iteration is extremely fast at producing a solution to the majority of the representative examples considered in Section 3. The importance of first identifying the nature of the problem in question and then matching to it a suitable multigrid sequence of operations is highlighted by those examples where the use of some relaxation procedures failed to readily provide a convergent solution.

Within the context of this study, the automatic prescription offers a robust and efficient multigrid method for solving typical reservoir simulation problems. For anisotropic cases, the choice of relaxation scheme has a decisive bearing on efficiency and speed. The algorithm is particularly effective in two dimensions. In three dimensions, the gain in speed is partly off-set by the increase in computational time needed to set up the multi-level operators on each grid and by the degree of anisotropy and discontinuity possessed by the problems.

Moreover, further work has shown that performance on two-dimensional problems is comparatively better than on three-dimensional examples and that discontinuities take on more significance in three dimensions by demanding additional computing time. Overall it has been observed from a comparison between 2-D and 3-D problems that a general trend is revealed whereby the algorithm used in this work is better suited to solving discontinuous and anisotropic problems in the plane rather than in space. It is believed however, that further investigation and development of the three-dimensional interpolation and especially the relaxation procedure should remedy this situation.

ACKNOWLEDGEMENTS

Discussions with Professor A. Brandt (Weizmann Institute of Science) and correspondence with Dr. J.E. Dendy, Jr. (Los Alamos National Laboratory) in the initial stages of this work are gratefully acknowledged.

The work reported in this paper has been supported by the United Kingdom Department of Energy.

REFERENCES

Alcouffe, R.E., Brandt, A., Dendy, J.E., Jr. and Painter, J.W. (1981) The Multigrid Method for the Diffusion Equation with Strongly Discontinuous Coefficients, *SIAM J. Sci. Stat. Comput.*, **2**, 430-454.

Aziz, K. and Settari, A. (1979) Petroleum Reservoir Simulation, Applied Science, London.

Brandt, A. (1977a) Multi-Level Adaptive Solutions to Boundary-Value Problems, *Math. Comp.*, **31**, 333-390.

Brandt, A. (1977b) Multi-Level Adaptive Techniques (MLAT) for Partial Differential Equations: Ideas and Software, Mathematical Software III, (J.R. Rice, Ed.), Academic Press, New York, 277-318.

Brandt, A. (1982) Guide to Multigrid Development, Proc. Multigrid Methods Conference, (W. Hackbusch and U. Trottenberg, Eds.), Lecture Notes in Mathematics, 960, Springer-Verlag, Berlin, 220-312.

Dendy, J.R., Jr. (1982) Black Box Multigrid, *J. Comput. Phys.*, **48**, 366-386.

Nicolaides, R.A. (1977) On the ℓ^2 Convergence of an Algorithm for Solving Finite Element Equations, *Math. Comp.*, **31**, 892-906.

A MULTIGRID ALGORITHM USING A HIERARCHICAL FINITE ELEMENT BASIS

A.W. Craig and O.C. Zienkiewicz

(University College Swansea, Wales)

1. INTRODUCTION

The introduction of multigrid methods has been, without doubt, one of the major advances in computational linear algebra in recent years. It has led to the development of optimal order algorithms for the solution of finite difference and finite element approximations to partial differential equations - that is numerical algorithms where the number of arithmetic operations required in order to obtain a solution to the resulting matrix equations is directly proportional to the dimension of the approximating subspace. The basic ideas of the multigrid method are explained fully elsewhere. For a more comprehensive introduction the reader is directed to Brandt (1977) and Stüben and Trottenberg (1982).

The basic ideas of multigrid, although simple to explain, have several computational difficulties associated with them. Firstly it is necessary to calculate and store discrete approximations to the differential operator and the data on each mesh used. This increases the time taken for generation of the discrete equations and the storage required by a factor of two for one-dimensional problems and a factor of 4/3 for two-dimensional problems. Secondly, due to the fact that the approximate solution is normally represented by completely different sets of basis functions on each mesh used it is necessary to calculate transfer operators to move data from mesh to mesh. These operators are only simple to calculate on regular meshes, and their employment counts for a substantial part of the number of operations required to solve the problem. There have been many attempts to use transfer operators which are of a simpler form than those strictly required by the algorithms, but unfortunately this has only led to a decrease in efficiency. Thirdly, although it is possible to show that the number of operations required is proportional to the number of degrees of freedom used, the constant of proportionality may be very large, therefore although the method is of optimal order the actual speed of execution may be slower than other methods of sub-optimal order for the meshes which are used. Lastly, a general multigrid algorithm is not a simple method to program. The complications involved would lead one to suspect that this is a major reason that the method has not been embraced by the engineering community at large.

In this short paper we describe a method of obtaining a finite element discretisation using hierarchical bases. The application of a standard multigrid algorithm to the resulting discrete equations then takes a particularly simple form. Indeed it may be described as a block iterative technique. This use of non-standard basis functions thus circumvents all of the problems catalogued above, producing an iterative algorithm which is simple to understand, easy to program, and efficient to run. We shall restrict ourselves to the description of the formulation and a statement concerning the convergence and optimality of a sub-class of our class of algorithms. A general convergence theorem and numerical results will be presented in a future paper.

We shall also describe briefly how to extend the results to finite difference approximations, and also how standard programs may be changed to hierarchical ones.

2. HIERARCHICAL FINITE ELEMENT BASES

Although these bases have been described elsewhere (e.g. Zienkiewicz et al. (1983)) we shall include a brief description here in order that the paper shall be self-contained. In order to produce the basis which represents the finite element subspace on each mesh we start with the lowest dimensional subspace, that is, the coarsest mesh. On this mesh we may represent functions by using the standard piecewise polynomial basis functions (for example the linear 'hat' functions). In order to produce the next mesh we would normally either subdivide each element (the h-method), or increase the order of the basis functions (the p-method). When performing this operation using standard bases the subdivision or augmentation of an element is accompanied by introducing entirely new basis functions (see Fig. 1a).

This could of course be remedied by introducing the new approximation hierarchically (see Fig. 1b), but here the traditional concept of elements, each formulated by an immutable procedure would have to be partially abandoned. The only 'element' in such a form would be the original one. Such an idea is of course possible although special integration rules may have to be designed for the interaction of 'senior' hierarchical functions (such as N_4) with 'juniors' (N_6). However, we shall see later that it is possible to introduce special procedures which allow us to retain the simplicity of standard bases with the computational advantages of hierarchical bases.

We emphasise at this point that these basis functions span precisely the same space as the standard basis, and therefore we have the standard approximation properties of the finite element method intact.

Hierarchical functions are introduced more simply in the context of the p-method and here their simplicity is evident. In Fig. 2 we show for instance the generation of hierarchical and non-hierarchical functions for a quadratic element of the serendipity family. It was in this environment that hierarchical functions were initially introduced by Zienkiewicz et al. (1971). At that time the objective was simply the introduction of conforming p-graded meshes in an a-priori chosen manner. Since that time use has been made of hierarchical subdivision to produce mixed order interpolation and indeed new and useful families of hierarchical p-type elements have been introduced by Peano (1976).

A short remark over the rates of convergence of the p-method is in order at this point. It has long been assumed, due to a misinterpretation of the standard finite element interpolation error, that it was inadvisable to use higher order elements where there was a singularity present in the domain. This has recently been proved false and indeed it is possible to prove that under certain mild conditions that higher

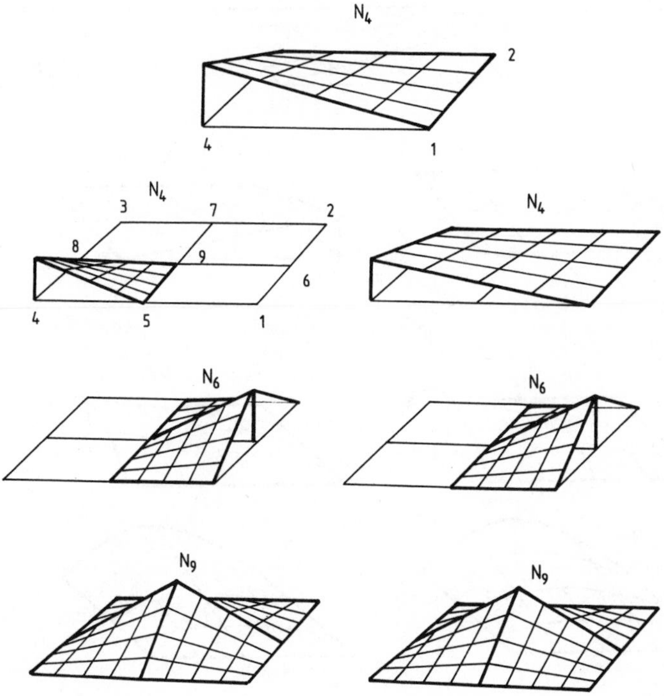

(a) Standard (b) Hierarchic

Fig. 1 Subdivision of an element

order elements give a more accurate approximation, for the same number of degrees of freedom, than lower order elements (see Babuska et al. (1981)).

It should now be obvious that we can further refine the mesh by introducing a further level of hierarchical approximation and proceeding to any degree of approximation which we desire. We may formally state:

Definition 2.1: Let V_h be a conforming finite element subspace of a Hilbert space V. We shall describe a basis $\{\phi_i\}_{i=1,n}$ of the space V_h to be hierarchical if we can write

$$V_h = V_1 \oplus V_2 \tag{2.1}$$

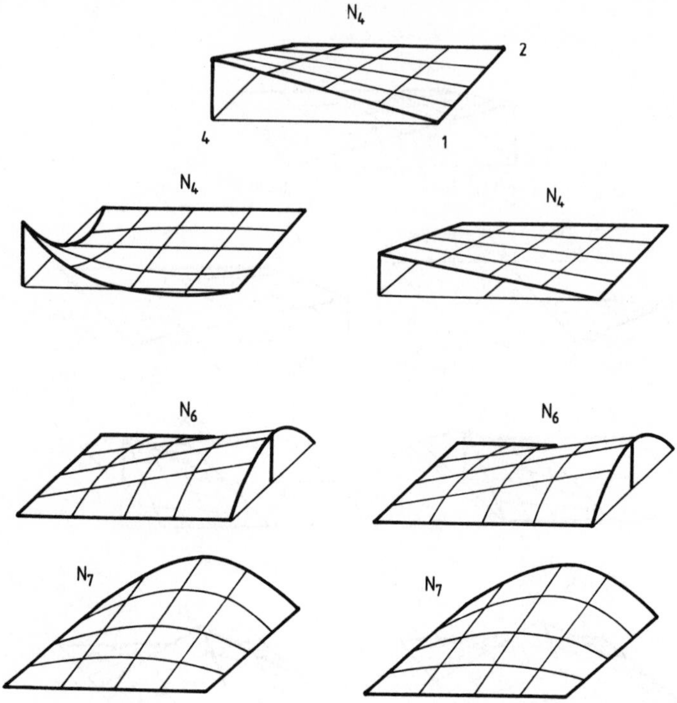

(a) Standard (b) Hierarchic

Fig. 2 Increase in order of polynomial over an element

where for some integer k, $1 \leq k < n$

$$V_1 = \underset{i=1,k}{\text{span}} \{\phi_i\} \tag{2.2}$$

$$V_2 = \underset{i=k+1,n}{\text{span}} \{\phi_i\} \tag{2.3}$$

and V_1 is a conforming finite element subspace of V. (Of course the bases representing V_1 and V_2 may in themselves be hierarchical.)

The term 'hierarchical' is not immediately apparent from this definition. It comes from the fact that if we have

$$V_h = \bigoplus_{i=1,m} V_i \tag{2.4}$$

then

$$V_1 \subset V_1 \oplus V_2 \subset V_1 \oplus V_2 \oplus V_3 \subset \quad \subset V_h. \tag{2.5}$$

The subspace V_2 need have no worthwhile approximation properties of its own, as we are interested in $V_1 \oplus V_2$, or more generally, in the hierarchies (2.5).

Hierarchical forms have many advantages which we have discussed in full detail elsewhere (Zienkiewicz and Craig (1983a)), in particular a-posteriori error estimation and adaptive mesh refinement. It is indeed in the context of adaptive theory that we have produced our multigrid algorithm, however for clarity we present it here on its own.

3. PROPERTIES OF THE HIERARCHICAL EQUATIONS

As we have already mentioned the hierarchical concept has many merits. Here we shall deal with those specific to multigrid algorithms.

3.1 Structure of equations

Let us consider a typical linear problem for which a discrete approximation is sought. Such a problem could well be the solution of elasticity equations in a typical stress analysis, but for generality we state it as:

Find $u \in V$:

$$a(u,v) = (f,v) \ \forall \ v \in V \tag{3.1}$$

where $a(.,.)$ is a continuous, bilinear, elliptic form on some Hilbert space V, and $f \in V'$ (the dual of V). Such a problem is known to possess a unique solution (Ciarlet (1978)). To obtain a finite element approximation to u on our coarsest grid we introduce the approximating subspace V_1, $\dim(V_1) = n$, and pose:

Find $u \in V_1$:

$$a(u_1,v_1) = (f,v_1) \ \forall \ v_1 \in V_1 \tag{3.2}$$

or writing this equation in matrix notation:

$$K_{(n)}u^{(n)'} = f^{(n)}. \tag{3.3}$$

When we now attempt to obtain the discretisation of (3.1) on our next (hierarchically created) subspace of, for example, dimension n + m, the resulting matrix equations are:

$$K_{(n+m)}\ u^{(n+m)} = f^{(n+m)} \tag{3.4}$$

or more precisely:

$$\begin{bmatrix} K_{(n)} & K_{(n,m)} \\ K_{(m,n)} & K_{(m)} \end{bmatrix} \begin{bmatrix} u^{(n)} \\ u^{(m)} \end{bmatrix} = \begin{bmatrix} f^{(n)} \\ f^{(m)} \end{bmatrix}. \tag{3.5}$$

Immediately we observe the first merit, the coarse grid approximation to the differential operator and the coarse grid representation of the data are contained within the corresponding fine grid discretisations. This process may be repeated up to the finest grid which we wish to use, creating a system of matrix equations of precisely the same order as a standard discretisation on the finest grid yet containing all of the necessary information about the problem posed on the coarser grids. When we come to discuss the multigrid algorithm the advantages of this structure will be apparent.

3.2 *Equation conditioning*

It is well known that numerical methods for inverting matrix equations are very sensitive to the condition number of the matrix (Johnson and Riess (1977)). It may be reasonable to expect the equations to have a more dominantly diagonal form due to the use of relative variables. This has important consequences of ensuring improved conditioning and a more accurate solution of the matrix system than would be possible with non-hierarchical forms.

In Fig. 3 we show in detail this improvement in conditioning for the Navier equations on a single cubic element and a patch of four cubic elements. In both cases the condition number of the matrix is improved by an order of magnitude and in larger assemblies more dramatic improvements are expected.

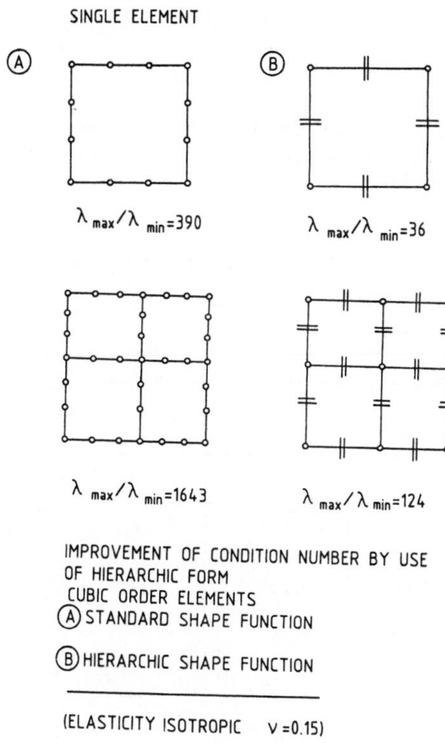

Fig. 3 Improvement in conditioning associated with hierarchical formulation for the Navier equation

This is clearly of importance in equation solution if a small number of digits is carried, as encountered in the increasing use of micro-computers. In Table 1, and Fig. 4 we show the simple example of a slender cantilever. This problem is well-known to produce ill-conditioned matrices. We remind the reader that the approximation spaces for hierarchical and non-hierarchical bases are precisely the same and that, if we could perform exact arithmetic, the solutions would be identical.

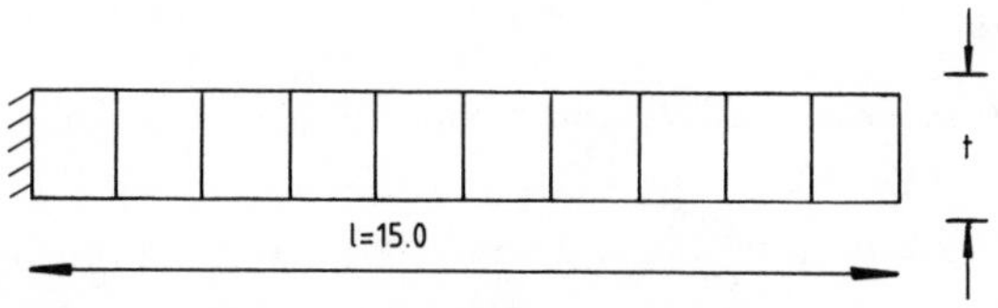

Fig. 4 Cantilever beam for conditioning studies

Table 1

Relative errors in cantilever computations

Pure traction example on cantilever beam with varying thickness t. Stress evaluation at point (7.7, 0.5t) $\sigma_{exact} = 1.0000$, tip displacement $\delta_{exact} = 15.0$

CASE A - 10 eight-noded isoparametric elements

CASE B - 10 eight-noded hierarchic elements

CASE I - relative error in computed σ_{xx} stress

CASE II - relative error in computed displacement

	I		II	
t	A	B	A	B
0.01000	0.0%	0.0%	0.0%	0.0%
0.00100	1.6%	0.1%	0.2%	0.0%
0.00010	15.7%	7.0%	17.4%	0.5%
0.00001	FAILS	FAILS		

These comparisons were found to be typical for a large range of physical parameters.

4. MULTIGRID METHODS

4.1 Description of algorithm

In this section we shall take the finite element multigrid algorithm of Bank and Dupont (1981) as our model. However, the results presented here are valid for any multigrid algorithm posed in terms of hierarchical basis functions and indeed can be extended simply to finite difference algorithms.

The algorithm may be stated as follows:

$\exists$ a sequence of subspaces of V such that

$$V_1 \subset V_1 \oplus V_2 \subset \ldots \subset \bigoplus_{i=1,k} V_i \subset V \qquad (4.1)$$

and we wish to find the function

$$u_k \in \bigoplus_{i=1,k} V_i \qquad a(u_k,v) = G(v) \quad \forall v \in \bigoplus_{i=1,k} V_i. \tag{4.2}$$

We choose parameters p,m and an initial guess $u^0 \in \bigoplus_{i=1,k} V_i$. Then for i=1, m

(1) $$(u^i - u^{i-1},v) = \Lambda_k^{-1} [G(v) - a(u^{i-1},v)] \quad \forall v \in \bigoplus_{i=1,k} V_i \tag{4.3}$$

(or if k = 1, solve directly)

(2) obtain $q \in \bigoplus_{i=1,k} V_i$ by applying the scheme p times to the equation

$$a(q,v) = G(v) - a(u^m,v) \quad \forall v \in \bigoplus_{i=1,k-1} V_i \tag{4.4}$$

(3) $u^{m+1} = u^m + q$

where Λ_k is the maximum eigenvalue associated with the space $\bigoplus_{i=1,k} V_i$.

Theorem

Under the conditions already stated in this paper, this algorithm is convergent. Furthermore for certain choices of the parameter p, the number of operations required to solve the system to acceptable accuracy is proportional to dim ($\bigoplus_{i=1,k} V_i$).

Proof: see Bank and Dupont (1981).

The previously mentioned difficulties associated with this and any other multigrid algorithm using standard bases are now obvious. Firstly, the matrices corresponding to a(.,.) in equations (4.3) and (4.4) are completely different. As we have seen in the previous section, this is not the case for hierarchical bases as the matrix takes a nested form. The same remarks also hold for the data.

More importantly, the calculation of the residual term G(v)-a(u,v) in equation (4.4) involves inner products of functions defined on different spaces. If these functions are expanded in terms of different bases then effectively the equation implicitly defines a transfer operator to move the discrete form of the functions from one mesh to another. Similar remarks hold true for the addition of the functions u^m and q in equation (4.5). This transfer operator may be difficult to calculate on the irregular meshes which are needed to solve practical problems accurately, and its operation contributes a large proportion of the solution time. However, let us re-examine the problem in terms of hierarchical bases.

Writing out the iteration applied to the second level equation (4.4) in full we obtain:

$$(q^i - q^{i-1}, v) = \Lambda_{k-1}^{-1}[G(v) - a(u^m, v) - a(q^{i-1}, v)] \forall v \in \bigoplus_{i=1,k-1} V_i \tag{4.6}$$

where u^m is obtained from the previous level. In principle we could update u^m directly instead of obtaining the correction vector q. However, in practice, for standard bases, this obviously requires the use of the transfer operator at each stage of the iteration. But for the hierarchical basis we may write,

$$u^m = \tilde{u}^m + \bar{u}^m \tag{4.7}$$

where

$$u^m \in \bigoplus_{i=1,k-1} V_i$$
$$\bar{u}^m \in V_k. \tag{4.8}$$

Now the addition

$$\tilde{u}^{m+i} = \tilde{u}^m + q^i \tag{4.9}$$

may be performed by pointwise addition of the vectors of the coefficients, and we may rewrite (4.6) as

$$(\tilde{u}^{m+i} - \tilde{u}^{m+i-1}, v) = \Lambda_{k-1}^{-1}[G(v) - a(\bar{u}^m, v) - a(\tilde{u}^{m+i-1}, v)]$$
$$\forall v \in \bigoplus_{i=1,k-1} V_i, \tag{4.10}$$

in other words, we update the coefficients of the lower order basis functions directly, and leave those corresponding to V_k unchanged. This not only removes the problem of transfer operators, but also does away with the need to calculate and store vectors corresponding to q. The full simplicity of this algorithm, as opposed to the normal multi-grid procedure becomes evident when we present an example in terms of matrix notation:

If we had an equation approximated on $V_1 \oplus V_2$, i.e.

$$\begin{bmatrix} k_{11} & k_{12} \\ k_{21} & k_{22} \end{bmatrix} \begin{bmatrix} u_1 \\ u_2 \end{bmatrix} = \begin{bmatrix} f_1 \\ f_2 \end{bmatrix} \tag{4.11}$$

where k_{11} is the 'coarse' stiffness matrix, then the algorithm would simply state

(1) choose $[u^o, u_2^o]^T$

(2) perform m iterations of a relaxed Jacobi algorithm to equations (4.11) to obtain $[u_1^m, u_2^m]^T$

(3) solve directly

$$k_{11}\, u_1^{m+1} = f_1 - k_{12}\, u_2^m$$

(4) let $\begin{bmatrix} u_1^{m+1} \\ u_2^m \end{bmatrix} \Rightarrow \begin{bmatrix} u_k^o \\ u_2^o \end{bmatrix}$ and return to (2)

that is, a simple block iteration scheme. It is obvious that the scheme is more efficient than normal multigrid due to the extreme simplicity of execution.

4.2 Implementation

The implementation of these algorithms is very simple for p-type hierarchical elements, and can be done in a straightforward manner. There are some problems however with h-type elements, due to the integration of the inner products of 'high order' basis functions with 'low order' basis functions (their supports may differ greatly), but it is of course possible to define a local linear mapping (on an element level), between the standard and hierarchical representations. This mapping is obviously a simple function of the transfer operators mentioned before. Therefore, although we would advise recasting the problems in terms of hierarchical bases to obtain the full power of the method, it is possible to take a standard program, calculate the element stiffness matrices and load vectors, then before assembly map these into their hierarchical counterpoints. We emphasise that this operation need only be performed once, at the beginning of the program, as opposed to many times in a standard multigrid procedure.

4.3 Extension to finite difference discretisation

This theory has been expounded in terms of finite element analysis as the concept of relative displacement variables is not normally used in finite differences. However, the analyst interested in implementing this theory in a finite difference context need only pose his difference equations for the finer mesh points in this form in order to transfer multigrid to block iteration. The precise form of the relative displacement variables is usually obvious from the context of the problem. However, if not, all that need be done is to pose the problem in a standard manner, and then using the transfer operators in the manner described in section 4.2, map into the appropriate matrix equations.

REFERENCES

Babuska, I. and Dorr, M. (1981) Error estimates for the combined h and p versions of the finite element method. *Num. Math.*, **25**, 257-277.

Bank, R. and Dupont, T. (1981) An optimal order process for solving elliptic finite element equations. *Math. Comp.*, **36**.

Brandt, A. (1977) Multi-level adaptive solutions to Boundary-Value Problems. *Math. Comp.*, **31**, 333-390.

Ciarlet, P.G. (1978) The finite element method for elliptic problems. North-Holland.

Johnson, L.W. and Riess, R.D. (1977) Numerical Analysis. Addison-Wesley.

Peano, A.G. (1976) Hierarchies of conforming finite elements for plane elasticity and plate bending. *Comp. Math. with Appl.*, **2**.

Stüben, K. and Trottenberg, U. (1982) Multigrid Methods. In: Hackbusch and Trottenberg, Multigrid Methods, Proceedings, Köln-Porz, 1981. Lecture Notes in Mathematics 960. Springer Verlag.

Zienkiewicz, O.C., Irons, B.M., Scott, F.C. and Campbell, J. (1971) Three dimensional stress analysis. Proc. IUTAM Symp. on High Speed Computing of Elastic Structures, Univ. of Liege Press, 413-433.

Zienkiewicz, O.C. and Craig, A.W. (1983a) Adaptive mesh refinement and a-posteriori error estimation for the p-version of the finite element method. In: Babuska, Chandra and Flaherty, Adaptive computational methods for partial differential equations, SIAM, 1983.

Zienkiewicz, O.C., de S.R. Gago, J.P. and Kelly, D.W. (1983b) The hierarchical concept in finite element analysis. *Comp. and Struct.*, **16**, 53-65.

SUBJECT INDEX

AUTHOR INDEX